Biosafety of Forest Transgenic Trees

FORESTRY SCIENCES

Volume 82

More information about this series at http://www.springer.com/series/5991

Cristina Vettori · Fernando Gallardo
Hely Häggman · Vassiliki Kazana
Fabio Migliacci · Gilles Pilate
Matthias Fladung
Editors

Biosafety of Forest Transgenic Trees

Improving the Scientific Basis for Safe Tree Development and Implementation of EU Policy Directives

Springer

Editors
Cristina Vettori
National Research Council (CNR)
Institute of Biosciences and Bioresources (IBBR)—UOS Florence
Sesto Fiorentino
Italy

Fabio Migliacci
National Research Council (CNR)
Institute of Chemistry of Organometallic Compound (ICCOM)
Sesto Fiorentino
Italy

Fernando Gallardo
Departamento de Biología Molecular y Bioquímica, Facultad de Ciencias
Universidad de Málaga
Malaga
Spain

Gilles Pilate
INRA
UR0588, Amélioration, Génétique et Physiologie Forestières
Orleans
France

Hely Häggman
Genetics and Physiology Department
University of Oulu
Oulu
Finland

Matthias Fladung
Thuenen Institute of Forest Genetics
Großhansdorf
Germany

Vassiliki Kazana
Department of Forestry and Natural Environment Management
Eastern Macedonia and Thrace Institute of Technology
Drama
Greece

ISSN 0924-5480 ISSN 1875-1334 (electronic)
Forestry Sciences
ISBN 978-94-017-7529-8 ISBN 978-94-017-7531-1 (eBook)
DOI 10.1007/978-94-017-7531-1

Library of Congress Control Number: 2015960219

Printed on acid-free paper

This Springer imprint is published by SpringerNature
The registered company is Springer Science+Business Media B.V. Dordrecht

Foreword

The studies presented in this book provide an essential perspective for the contemporary evaluation of governmental regulation and industrial deployment of transgenic trees. The first transgenic tree was a poplar hybrid, modified by the introduction of a gene conferring resistance to the herbicide glyphosate, only three years after the first expression of foreign genes in plant cells (tobacco) and before the development of the first transgenic soybeans or maize. Over the past 30 years, crop varieties created using gene splicing have been widely adopted increasing at a rate of 3 % per year since 1996. By 2014, the global acreage of transgenic maize and soybeans reached 181 million hectares. Gene splicing has been a more limited success with consumers than with agrobusiness. A substantial number of consumers are hostile to genetically modified (GM) crops, and nongovernmental organizations such as Greenpeace have opposed deployment of GM crops, worldwide.

GM technology should have substantial benefits for forest trees, potentially greater than for agricultural crops. Genetic improvement of forest trees by breeding is limited by the long generation times and by the space required to maintain large populations of forest trees. Generation times from seed to seed can vary from a few years to decades. GM technology has the potential to produce large genetic effects based on the introduction of small numbers of genes in a single step through an asexual process. In contrast, selection for quantitative variation may require several generations. The pace of tree improvement through traditional breeding, for example, for loblolly pine in the southern US, is such that four generations of selection has required about half a century.

Accelerated tree improvement by GM technology could produce more wood on less land, improve quality, increase resistance to pests and pathogens, and increase stress tolerance in the face of climate change, a major challenge to biodiversity. GM technology has potential for long-term environmental benefits. An example of a potential long-term environmental benefit is the engineering of resistance to blight in the American chestnut. Chestnut was a highly valued and beneficial forest tree species both to human communities and to the Appalachian ecosystem in the mid-Atlantic forests of North America. American chestnut is moribund and will become extinct if it is not rescued by genetic development of resistance. Hundreds of forest tree species around the world are currently threatened by epidemics of pests and pathogens. Some of the epidemics are associated with climate change either because of reduced stress on the pests or increased stress on the hosts.

It might have been expected that application of gene splicing to tree species would be rapid and enthusiastic. What actually happened was that deployment of transgenic trees has been quite limited and has lagged far behind the major commercial agricultural crops. Deployment of transgenic trees has been highly restricted in Europe and limited to a few test plantations in the USA. The most extensive transgenic tree plantings have been in China where hundreds of hectares of transgenic poplars are growing. Why has the deployment of transgenic crops been so widespread while transgenic trees have lagged so far behind?

This volume describes many of the reasons for the delayed application of transgenic trees, based in part on the lack of technology development, but primarily on the barriers of government regulation and the public opposition to GM plants and trees. Technology development has been a significant factor. Many of the methods developed for crop plants, particularly those of Agrobacterium-based transformation can be and have been applied to many tree species, where cell and tissue culture methods had been adequately developed. The development of culture and regeneration technology has been a greater barrier than the method of gene transfer.

As a research tool, transgenic technology has been used extensively for genetic analysis of forest trees. Major successes have been achieved for specific gene expression, for control of flowering, and particularly for the modification of metabolic pathways. Some large-scale mutation screens have been established using transposon tagging, and new genes affecting a spectrum of phenotypes have been identified. Important new methods of genome editing can create homozygotes for an introduced mutation in a single generation, and such genetic drive technology may be able to spread a mutation through a tree population to counteract attacks by pests and pathogens. Climate change is likely to increase the number of cases where intervention using GM technology may be needed. GM trees may be able to prevent the loss of entire species and the loss of a dominant tree species may threaten an entire ecosystem.

The process of gene splicing is closely related to the genetic changes that occur naturally in trees through insertional transposition and mobile DNA elements. The process of gene splicing should not be the basis for regulation of a crop or a tree, it should be the nature of the modified plant itself. Based on this principle, it should be possible to create a regulatory system that provides a realistic evaluation of costs, risks, and benefits. Much work needs to be done to rationalize the regulatory process to bring restrictions and benefits into a more consistent and logical framework. A "regulatory trigger" of gene splicing technology is still widely used, even though the safety of the gene splicing process per se, has been unequivocally established. Gene splicing technology has been tested in thousands of laboratory experiments, deployed over hundreds of millions of hectares, and fed to millions of animals, without a single valid example of harmful effects. The regulatory burden of unnecessary regulation has cost society on the order of a trillion of euros/dollars.

Deployment of exotic species and interspecies or even intergeneric hybrids of forest trees are considered safe, while the directed introduction of a highly defined and precisely controlled specific gene is still considered a substantial health or environmental hazard. In spite of these limitations, a GM loblolly pine has recently been classified by the USDA as "unregulated", paving the way for deployment, while a transgenic *Eucalyptus* has been approved in Brazil shortly after a major tree nursery containing transgenic trees was destroyed by anti-GM vandalism.

There is a great need for long-term efforts for public involvement and education. The general public cares deeply about forest trees. However, there is limited understanding of the dynamic nature of tree genomes and the mechanisms of genetic changes that have occurred in our trees and forests in recent times. The application of GM technology to solving problems of our forests is just beginning and the future of these applications should be important and exciting. A major impact of this book will be that it advances a rational basis for the path forward.

Ronald Sederoff
Department of Forestry and Environmental Resources
North Carolina State University Raleigh
NC, USA

Preface

Since the first report on transgenic tobacco published in 1983, a large but diverse and fragmented body of knowledge on the environmental interactions and safety of transgenic plants and in particular trees has been acquired over the past more than 30 years. The main objective of the EU-COST Action FP0905 (Biosafety of forest transgenic trees: improving the scientific basis for safe tree development and implementation of EU policy directives) has been to evaluate and substantiate scientific data relevant to the biosafety of Genetically Modified Trees (GMT). Therefore, the idea of this book arises from the need to report the outcome of the scientific discussion and synthesis from the four years of activity of this COST Action.

COST—Cooperation in the field of Scientific and Technical Research—is one of the longest running European instruments supporting cooperation among scientists and researchers across Europe. COST is also the first and widest European intergovernmental network for coordination of nationally funded research activities. Therefore, this EU instrument has permitted the establishment of an unique platform on biosafety of GMTs like this COST Action is.

The Action started on April 12, 2010 and ended on April 11, 2014. In total, 27 COST countries (Austria, Belgium, Bosnia and Herzegovina, Bulgaria, Croatia, Czech Republic, Denmark, Estonia, Finland, France, Germany, Greece, Israel, Italy, Latvia, Netherlands, Norway, Poland, Portugal, Romania, Serbia, Slovak Republic, Slovenia, Spain, Sweden, The Former Yugoslav Republic of Macedonia, and United Kingdom) and seven NON-COST countries (Albania, Australia, Canada, China, New Zealand, South Africa, USA) have participated in the Action.

To reach its objectives, the Action work plan organized into four Working Groups (WGs) focuses on: (i) the biological characterization of GMTs aiming to evaluate existing knowledge, including the experience from expert scientists actively working in the field of forest GMTs (WG 1), (ii) the assessment of possible environmental impacts and monitoring of GMTs in the whole production chain from plantation to final products (WG 2), (iii) the socio-economic implications and public acceptance and concerns of potential use of GM forest trees and R&D

investments in the framework of Cost–Benefit Analysis (WG 3), and (iv) increasing public awareness and understanding of GM tree plantations by providing science based information through management of the www.cost-action-fp0905.eu dynamic website (WG 4).

This book is subdivided into four sections each representing a WG, and the authors of the chapters are experts of the COST Action as well as colleagues in the field of GMT, all researchers and scientists with proven international reputation in the fields of tree biotechnology, risk assessment, monitoring, socio-economic, and dissemination.

The scientific outputs of this Action can be useful to bodies performing environmental risk assessment of GM plants (at European and local level but also at the worldwide level) and to authorities involved in risk management (e.g., commercial release authorization). Therefore, several end users will be interested in the content of this book, including public bodies (e.g., non-governmental organizations (NGOs) as well as organizations at regional and municipal levels) responsible for policy regulation at the scientific, ethic, and environmental levels, and private industries (e.g., paper, timber).

In particular, target groups are:

- Research organizations working with transgenic organisms and in related fields, such as ecology, biochemistry, physiology, economy, society, and policy
- Relevant national ministries responsible for GM plants
- European regional authorities (via relevant regional networks like AREPO—Association des Région Européenne des Produits d'Origine, GMO-free European Regions Network, CRPM—Conférence des Régions Périphériques Maritimes d'Europe, AER—Assembly of European Regions)
- European Commission services (particularly DG AGRI and DG ENV)
- GMO panel of the European Food Safety Authority (EFSA)
- Members of the European COEXNET initiative.

In summary, this book can be useful to public/private organizations as well as to research private and public bodies and universities worldwide since the status of GMT research and policy is reported at the world level.

Cristina Vettori
Matthias Fladung

Acknowledgments

This publication is based upon work from COST Action FP0905 "Biosafety of forest transgenic trees: improving the scientific basis for safe tree development and implementation of EU policy directives", supported by COST (European Cooperation in Science and Technology).

EUROPEAN COOPERATION
IN SCIENCE AND TECHNOLOGY

COST (European Cooperation in Science and Technology) is a pan-European intergovernmental framework. Its mission is to enable breakthrough scientific and technological developments leading to new concepts and products and thereby contribute to strengthening Europe's research and innovation capacities.

It allows researchers, engineers, and scholars to jointly develop their own ideas and take new initiatives across all fields of science and technology, while promoting multi- and interdisciplinary approaches. COST aims at fostering better integration of less research intensive countries to the knowledge hubs of the European Research Area. The COST Association, an International not-for-profit Association under Belgian Law, integrates all management, governing, and administrative functions necessary for the operation of the framework. The COST Association has currently 36 member countries (www.cost.eu).

COST is supported by
the EU Framework Programme
Horizon 2020

We are grateful to all participants of the COST Action FP0905 for their contributions during the four years of activity. We thank Dr. Lise Jouanin (INRA Versailles, France) for coordinating the STSM (Short Term Scientific Mission) activity within this COST Action. In particular, we thank the Early Stage Researchers (ESR), who greatly contributed in collecting the data for the realization of the GMT database which was possible through STSMs financed by this Action: Ms. Maja Popovic (Division of Genetics, Forest Tree Breeding and Seed Husbandry, Croatian Forest Research Institute, Jastrebarsko, Croatia), and Mr. José Manuel Ramos Sánchez (Universidad Politécnica De Madrid, Madrid, Spain).

We deeply thank the COST office staff of Forests, their Products and Services (FPS), and in particular our Science Officers, Ms. Melae Langbein and Dr. Fatima Bouchama, the Administrative Officer, Ms. Cassia Azevedo, in supporting our Action and also providing helpful suggestions for the correct management of the Action. Moreover, we are grateful to the Communication staff of COST Office for support in preparing a video communicating the reasons why this COST Action was established, in response to anti-GMO hacker attacks on the Action website.

We thank all members of the COST Action FP0905 focus group participating in the Cost–Benefit Analysis assessment (chapter "Approaches and Tools for a Socioe-conomic Assessment of GM Forest Tree Crops: Factors for Consideration in Cost–Benefit Analyses").

We thank all the colleagues at the universities for the support given in collecting the information/opinion of the students by circulating the WG 3 questionnaire during the class lessons (chapter "Public Knowledge and Perceptions of Safety Issues Toward the Use of Genetically Modified Forest Trees: A Cross-Country Pilot Survey"): Dr. Helena Almeida (ISA-UL, Portugal), Dr. Margarida Ataíde (IPCB, Portugal), Dr. Dirk Becker (University of Hamburg, Germany), Dr. Alžbeta Blehová (Comenius University in Bratislava, Slovakia), Dr. G. Covelo (Univ. Santiago de Compostela), Alfredo Cravador (University of Algarve, Portugal), Prof. Dr. Thomas Debener (University of Hannover, Germany), Dr. Želmíra Gregáňová (Slovak University of Agriculture in Nitra, Slovakia), Prof. Stefano Leonardi (Dipartimento di Bioscienze, Università degli Studi di Parma, Italy), Dr. J.L. Fernandez-Lorenzo (Univ. Santiago de Compostela, Spain), Dr. Patrik Meszároš (Constantine the Phylosopher University in Nitra, Slovakia), Dr. Leonor Morais (ISA-UL, Portugal), Prof. Susanna Nocentini (Dipartimento di Gestione Sistemi Agrari, Alimentari e Forestali—GESAAF, Università degli Studi di Firenze, Italy), Prof. Dr. Thomas Schmülling (University of Berlin, Germany), Dr. Thomas Teichmann (University of Göttingen, Germany), Prof. Davide Travaglini (Dipartimento di Gestione Sistemi Agrari, Alimentari e Forestali—GESAAF, Università degli Studi di Firenze, Italy).

The authors of the chapter "Public Knowledge and Perceptions of Safety Issues Toward the Use of Genetically Modified Forest Trees: A Cross-Country Pilot Survey" thank all the students of the following universities who participated in the survey: Agricultural University of Tirana (Albania), University Nacional de la Plata (Argentina), University of Melbourne (Australia), University of Banjaluca (Bosnia and Herzegovina), Forestry University of Sofia (Bulgaria), University of Zagreb (Croatia), Czech University of Life Sciences (Czech Republic), Charles University (Czech Republic), University of Economics (Czech Republic), University of Goettingen (Germany), University of Hamburg (Germany), Thuenen Institute (Germany), University of Berlin (Germany), University of Hannover (Germany), Eastern Macedonia and Thrace Institute of Technology (Greece), Hebrew

University (Israel), Ben Gurion University (Israel), University of Florence (Italy), University of Parma (Italy), Instituto Superior de Agronomia—University of Lisbon (Portugal), University of Algarve (Portugal), Instituto Politecnico de Castelo Branco (Portugal), University of Belgrade (Serbia), University of Singidunum (Serbia), University of Novi Sad (Serbia), Constantine the Phylosopher University in Nitra (Slovakia), Slovak University of Agriculture in Nitra (Slovakia), Comenius University in Bratislava (Slovakia), University of Lubljana (Slovenia), Universidade de Santiago de Compostela (Spain).

Finally, in relation to the chapter "Public Knowledge and Perceptions of Safety Issues Toward the Use of Genetically Modified Forest Trees: A Cross-Country Pilot Survey": Susana Araújo and Jorge A.P. Paiva acknowledge Fundação para a Ciência e a Tecnologia (FCT, Portugal) for the research contract on the frame of Programa Ciência 2008 funded by POPH (QREN), Gregor Božič acknowledge the Slovenian Research Agency for the research contract P4-0107, and Jana Malá, Helena Cvrčková and Pavlina Máchová acknowledge the Ministry of Education, Youth and Sports of Czech Republic for the research project COST No. LD13008.

Moreover, we are very grateful for the reviewing work of the chapters of the book done by: Dr. Detlef Bartsch (Federal Office of Consumer Protection and Food Safety, Germany), Dr. Dirk Becker (University of Hamburg, Germany), Dr. Wout Boerjan (VIB, Department of Plant Systems Biology, Belgium), Prof. Dr. Gerd Bossinger (The University of Melbourne, Australia), Dr. Didier Breyer (Scientific Institute of Public Health, Biosafety and Biotechnology Unit, Belgium), Dr. Rene Custers (VIB headquarters, 9052 Ghent, Belgium), Dr. Matthias Fladung (Thünen Institute of Forest Genetics, Germany), Dr. Fernando Gallardo (Departamento de Biología Molecular y Bioquímica, Facultad de Ciencias, Universidad de Málaga, Spain), Dr. Helmut Gaugitsch (Umweltbundesamt [Environment Agency Austria], Unit Land Use and Biosafety, Austria), Dr. Boet Glandorf (GMO Office, National Institute of Public Health and the Environment, the Netherlands), Dr. Lise Jouanin (INRA Versailles, France), Prof. Katri Kärkkäinen (Natural Resources Institute Finland, The Finnish Forest Research Institute, Oulu Unit, Finland), Prof. Gabi Krczal (RLP AgroScience GmbH, AlPlanta—Institute for Plant Research, Germany), Dr. Efstratios Loizou (Division of Agricultural Economics—Department of Agricultural Technology, Technological Education Institute (TEI) of Western Macedonia, Greece), Dr. Klaus Minol (Genius, Germany), Dr. Marina Nonić (University of Belgrade-Faculty of Forestry, Serbia), Prof. Maria Partalidou (Aristotle University of Thessaloniki, School of Agriculture Greece), Dr. Marja Ruohonen-Lehto (Finnish Environment Institute, Finland), Prof. Irma Saloniemi (Turku University, Finland), Dr. Ralph Scorza (USDA, USA), Dr. Susanne Stirn (University of Hamburg, Germany), Prof. Steven Strauss (Oregon State University), Dr. Jeremy Sweet (Sweet Environmental Consultants, UK), Prof. Dr. Christoph Tebbe (Thünen Institute for Biodiversity, Germany).

We also thank Susann Harnik and all staff of the "Reisekostenstelle" (Thünen Institute, Braunschweig, Germany) for administrative support.

We are particularly grateful to Simonetta Bianchi, Sandro Capaccioli, and Catia Boggi of the Institute of Biosciences and Bioresources (CNR, Italy) for the support given in the administration and management as Grant Holder of the COST Action FP0905.

Cristina Vettori
Matthias Fladung

Contents

Introduction

Cristina Vettori and Matthias Fladung

Abstract In the last 25 years, various genetically modified forest trees (GMTs) with modified or novel characteristics have been produced. Economic and ecological benefits like reduction of product costs and less pressure on native forests are expected but also risks like unintended spread of GMTs and pleiotropic effects of transferred genes are speculated by the opponents. This book summarizes a 4-years lasting COST Action on the biosafety of GMTs and focuses on the following four key aspects: (a) biological characterization of GMTs aiming to evaluate existing knowledge including the experience from expert scientists in the field of forest GMTs, (b) evaluation of the environmental impacts of the GMTs already developed, (c) assessments of the efficiency of existing transgene containment strategies, and (d) conduction of socio-economic and cost/benefit analyses in relation to the use of GMTs in plantations. This book combines multidisciplinary knowledge generated with transgenic lines of forest trees (such as, *Populus* spp., *Pinus* spp., *Eucalyptus* spp., *Betula* spp., *Castanea* spp., *Picea* spp., etc.) as well as extensive expertises in correlated topics. Moreover, considering the increase of public awareness on the use of GM forest plantations and environmental protection, the final part of the book focuses on communication about GMTs through science based information management of the http://www.cost-action-fp0905.eu dynamic website. This book aims to present useful information for the scientific community that is interested or engaged in socio-economic implications and biosafety issues of GMTs, but also to provide scientifically based support for decision processes of policy-makers and regulators.

Forests are important for numerous aspects of human health and the environment. For humans, forests play a significant role by provision of many economic, leisure

---

C. Vettori (✉)
National Research Council (CNR), Institute of Biosciences and Bioresources (IBBR)—UOS Firenze, Via Madonna Del Piano 10, 50019 Florence, Sesto Fiorentino (FI), Italy
e-mail: cristina.vettori@cnr.it

M. Fladung
Thünen Institute of Forest Genetics, Sieker Landstr. 2, 22927 Grosshansdorf, Germany
e-mail: matthias.fladung@ti.bund.de

C. Vettori et al. (eds.), *Biosafety of Forest Transgenic Trees*,
Forestry Sciences 82, DOI 10.1007/978-94-017-7531-1_1

and spiritual benefits. Trees convert the energy of the sun into woody biomass and release oxygen through the process of photosynthesis. At the same time, forests are highly important as sinks for CO_2, thus moderating climate change. There is a growing worldwide demand for wood and biomass to provide fuels, chemicals and industrial products. Current wood fibre usage is about 3 billion m^3 per annum, and 50 % of the wood required is supplied by forest plantations that comprise only 7 % of the world forest area (Hirsch 2015). For 2050, a need of about 10 billion m^3 per annum is predicted, exceeding the productivity of natural forests and also "classical" forest plantations. Biotechnologies provide new tools to help increase the productivity of forest plantations to the required levels (e.g. Klocko et al. 2013).

Biotechnology can meet the demands and needs of society by reducing pressure for harvest of forests that are important for support of biodiversity and other ecosystem services. Like any technology and human intervention made by man, it must be applied prudently, following principles of ecosystem stewardship (Institute of Forest Biotechnology 2011). It will be a great challenge to increase productivity of forest plantations while at the same time ensuring sustainable practices. One of the greatest challenges to the human race today is to secure the material, food and energy supplies for our children and grandchildren while keeping the planet viable.

For transgenic plants in general, environmental impact assessment policies have been developed and regulated at both the European and international levels. Risk characterization is described in detail by the European Food Safety Authority (EFSA; http://www.efsa.europa.eu/) in its guidance document on GM plants. The EFSA provides independent scientific advice on the safety of GMOs (Directive 2001/18/EC: deliberate release into the environment of GMOs; Recommendation EC/553/2003: coexistence), and on genetically modified food and feed (Regulation EC/1829/2003). In particular, the EFSA GMO Panel carries out risk assessments to evaluate the safety of a given GMO by reviewing scientific data and information.

At the international level, several documents are available from the US Department of Agriculture (APHIS guidance documents, field database of field trials for more than 40 woody species), the Canadian Food Inspections Agency and other organizations. Relevant documents and research reports have been produced by OECD in the field of the environmental and food/feed safety. A Biosafety Clearing House database has also been created to facilitate international coordination. The recommendations of the Cartagena protocol on biosafety, part of the international Convention on Biological Diversity, can be considered in risk assessments.

A number of genetically modified trees (GMTs) have been developed in the past 25 years (see chapter by Häggman et al. in this book). In many countries, including the USA, France, Belgium, Germany and many others, field trials with transgenic poplar (*Populus* spp.) have been conducted to study the behaviour and growth of GMTs (Walter et al. 2010, and see chapters by Pilate et al. and Strauss et al. in this book). These GMTs carry novel or modified traits such as herbicide and insect resistance, modified lignin and cellulose for pulp and energy production, heavy metal accumulation for phytoremediation, and fast-growing varieties for biomass production (Walter et al. 2010; Robischon 2006; Busov et al. 2005). New transformation technologies have been developed for forest trees (see chapter by

Alburquerque et al. in this book) to help overcome the common recalcitrance to transformation/regeneration by woody plants. China initiated commercial plantations of insect resistant varieties more than a decade ago (Ewald et al. 2006). A genetically modified *Eucalyptus* with yield enhancement characteristics has been recently approved for commercialization in Brazil (Ulian and Abrahão 2015).

Because of the long life cycles of trees and their particular significance in many terrestrial ecosystems, concerns of potential impacts of GMTs have been raised, leading to attempts to ban GMTs (Strauss et al. 2009). These alleged impacts include damage to wild ecosystems by uncontrolled spread of transgenes via vertical or horizontal transfer, as well as pleiotropic and unintended effects of the transgenes on target and non-target species. International groups of scientists, in publications (e.g. Walter et al. 2010) and by direct participation in Convention on Biological Diversity conferences (e.g., Strauss et al. 2009), have examined and rebutted the many reckless and disproportionate claims.

Nonetheless, to further understand and improve the biosafety of GMTs, research on biosafety has been initiated in a number of European countries (e.g. Germany, Finland, France, Spain, etc.) and non-European countries (Häggman et al. 2014; Fladung and Hoenicka 2012; El-Kassaby and Prado 2010). This research included the potential release and fate of GMT pollen, the development of containment strategies, the integration of the transgene and its inheritance, and the relationship between the genomic position of the transgene and genomic as well as expression stability (e.g. Ahuja and Fladung 2014; Fladung et al. 2013; Brunner et al. 2007; Valenzuela et al. 2006; Hoenicka and Fladung 2006a; Strauss et al. 2004; Pilate et al. 2002). Other studies focused on the possibility of pleiotropic, often undesirable alterations in plant metabolism and physiology due to the recombination events in the transgenic host plant genome (for discussion on this point see the chapter by Biricolti et al. in this book), effects on composition and activities in the soil microbial community and the fate of the transgene and proteins in soil (e.g. Pasonen et al. 2005; Kaldorf et al. 2002, and see chapters by Gallardo et al. and Robischon in this book).

In principle, three main technological issues need to be considered in GMT development and regulatory assessment: (i) the establishment of efficient containment strategies to avoid possible vertical and horizontal gene transfer, when these raise significant ecological or social problems, (ii) the analysis of the transgene insertion behaviour, including number and orientation of copies, and the presence of vector backbone sequences to help ensure predictable and reproducible trait expression and (iii) selective inclusion of new "omics"-technologies when they are needed to study cellular effects following the expression of novel gene(s).

Containment strategies have already been proposed, including the establishment of male and/or female sterility aand inhibition of vegetative propagation (Brunner et al. 2007; Strauss et al. 2004; Hoenicka and Fladung 2006b). It has been shown that the integration site of the transgene affects the stability of the expression, which is partly due to methylation and silencing events. In addition, depending on the molecular layout of the integration locus as well as possible alternative splicing of the foreign gene, the production of "variants" of the intended protein is theoretically possible. In this context, the targeted integration into known genomic positions appears to be a valuable approach (Fladung et al. 2013; Li et al. 2009). The new mutagenesis technologies, particularly CRISPR/Cas-systems, now makes gene targeting efficient in plants (Gaj et al. 2013) though it is yet to be used commercially in GMTs. A major activist concern about GM technology is that the expression of novel gene(s) may cause alterations in cell homeostasis, which could be considered in environmental risk assessments. However, all information available regarding transcriptomics, proteomics and metabolomics ("omics") of genetically modified plants indicate that unintended effects are considerably less than for conventional breeding (Ricroch 2013; Gao et al. 2013) and could not be detected to be caused by the transformation methods (Metzdorff et al. 2006).

A large body of knowledge on environmental effects and biosafety issues of (GMTs) has been collected in many countries over the past 25 years. Therefore, it was important to compile this scattered information to build up a European knowledge platform. This was the idea behind the 4-year EU-COST Action FP0905 project—Biosafety of forest transgenic trees: improving the scientific basis for safe tree development and implementation of EU policy directives. A main objective of this ACTION was to evaluate and substantiate the scientific knowledge relevant for GM forest tree biosafety, which is important for future EU policy and regulation of GM forest trees. This is particularly important because the majority of the well-documented knowledge present in public accessible websites and regulations are mostly related to crops. In respect to risk assessment of GM trees, forest trees differ in a number of important characteristics (i.e. complex ecosystem, longevity, etc.) from annual agricultural crops. The goal was to collect the scattered information on transgenic forest trees and make it available for those organizations (as EFSA) and institutions (as state department or Ministries of the Environment, etc.) that have to evaluate and regulate any introduction of transgenic tree to the market.

Yet, we are not aware of network collaborations on GMTs, similar to this COST Action, under consideration in the EU Framework Programme or in other European organizations, such as EUREKA, ESF, etc. A COST Action on this subject is therefore relevant, and innovative, especially in the context of the existing debate in the EU on the cultivation and commercialization of GM-plants, as the 2912th European Council on Environment showed with the adoption of EU Council Conclusions on GMOs.

This ACTION was aimed both at EU economic/societal needs and scientific/technological advance. From an economical perspective, this ACTION allowed information exchanges to improve European knowledge about the importance of forestry and forest trees productions in different countries and constraint factors in this sector. This is highly important to improve R&D efforts on biotech development in trees. From a societal perspective, this ACTION has explored and identified the real and perceived importance/utility of forests and plantations in different Europe societies, with a strong accent on how some types of GMTs could either ameliorate or impair forest sustainability from a citizen's perspective. Thus is very important to understand the kind of policies needed to meet the concerns of society, which are highly diverse among and within Europe countries.

Summarizing the results from experience with GMTs in this book is expected to help policy-makers and regulators to respond to public concerns in a serious and responsible way, particularly concerning socio-economic implications and biosafety issues of transgenic tree plantations.

This book is divided into four sections and summarizes the activities of the four working groups of the COST Action FP0905. The different sections are interconnected and the chapters report the state of knowledge on GMT research, risk assessment, biosafety and social perception. The book aims to present useful information for the scientific community that is interested or engaged in socio-economic implications and biosafety issues of GMTs, but also to provide scientifically based support for decisions of policy-makers and regulators.

Part I of the book addresses the topic of genetic engineering and transformation technologies and how these can be useful to improve important productive characteristics of forest trees for specific purposes. The global experience with GMT field trials is presented, and a database that gathers the current knowledge on field trials and greenhouse experiments with transgenic trees in EU and non-EU countries is provided (Part IV). The database is freely available to the scientific community and Europe organizations at the website of the COST Action FP0905 (http://www.cost-action-fp0905.eu/). The different chapters of this section give a factual overview of the status of GMTs in European and non-European countries in order to provide sound scientific data for risk assessment, as is further addressed in Part II.

Part II focuses on environmental impact assessment and monitoring of GMTs. It includes environmental impacts of the GMTs already developed and guidance on their risk assessment and monitoring. This information will help regulators and politicians to understand the kind of policies needed to meet the concerns of EU citizens in relation to the possible use of known types of GMTs.

Part III is strictly related to environmental and socio-economic indicators that may be included in cost–benefit analyses. It includes data collected in different EU countries, and a survey to explore public attitude towards adoption of transgenic forest trees. The cross-country results of the surveys are expected to provide policy support to the European Commission with regard to public acceptance of transgenic forest trees and the associated conflicts among citizen values (e.g. economic efficiency *vs.* perceived natural integrity). One of the main problems discussed is how to transfer knowledge to citizens in an objective and unbiased, but understandable manner.

Part IV is focused solely on communication about GMTs. Two chapters describe the knowledge needed to communicate effectively in the field of agri-food biotechnology and the experience of this COST Action in this regard for GMTs. The advantages of using thoughtful, audience-appropriate communication tools to explain technical, socio-economic and environmental aspects of GMTs to the general public cannot be over-emphasized.

We thank all the COST Action FP0905 participants and collaborators for their active engagement in this ACTION, providing data for some of its analysis, and taking part in education events. It provides a foundation to guide further research, policy development and regulatory decisions.

References

Ahuja MR, Fladung M (2014) Integration and inheritance of transgenes in crop plants and trees. Tree Genet Genomes 10:779–790

Brunner AM, Li J, DiFazio SP, Shevchenko O, Mohamed R, Montgomery B, Elias A, Van Wormer K, Strauss SH (2007) Genetic containment of forest plantations. Tree Genet Genomes 3:75–100

Busov VB, Brunner AM, Meilan R, Filichkin S, Ganio L, Gandhi S, Strauss SH (2005) Genetic transformation: a powerful tool for dissection of adaptive traits in trees. New Phytol 167:9–18

El-Kassaby YA, Prado JA (2010) Forests and genetically modified trees. Food and Agriculture Organisation of the United Nations, Rome 235 pp

Ewald D, Hu JJ, Yang MS (2006) Transgenic forest trees in China. In: Fladung M, Ewald D (eds) Tree transgenesis: recent developments. Springer, Heidelberg, pp 25–45

Fladung M, Hoenicka H (2012) Fifteen years of forest tree biosafety research in Germany. iForest 5:126–130

Fladung M, Hoenicka H, Ahuja MR (2013) Genomic stability and long-term transgene expression in poplar. Transgenic Res 22:1167–1178

Gaj T, Gersbach CA, Barbas CF (2013) ZFN, TALEN, and CRISPR/Cas-based methods for genome engineering. Trends Biotechnol 31:397–405

Gao L, Cao Y, Xia Z, Jiang G, Liu G, Zhang W, Zhai W (2013) Do transgenesis and marker-assisted backcross breeding produce substantially equivalent plants?-A comparative study of transgenic and backcross rice carrying bacterial blight resistant gene Xa21. BMC Genom 14:738

Häggman H, Sutela S, Walter C, Fladung M (2014) Biosafety considerations in the context of deployment of GE trees. In: Fenning T (ed) Challenges and opportunities for the world's forests in the 21st century, Forestry sciences 81, Springer Science + Business Media Dordrecht, pp 491–524

Hirsch S (2015) Forest biotechnology and sustainability. In: Communication presented at the IUFRO tree biotechnology 2015 conference. Florence, Italy, 8–12 June 2015

Hoenicka H, Fladung M (2006a) Biosafety in *Populus* spp. and other forest trees: from non-native species to taxa derived from traditional breeding and genetic engineering. Trees 20:131–144

Hoenicka H, Fladung M (2006b) Faster evaluation of sterility strategies in transgenic early flowering poplar. Silvae Genetica 55:285–291

Institute of Forest Biotechnology (2011) Responsible use: biotech tree principles. A publication by the Institute of Forest Biotechnology. http://forestbio.org/wp-content/uploads/2013/02/Responsible_Use_Biotech_Tree_Principles_v.P.1.0.B.pdf

Kaldorf M, Fladung M, Muhs HJ, Buscot F (2002) Mycorrhizal colonization of transgenic aspen in a field trial. Planta 214:653–660

Klocko A, Stanton B, van Oosten C, Strauss SH (2013) Green revolution plantations: could short trees be a big thing? ISB news report. May 2013, pp 7–9

Li J, Brunner AM, Meilan R, Strauss SH (2009) Stability of transgenes in trees: expression of two reporter genes in poplar over three field seasons. Tree Physiol 29:299–312

Metzdorff SB, Kok EJ, Knuthsen P, Pedersen J (2006) Evaluation of a non-targeted "omic" approach in the safety assessment of genetically modified plants. Plant Biol 8:662–672

Pasonen HL, Degefu Y, Brumós J, Lohtander K, Pappinen A, Timonen S, Seppänen SK (2005) Transgenic *Betula pendula* expressing sugar beet chitinase IV forms normal ectomycorrhizae with *Paxillus involutus in vitro*. Scand J Forest Res 20:385–392

Pilate G, Guiney E, Holt K, Petit-Conil M, Lapierre C, Leplé JC, Halpin C (2002) Field and pulping performances of transgenic trees with altered lignification. Nat Biotechnol 20:607–612

Ricroch AE (2013) Assessment of GE food safety using '-omics' techniques and long-term animal feeding studies. New Biotechnol 30:349–354

Robischon M (2006) Field trials with transgenic trees. In: Fladung M, Ewald D (eds) Tree transgenesis: recent developments. Springer, Heidelberg, pp 3–23

Strauss SH, Brunner AM, Busov V, Ma C, Meilan R (2004) Ten lessons from 15 years of transgenic *Populus* research. Forestry 77:455–465

Strauss SH, Tan H, Boerjan W, Sedjo RA (2009) Strangled at birth? Forest biotech and the convention on biological diversity. Nat Biotechnol 27:519–527

Ulian E, Abrahão O (2015) Biosafety of a genetically modified *Eucalyptus* with yield enhancement characteristics. In: Communication presented at the IUFRO Tree Biotechnology 2015 conference. Florence, Italy, 8–12 June 2015

Valenzuela S, Balocchi C, Rodríguez J (2006) Transgenic trees and forestry biosafety. Electron J Biotechnol (online), 9

Walter C, Fladung M, Boerjan W (2010) The 20-year environmental safety record of GM trees. Nat Biotechnol 28:656–658

Part I
Biological Characterization of Genetically Modified Trees (GMTs)

Genetic Engineering Contribution to Forest Tree Breeding Efforts

Hely Häggman, Suvi Sutela and Matthias Fladung

Abstract Forests have multiple roles, including traditional timber and fibre production, to sequestering global carbon emissions, preserving water systems and providing ecosystem services and shelter for a variety of plant and animal species. Genetically engineered (GE) forest trees can play a significant role in highly productive plantation culture for commercial products, where they would be beneficial for specific purposes. We review the knowledge of transformation technologies, and compare the role of GE technology in forest tree breeding to traditional tree breeding approaches. The role of genetic engineering in the climate change context is also covered. Finally, we speculate on the importance of tree genomics for future forestry, genetic engineering and forest tree breeding.

1 Introduction

Looking back at the history of mankind, domestication of the most important crop species has played an enormous role. Around 12,000 years ago, ancient people began a plant-breeding programme by transforming a number of wild plant species into domesticated crops by means of selective breeding. These crops also included the most highly productive crops, rice, wheat and maize, on which human survival is still dependent today (Doebley et al. 2006; Meyer and Purugganan 2013). Food grain production reached 1 billion tons in 1960 and it took almost 10,000 years, but to reach the additional 1 billion tons produced in the year 2000, it took only

H. Häggman (✉) · S. Sutela
Genetics and Physiology Department, University of Oulu, P.O. Box 3000,
90014 Oulu, Finland
e-mail: hely.haggman@oulu.fi

S. Sutela
e-mail: suvi.sutela@oulu.fi

M. Fladung
Thünen Institute of Forest Genetics (TI-FG), 22927 Grosshansdorf, Germany
e-mail: matthias.fladung@ti.bund.de

C. Vettori et al. (eds.), *Biosafety of Forest Transgenic Trees*,
Forestry Sciences 82, DOI 10.1007/978-94-017-7531-1_2

40 years. The recent increase in productivity was due to Dr. Norman Borlaug the 'Father of the Green Revolution' and recipient of the Nobel Peace Prize in 1970, and his many partners. In the Green Revolution, the genetically improved crop varieties were combined with the application of improved agronomic practices. The Green Revolution boosted crop yields especially in developing nations by introducing semi-dwarf genotypes of wheat and rice capable of responding to fertilization without lodging. However, population growth, increased standards of living, ongoing soil degradation and increasing costs of chemical fertilizer will make the second Green Revolution a priority for plant biology in the twenty-first century (Khush 2001).

Today, the global cultivation area of genetically engineered (GE) crops is already 175.2 million ha, which is 3 % (i.e. 5 million ha) more than in 2012, in 27 countries by 18 million farmers. Biotech (GE) crops, the fastest adopted crop technology, are considered to deliver substantial socioeconomic and environmental benefits (Clive 2013; Klümper and Qaim 2014). In the future, global growth of biotech crop hectares may be more modest due to the already high rate of adoption in all the principal crops in mature markets in both developing and industrial countries (Clive 2013); however, a much greater diversity of genes and traits is expected to be used.

Concerning forest trees and taking into account their specific biological properties such as the long generation interval/breeding cycles and lengthy juvenile period compared to crop species, it is obvious that breeding programmes are today only ongoing with commercially important species. Generally, forest tree domestication and breeding programmes mostly started at the earliest in the last half century and are, therefore, still in its infancy (reviewed by Campbell et al. 2003). Thus, the application of biotechnology/GE technology to forest trees offers a great potential to hasten the pace of tree improvement for desirable end uses. However, the challenges involved in predicted population growth and climate change including weather extremes (Nellemann et al. 2009) may increase the need for new cultivation areas and/or more effective food production. This means also increased pressure to accelerate forest tree domestication in a sustainable way to fulfil the future needs of wood, biomass and other bio-based products.

In this context, the traditional genetic improvement protocols, due to the long breeding cycles, alone may be too slow to achieve rapid domestication of forest trees. In these cases, both the lessons from the almost 25-year history of genetic engineering of forest trees combined with the experience on conventional breeding programmes as well as the information available and applications depending on the genetic and genomic knowledge, could speed up tree domestication. The cost-effectiveness, as well as the regulatory approval of the technology/end-products by the authorities, and last but not least, public acceptance is vital for the practice, as for the adoption of the technologies. However, even if the economic reasons favour the use of GE trees in plantation/production forests, it is also good to remember that the technology can be utilized also for conservation purposes of the species threatened with extinction (e.g. chestnut in USA or European elms).

In this chapter, the possibilities of GE tree exploitation are discussed in relation to the tree breeding efforts. This is done at the time when the European Council made an important step by reaching political agreement on 12 June 2014 towards allowing Member States to restrict or ban GMO cultivation in their territory and when simultaneously there are commercial cultivations of *Populus* species in China (Ewald et al. 2006; Hu et al. 2014) and ongoing efforts for deregulation of GE forest trees in South and North America. Transgenic trees are on the way from lab research to trait oriented uses in the next 25 years.

2 Tree Breeding from Its Infancy Towards Future Innovations

The breeding process for forest trees differs fundamentally from that of annual crops (Fladung 2008; Flachowsky et al. 2009). The reason simply lies in the special characteristics of trees, as these are predominantly out-crossing and have extensive vegetative phases as well as a high, individual age. Unlike crop plants, which have a high degree of domestication and often no wild relatives in Europe, many "bred" varieties of tree species are comparatively still with "wild plants" that are characterized by a high genetic diversity.

The breeding process of an agricultural crop is pure cross-breeding, i.e. after an initial cross between a wild plant (carrying e.g. a resistance gene) with an elite line (deficient in the resistance gene), repeated back-crosses and field testing ensures the presence of the resistance gene together with the accumulation of the genomic background of the elite line. On the contrary, in trees, the establishment of a second or third filial generation is practically excluded due to the long generation times of most forest tree species. In consequence, the concept of selective breeding has been coined in forest tree breeding. This means the inspection and selection of natural occurring forest trees (termed as elite or plus trees), followed either by simply collection of seeds and using them for the establishment of seed orchards (Langner 1957/58) or by an initial cross between elite or plus trees and evaluation of a hybrid F1 progeny in terms of forestry-relevant features, e.g. straight stem or vigour growth (hybrid selection; von Lochow 1929; Wolf 2003).

The forest tree breeding efforts were initiated between 1920 and the 1940s in many European countries including Finland, Germany, Norway, Portugal and Sweden. Since then, most of the European countries have had breeding programmes for a variety of hard- and softwood species, in which the general aim has been the improvement of traits considered economically or ecologically important (White et al. 2007; Mullin and Lee 2013; Pâques 2013). For instance in Germany, selective breeding of forest trees has a long tradition (Langner 1957/58; Wolf 2003). Examples are the breeding of fast growing hybrid poplar clones, micro-vegetative propagation of wild cherry clones, and selection of phenotypically superior hybrid trees in sessile oak, pine, alder or larch. However, many of the European tree breeding programmes have already been suspended (Mullin and Lee 2013; Pâques 2013). Globally, the present breeding programmes include the following forest tree species: Douglas-fir (*Pseudotsuga menziesii*), *Eucalyptus* species, loblolly pine (*Pinus taeda*), Monterey pine (*Pinus radiata*), Norway spruce (*Picea abies*), *Populus* species, and Scots pine (*Pinus sylvestris*).

As reviewed by Neale and Kremer (2011) research in forest tree genomics lagged behind that of model and agricultural species. The ongoing next-generation sequencing efforts with several forest tree species is proceeding fast but the draft genomes of tree species still represent only 6 of the more than 100 plant families containing tree species, and thus, the possibilities for comparative genomic and genomic selection approaches are still limited (Neale et al. 2013). Before utilization in tree breeding or conservation or restoration purposes the reference genomes should be functionally annotated with omics approaches and preferably mapped with high density. This, together with feasible genotyping platforms would potentially in future enable marker-assisted selection (MAS) and genomic selection in forest tree species (Nieminen et al. 2012; Isik 2014; Zhang et al. 2014).

3 Genetic Transformation Technologies

Genetically modified organisms (GMOs) are officially defined in the EU legislation as organisms in which the genetic material (DNA) has been altered in a way that does not occur naturally by mating and/or natural recombination (Plan and Van den Eede 2010). In the Organisation for Economic Cooperation and Development (OECD), genetic modification is defining an organism whose genetic material has been altered by whatsoever method including, e.g. traditional methods such as selection breeding.

The first scientific peer-reviewed paper on GE forest tree was published by Fillatti et al. (1987). They introduced *AroA* gene coding 5-enol-puryvate 3-phosphate and conferred herbicide resistance via *Agrobacterium tumefaciens*-mediated gene transfer to *Populus* hybrid (*Populus alba* × *grandidentata*). In the early days, the efforts were focused on the development of transformation technologies rather than trait-oriented transformation. Back in the days and actually still today, *Agrobacterium* (*A. tumefaciens* and *A. rhizogenes*) species-mediated gene transfer (Chilton et al. 1977; Schell et al. 1979; Zambryski et al. 1983; Bevan 1984; Wood et al. 2001) is the most common transformation technology. Biolistic transformation (particle/microprojectile bombardment) (Sanford 1988, 1990), in contrast to *Agrobacterium*-mediated transformation, is a physical process based on the delivery of DNA coated gold particles into the plant cells. Moreover, during the years, also other genetic transformation technologies were developed/used, such as electrophoretic transfection (Ahokas 1989; Vik et al. 2001) and microinjection (Miki et al. 1989). In addition, DNA delivery to plant protoplasts was performed by electrophoresis and in some cases also via liposomes (as reviewed by Dan and Ow 2011). Furthermore, there are several so-called combination technologies where particle bombardment or silicon carbide whiskers (Nagatani et al. 1997) were used prior to *Agrobacterium* transformation to wound plant tissue/cells to improve the transformation efficiency. Similarly, Trick and Finer (1997) used sonication before *Agrobacterium*-mediated transformation. Nevertheless, in spite of several possibilities, the most common transformation technologies within forest tree species are biolistic transformation and *Agrobacterium*-mediated transformation. For instance, Trontin et al. (2007) list in their review 30 *Pinus* species, including the important forest and plantation species *P. pinaster* in Europe, *P. radiata* in New Zealand, Australia and South America as well as *P. taeda* and *P. strobus* in North America, which have been targets of extensive research to improve the transformation methods. In the case of 17 pine species, foreign gene transfer and stable expression have been achieved either by co-cultivation with disarmed *A. tumefaciens* strains and/or biolistic transformation (Trontin et al. 2007).

The viral vectors have also been an advantageous alternative to the transgenic systems for the expression of foreign proteins in plants. There are two basic types of viral systems that have been developed for production of immunogenic peptides and proteins in plants, namely (a) epitope presentation where the sites of insertion are chosen so that the peptide is displayed on the surface of the virus particles and (b) polypeptide expression systems (Cañizares et al. 2005 and references therein). Plant viral vectors delivered by *Agrobacterium* are the bases of several manufacturing processes that are currently in use for producing a wide range of proteins for multiple applications, including vaccine antigens, antibodies, protein nanoparticles such as virus-like particles (VLPs), and other protein and protein-RNA scaffolds, but they are also important tools in research. Among the most often used viral backbones are those of the RNA viruses tobacco mosaic virus (TMV), potato virus X (PVX) and cowpea mosaic virus (CPMV), and the DNA geminivirus bean yellow dwarf virus (Gleba et al. 2014).

Virus-induced gene silencing (VIGS) is a transient RNAi-mediated gene silencing approach that facilitates rapid gene function assessment without the requirement of generating stable transgenes (Burch-Smith et al. 2004). Zhu et al. (2010) utilized *Nicotiana benthamiana* a well-proven platform for VIGS when studying plant cell wall formation. Sasaki et al. (2011) presented an effective VIGS system in apple, pear and Japanese pear using *Apple latent spherical virus* (ALSV) vectors. This was the first time when VIGS vectors were reported for *Rosaceae* fruit trees. Inoculation of ALSV vectors carrying a partial sequence of endogenous genes from apple [ribulose-1,5-bisphosphate carboxylase small subunit (*rbcS*), alpha subunit of chloroplast chaperonin (*CPN60a*), elongation factor 1 alpha (*EF-1a*), or actin] to the cotyledons of seeds by a particle bombardment induced highly uniform knock-down phenotypes of each gene from 2 to 3 weeks after inoculation.

In addition to transformation technologies, also a lot of experience has been gained concerning gene constructs used in transformation. Generally, the gene constructs include the gene of interest, promoter sequences (constitutive, inducible or synthetic, Venter 2007) guiding the expression or down-regulation of the target gene and selectable marker gene, which in most of the cases is *nptII* (neomycin phosphotransferase II) providing resistance to kanamycin antibiotics or for instance *bar*, which encodes for phosphinothricin acetyltransferase and which allow the selection of transgenic cells. With regard to the safety use of antibiotics in gene constructs EFSA (European Food Safety Authority) expressed already in 2004 and further confirmed in 2009, there is no rationale for inhibiting or restricting the use of category I antibiotic genes including *nptII* either for field experimentation or for placing on the market. Concerning the traits (genes of interest) in the research areas, multiple traits and species have been used during the 25 years, but looking at the existing field trials the traits could be grouped as recently presented by Häggman et al. (2013, 2014) marker genes, disease and insect resistance, abiotic stress tolerance, growth and wood properties and biosafety-related issues.

The new technologies to come such as zinc finger nucleases (ZFN), transcription activator-like effector nuclease (TALEN), LAGLIDADG homing endonuclease (LHE) (Curtin et al. 2012) or CRISPR-Cas systems where gRNA-directed Cas9 nuclease can induce indel mutations or specific sequence replacement or insertion (Sander and Joung 2014). The CRISPR-Cas systems can edit, regulate or target genomes with very specific outcomes. These new technologies may also cause very small changes in the genomes. Changes up to 20 bp are not considered as genetic modification as these changes cannot be distinguished from conventional mutagenesis or natural mutations (Cao et al. 2011). Therefore, it is at present under discussion whether or not these small changes will be regulated under the European gene technology directives. The result of this discussion will certainly guide the future technology choices in plant including tree breeding.

4 Genetic Engineering Can Accelerate Tree Breeding

The juvenility period of most of the forest trees are long, which also means long breeding cycles (White et al. 2007), usually >15 years for gymnosperms and for angiosperm species 7 or more years. Traditional tree breeding has principally focused on improving economically and ecologically important traits, including early stem growth and stem form, pathogen and disease resistance, adaptation and more recently wood properties (as reviewed by Häggman et al. 2013). These traits also show complex patterns of inheritance. The increase in genetic gain achieved by breeding cycle may also mean that applications for vegetative propagation are needed at the state of field testing. During the years, also MAS and discovery and use of quantitative trait loci (QTL) in breeding (as reviewed by Isik 2014) have been substantial and this has been done to reduce the long breeding cycles. However, MAS has not been successful in forest trees which generally lies in the fact that the QTLs discovered in experimental populations explain only a small percent of phenotypic variation and also the genotype by environment interactions make the MAS application more complicated (Isik 2014). The situation might, however, change together with the available high throughput genotyping platforms providing numerous markers to be utilized in high-density genetic maps.

By means of genetic engineering, the characteristics of forest trees could be improved faster by using "transgenesis" approaches. The tree characteristics of interested, including enhanced abiotic/biotic resistance, increased biomass production or modified/altered wood, could be achieved by altering the expression of specific (endogenous) gene/s or by introducing novel traits to trees. Another possibility could be reducing the vegetative period of trees from a few years down to one year, allowing breeding activities in a similar manner as in annual crop species. The shortening of the juvenile phase can be achieved by transgenesis with the transfer of "early flowering" genes leading to precocious flowering (Flachowsky et al. 2009) as described in detail in Sect. 4.2.

Besides being slow, traditional tree breeding can cause notable changes in the tree genome. In contrast, the mutagenesis based, genetic engineering technologies under development affect very moderately the genome organization—it may be that the changes are only present in a few bases. The new methodologies (ZFN, TALEN, LHE, CRISPR-Cas) hold promise that the genetic engineering of forest trees can be more precise in the near future in comparison to the biolistic and *Agrobacterium*-mediated transformation methods. However, the almost 25-year history of GE forest trees produced with these "old-school" transformation methods has proved that after the initial greenhouse characterization of GE forest trees, the foreign gene/s have stable expression under natural conditions causing no/minor unintended effects to the environment/other organisms (Häggman et al. 2013, 2014). Yet, we are lacking data on the performance of GE forest trees under long-term field trials crossing the border of juvenility and maturity.

We have a long experience on the genetic and phenotypic characteristics of the main forest tree species in conventional breeding programmes. The accumulation of genome sequence data is premise for better genetic maps needed for MAS. However, high-density genetic maps are not yet available and it will take time before they are for all the forest tree species involved in breeding programmes. The tree domestication can be speeded up bymeans of genetic engineering but the realization by genetic engineering requires public acceptance.

4.1 Genetic Engineering Versus Natural Genetic Variation

Natural genetic variation is the driving force of the evolution of all living individuals. Each spontaneously occurred mutation is checked for its value for the fitness and survival of each living individual. This is the principle of Darwin's theory of selection and evolution. Along with the rapid development of new methods for genome sequencing, numerous genomes of different individuals already have been or will be totally sequenced and uncountable genetic variations are being or will be detected. Such genetic differences can be classified in different sections, small ones, i.e. single nucleotide variations (called single nucleotide polymorphisms = SNPs) including insertion/deletions ("Indels") of up to 20 bp, rearrangement of intragenomic sequences or transposition events, or uptake of foreign DNA (Arber 2010). Genetic variations can in consequence be beneficial or disadvantageous for living organisms, depending on the present and future environmental factors. When transmitted to the progeny, according to Darwin's theory, in the long-term end survival of the fittest will value the "best" genetic variation. Genetic variations can also be independent of environmental factors, e.g. when such variations inhibit necessary life processes or are even lethal in extreme cases (Arber 2010).

Molecular mechanisms behind genetic variations are manifold. Single nucleotide variations can occur spontaneously as in textbooks given at 10^{-6} frequency. However, the frequency of this type of mutation can be increased by induced mutagenesis by applying chemical mutagens or UV- or gamma-irradiation. Single nucleotide variations, however, can also be induced by the molecular machinery in each living cell, e.g. DNA replication mechanism. The larger genetic variations, i.e. DNA rearrangements or transposition events, can be triggered through homologous or site-specific recombination and mobile genetic elements (transposons, retrotransposons), respectively. Uptake of DNA is known as natural horizontal gene transfer or along with modern biotechnological methods such as genetic engineering.

Genetic variations up to 20 bp can occur when living cells are underlying all the mechanisms leading to small genetic variations as described above (Cao et al. 2011), thus these changes cannot be distinguished from genetic variations induced by genetic engineering (Arber 2010). Nevertheless, the larger genetic variations such as recombination and transposition leading to natural, spontaneous generation of genetic variants can also be observed along with genetic engineering. No differences in the amounts of nucleotides, but also in the molecular mechanisms are obvious, thus there is no scientific reason to assume that genetic engineering bears particular conjectural evolutionary risks (Arber 2010).

With transgenesis, the goal is to alter the expression of specific gene/s or to introduce novel trait to organisms in which case the introduced gene/genes are possibly absent from the genome of the recipient. If the introduced gene originates from a close, sexually compatible relative of the recipient or from the same species and the gene (or flanking regions) have not been modified, but include the T-DNA borders of *Agrobacterium*, term cisgenesis is applicable. Similarly, intragenesis refers to utilization of gene retrieved from a close relative or the same species of recipient organism but in this case, the gene is modified, that is, the recombinant version of the original gene. In contrast to transgenesis, intragenesis could, in theory, rise in plants naturally. Hence, EFSA (2012) concluded that the hazards associated to cisgenesis are the same as in conventionally bred plants. Thus, forest trees could be modified with constructs not dissimilar to the genetic content of the species; however, at present, the limited gene delivery methods usable for forest tree species hinder the accuracy of the approaches as the integration of constructs to the genome is random in biolistic and *Agrobacterium*-mediated transformation (Ahuja and Fladung 2014).

4.2 The Use of Early Flowering Genes to Accelerate Tree Breeding

The main factor hindering effective breeding in forest trees is the prolonged vegetative phase, variable lasting in some tree species until 40 years (e.g. *Fagus sylvatica* L.) (Meilan 1997). The use of a transgenic early flowering plant can accelerate the breeding of forest trees significantly (Flachowsky et al. 2009). The initial step in the breeding process is based on the cross of a transgenic early flowering plus tree with a tree carrying an interesting gene ("gene-of-interest"), e.g. for disease resistance. In the progeny of this initial cross, the early-flowering gene and the "gene-of-interest" will segregate following the Mendelian rules. By employing molecular marker, both genes can be detected in each plant in the progeny. The plants carrying the two genes will be submitted to back-crosses to increase the genomic background of the plus tree. In the next generation, again molecular marker can be used to select progeny plants carrying both genes. This cycle can be repeated several times. At the end, the early-flowering gene will be out-crossed, resulting in a disease-resistant plus tree that is not transgenic (Flachowsky et al. 2009). In this way, new, permanently improved tree varieties can be created in a reasonable time frame, which are also free of any "foreign" DNA sequences.

Very early, the idea came up to employ early flowering trees, either natural or transgenic ones, to accelerate tree breeding (Häggman et al. 1996; Meilan 1997; Fladung et al. 2000; Fladung 2008; Flachowsky et al. 2009). Most genes involved in regulation of flowering have been discovered in *Arabidopsis* (Theissen et al. 2000; Liu et al. 2009). Some of these genes have been tested for early flowering, mainly in poplar (Rottmann et al. 2000; Hoenicka et al. 2006; Hsu et al. 2006; Böhlenius et al. 2006; Hoenicka et al. 2012a), but also in apple (Wada et al. 2002; Flachowsky et al. 2007; Tränkner et al. 2011) and birch (Elo et al. 2001, 2007).

The first report on successful application of a flowering-inducing gene from *Arabidopsis* was the constitutive expression of the meristem identification gene LEAFY in *Populus* (Weigel and Nilsson 1995). The *Arabidopsis* gene under the control of the 35S promoter but also the poplar native gene reduces flowering time in poplars very effectively (Weigel and Nilsson 1995; Rottman et al. 2000). However, this early system had some disadvantages because of the dwarf phenotype of transgenic plants and the formation of single flowers instead of catkins (Hoenicka et al. 2006; Hoenicka et al. 2012b). Later, a very effective early-flowering system based on the *FLOWERING LOCUS T* (*FT*) could be developed (Zhang et al. 2010), however, also with some disadvantages, e.g. not inducible in tissue culture or in very young greenhouse-grown plants. An improvement of the system could be achieved by combining the *FT* with the *SUPPRESSOR OF CONSTANS* (*SOC*) gene (Fladung and Hoenicka, unpublished results).

The expression of the 35S::*BpMADS4* gene construct caused an acceleration of flowering time in birch, tobacco and apple (Elo et al. 2001; Flachowsky et al. 2007), but not in poplar (Hoenicka et al. 2008). Similarly, genes inducing early flowering in *Arabidopsis* including *FLOWERING PROMOTING FACTOR 1* (*FPF1*) and *FUL* showed no effect on flowering time in poplar (Hoenicka et al. 2008, 2012a). Induced early flowering was achieved using the genes located on the Ri plasmid of *Agrobacterium rhizogenes* in tobacco (*rolD*, Mauro et al. 1996) and poplar (*rolC*, Fladung et al. 2003). Unfortunately, constitutive *rolD* expression in transgenic poplars did not lead to precocious flowering.

5 Towards Trait-Oriented Approach

Today, there is a lot of information from genetic transformation technology and gene constructs both at the laboratory level and at greenhouse, the focus being on basic studies and biosafety issues. The European Commission has spent between 1982 and 2010 over 300 million € to study the biosafety of GMOs (EUR 24473 2010) and altogether 50 research projects were funded between 2001 and 2010 the specific focus of the projects being in biosafety issues. The main conclusion which was drawn from the efforts of more than 130 research projects, covering a period of more than 25 years of research, and involving more than 500 independent research groups, is that biotechnology, and in particular GMOs, are not per se, more risky than conventional plant breeding technologies (EUR 24473 2010).

Moreover, we have considerable experience on approved confined field trials from the United States (500 field trials), China (78), Brazil (68), Canada (45), EU (44) Japan (9) and New Zealand (5) according to Häggman et al. (2013). The following forest tree species have been approved for the trials: the hardwood species being *Betula pendula*, *Castanea dentata*, *Eucalyptus* spp., *Liquidambar styraciflua*, *Populus* spp., *Robinia pseudoacacia*, *Paulownia and Ulmus americana* as well as the following coniferous species *Larix* spp. *P. abies*, *Picea glauca*, *Picea mariana*, *Pinus* spp. *P. radiata, P. sylvestris* (reviewed by Häggman et al. 2013).

Genome resources are developing fast in forest trees but the completed and ongoing genome-sequencing projects in forest trees are today still limited to around 25 species representing mostly members of 4 families, i.e. *Pinaceae*, *Salicacae*, *Myrtaceae and Fagaceae* (Neale et al. 2013) including species that have also been used in genetic transformation approaches. Comparative genomics tools may also be helpful in planning the genetic transformation approach and/or by comparing the GE tree functioning with the non-transgenic counterparts.

All the cumulative information on laboratory and greenhouse experiments as well as field trials and increasing amount of genome resources will support genetic engineering approaches to move from laboratory research to trait-oriented applications during the next 25 years. It will also influence and fasten the tree breeding efforts. However, also the general challenges of the society (population increase, food scarcity, deforestation and forest degradation, climate change) may emphasize the need for high yield fast growing GE production forests as one solution, but also the conservation of the species threatened with extinction such as chestnut in USA (or the elms in Europe) might be good candidates for transgenic approach. Thus, the next steps will be to consider deregulation of forest trees in specific cases and from the legislation point of view—in all cases anyhow the changes in the forest tree genome will be minor compared to the traditional breeding or hybridization.

6 The Global Climate Change and Need for Wood

The sufficient production of wood without disturbing primary forests will be one of the most important issues for the near future. During the last 100 years, an increase in the CO_2 concentration in the atmosphere from 290 to 400 ppm has been recorded in middle Europe. During the same period, Europe's average temperature increased by 0.95 °C and the global average sea site level has risen by 10–20 cm. These are strong indicators for the present ongoing climate change worldwide.

Besides increase in temperature and CO_2 concentration, other predicted consequences are an increase in the number of storms and other extremes as well as changes in precipitation patterns. For Europe, the expected consequences of the predicted climate change are dramatic. The melting of alpine glaciers and permafrost will lead to increased risks of flooding in central and northern Europe. Growth zones are shifting north, loss of important habitats such as wetlands and tundra will be the result (e.g. IPCC 2007).

6.1 Wood Production Should Be Ensured

The long-term goal for world forestry, under the scenario of climate change, is to ensure a sufficient wood production. This applies both for timber production in forests but also in plantations. However, the above described situations of increased number of storms, floods and droughts, heat waves and rising CO_2 concentrations towards 600 and 1000 ppm but also the spread of tropical diseases and parasites are severe threats for world forestry. Questions raised and to be answered are which tree species cope with climate change, which do not, and do native forest tree populations have a sufficient genetic variability to adapt to climate change? Is the existing natural adaptability of forests sufficient, since climate change is expected to take place very quickly? Or, is a forest conversion needed in a way that native tree species would partly be substituted with drought-tolerant exotic tree species? For the use of trees in plantation forestry, the question arises whether forest tree breeding is "fast" enough to meet a fast progressing climate change?

Wood production must be ensured without disturbing primary forests. Of course, stop of climate change by avoiding or reducing causing factors should have eminence. However, simply due to social responsibility and due to the possibility that climate change inexorably progresses, measures have to be taken to ensure survival of the world's forests in future, also in the sense of an adequately secured wood production. Such measures should consider all kinds of approaches to accelerate tree breeding, and could also include the use of GE trees. From today's perspective, no possibility can be excluded per se to assure stable forests for an ecological-driven forestry and sustainable wood production.

A large number of foreign tree species from northern America and Asia have been introduced to European countries (Hoenicka and Fladung 2006). As many of the exotic species are naturally hybridizing with European tree species, they have been successfully included in classical tree breeding programmes (Wolf 2003). Unfortunately, due to the predicted changing climatic conditions, the appearance of new biotic pests is expected in a relatively short time span. To accelerate tree breeding, in the last 25 years, biotechnological techniques already widely in use for annual crops have been adopted to various tree species, e.g. tissue culture, transformation (gene technology) and genome analysis (Taylor 2002; Wullschleger et al. 2002).

6.2 How GE Trees Can Help

The use of GE technology offers a unique opportunity to specifically improve desired traits of plants in a very sophisticated manner. Breeding objectives of tree species as well as acceleration of the tree breeding process to ensure reliable wood production also in future, in particular in regard to the predicted climate change, have been discussed some years ago in combination with domestication of poplar, and multiple targets were formulated (Bradshaw and Strauss 2001; Fladung 2008; Flachowsky et al. 2009). As the first important objective, maintenance and optimization of wood yield has to be considered. This goal could be achieved by increasing the fixation rate of atmospheric CO_2, enhancing the deposition of carbon in the stem or by diversion of photoassimilates into wood formation.

6.3 Efforts to Make Transgenic Trees Safer

6.3.1 Induction of Flower Sterility

To avoid uncontrolled spread of the mutated gene(s) within or outside the gene pool of the improved tree species should be avoided. Here, different molecular approaches have been developed to achieve gene containment in transgenic crops including trees, such as male and/or female sterility, and selection of non-GM plants by using recombination systems. The incorporation of sterility genes into transgenic lines of trees has been proposed early as a strategy to reduce or even avoid gene flow of transgenes into non-transgenic relatives (Brunner et al. 1998). However, evaluation of success concerning sterility in transgenic forest trees is difficult due to the long vegetative periods (Hoenicka et al. 2006). Strategies to shorten reproductive phases in forest trees by using flowering-induction genes have been developed (Weigel and Nilsson 1995; Rottmann et al. 2000; Hoenicka et al. 2006; Zhang et al. 2010), and their usefulness to study fertility/sterility of flowers has been summarized in Hoenicka et al. (2012b).

Publications dealing with sterility induction in forest trees are still very rare (Meilan et al. 2001; Skinner et al. 2003; Hoenicka et al. 2006; Lemmetyinen et al. 2004; Wei et al. 2007). In most cases, sterility constructs successfully used in crops were simply tested in trees. As disadvantage, the heterologous floral-specific promoters often revealed "leaky" expression resulting in activity of cytotoxic gene in non-target, vegetative tissues, leading sometimes to a lower performance of transgenic plants (Meilan et al. 2001; Skinner et al. 2003; Lemmetyinen et al. 2004).

In an early study by Hoenicka et al. (2006), early flowering 35S::*LEAFY* poplar lines were used for evaluation of two different sterility constructs, TA29::*Barnase* and C-GPDHC::*Vst1*, allowing evaluation of the sterility strategies only 2 years after transformation. It could be shown that CGPDHC::*Vst1* promoted sterility in poplar (Hoenicka et al. 2006). A similar sterility system based on the *STILBENE SYNTHASE* (*STS*) gene under control of the gymnosperm promoter PrMALE1 led to reduced pollen number and very low pollen germination in tobacco (Höfig et al. 2001). In poplar, pollen development was disturbed in 68 % of catkins (Hoenicka et al. 2012b).

6.3.2 Selection of Non-GM Trees

To overcome the random integration of foreign gene(s) related to biolistic and *Agrobacterium*-mediated transformation (Kumar and Fladung 2001, 2003), site-specific recombination systems can be used (Kumar and Fladung 2001; Ow 2002). For the first time, Fladung et al. (2005) indicated the usefulness of two recombination systems in the tree species *Populus*. The idea of gene targeting is based on the action of the site-specific recombinase catalyzing an intragenomic mobilization of the randomly inserted gene(s) into a desired, previously characterized recipient locus (Fladung and Becker 2010).

An example of such successful cassette exchange has been provided by Fladung et al. (2010) by using a two-constructs approach. The first construct contained the FLP gene controlled by a heat-inducible promoter as well as a marker gene under control of the 35S promoter, while the second carried a promoterless second marker gene. Poplar plants were double transformed with the two constructs, and following heat induction of FLP, the promoterless second marker gene became active by exactly replacing the FLP gene/first marker gene and getting under control of the 35S promoter (Fladung et al. 2010). In the long-term, however, direct targeting of the new gene to a defined position is sought that should allow one to produce transgenic trees in which the incoming gene can be expressed in a predictable pattern.

The same system can be applied to remove the herbicide or antibiotic marker genes or even to select non-GM plants (Endo et al. 2002). A combination of two site-specific recombination systems in one gene cassette has been successfully applied to remove the foreign genes from the pollen genome by having them still expressed in the plant body (Hoenicka et al. unpublished results). Here, the first recombination system was under the regulation of a heat inducible promoter, while the second recombinase was promoterless. A flower-specific promoter was also present but without a subsequent gene. Activation of the first recombination system led to in vivo assembling of the flower-specific second recombinase system. In consequence, when the transgenic tree starts to flower, the second recombinase becomes active and removes all transgenic sequences from the pollen genome.

Another possibility to reposition transgenes in the genome is based on transposon action (Cotsaftis et al. 2002). The functionality of transposons was first shown by Fladung and Ahuja (1997) and Kumar and Fladung (2003). Later, mobility of transposons across chromosomal barriers was confirmed by Fladung (2011) and Fladung and Polak (2012). A relocated transposon-borne transgene is free from unwanted T-DNA sequences and may be less prone to gene silencing than at the original integration site. Such a method could be useful for quickly generating large populations of T-DNA site-free transgenic plants.

7 Concluding Remarks

The predicted increase in population size together with accelerated climate change including weather extremes, droughts, floods, etc., will also increase the need for food production. This will definitely also affect the forests of the world. Tree breeding efforts have so far been quite slow and, generally, forest trees can still be considered as undomesticated compared to crop plants. For crop plants, all of the most important species are already genetically engineered and include transgenes providing insect or herbicide resistance or both. In mature markets the adoption rate of some GE crops is also high (>80 %).

Concerning forest trees, there is also a need to improve growth, increase resistance to biotic and abiotic stresses, and other traits such as wood properties or chemical composition. The latter might serve the future bioeconomy-oriented society. To reach all these goals, also non-transgenic alternatives and new breeding technologies should be critically considered and compared to GE technology, keeping in mind the importance to hasten the breeding efforts. Today, we already have a lot of information about GE technology and it is an everyday research work in molecular biology laboratories. In the case of forest tree species, we have a long experience on GE forest trees at laboratory, greenhouse and field trial level and therefore a lot of information has been gathered during the last 25 years. Quite recently, also other so-called mutagenesis-based methods have been developed/under development and also information on tree genome sequences is cumulating fast. All these may provide interesting breeding possibilities for the future, which could include utilization of both new breeding approaches (phenotype–genotype correlation, MAS) and transgenesis. Transgenic trees are on the way from lab research to trait-oriented uses in the next 25 years.

References

Ahokas H (1989) Transfection of germinating barley seed electrophoretically by exogenous DNA. Theor Appl Genet 77:469–472

Ahuja MR, Fladung M (2014) Integration and inheritance of transgenes in crop plants and trees. Tree Gen Genom 10:779–790

Arber W (2010) Genetic engineering compared to natural genetic variations. New Biotechnol 27:517–521

Bevan M (1984) Binary *Agrobacterium* vectors for plant transformation. Nucl Acids Res 12:8711–8721

Böhlenius H, Huang T, Charbonnel-Campaa L, Brunner AM, Jansson S, Strauss SH, Nilsson O (2006) *CO/FT* regulatory module controls timing of flowering and seasonal growth cessation in trees. Science 312:1040–1043

Bradshaw HD Jr, Strauss SH (2001) Breeding strategies for the 21th century: domestication of poplar. In: Dickmann DI, Isebrands JG, Eckenwalder JH, Richardson J (eds) Poplar culture in North America. Part B, Chapter 14. NRC Research Press, National Research Council of Canada, Ottawa

Brunner AM, Mohamed R, Meilan R, Sheppard LA, Rottman WH, Strauss S (1998) Genetic engineering of sexual sterility in shade trees. J Arboriculture 24:263–271

Burch-Smith TM, Anderson JC, Martin GB, Dinesh-Kumar SP (2004) Applications and advantages of virus-induced gene silencing for gene function studies in plants. Plant J 39:734–746

Campbell MM, Brunner AM, Jones HM, Strauss SH (2003) Forestry's fertile crescent: the application of biotechnology to forest trees. Plant Biotechnol J 1:141–154

Cañizares CM, Nicholson L, Lomonossoff GP (2005) Use of viral vectors for vaccine production in plants. Immunol Cell Biol 83:263–270

Cao J, Schneeberger K, Ossowski S, Günther T, Bender S, Fitz J, Koenig D, Lanz C, Stegle O, Lippert C et al (2011) Whole-genome sequencing of multiple *Arabidopsis thaliana* populations. Nat Gen 43:956–963

Chilton MD, Drummond MH, Merlo DJ, Sciaky D, Montoya AL, Gordon MP, Nester EW (1977) Stable incorporation of plasmid DNA into higher plant cells: the molecular basis of crown gall tumorigenesis. Cell 11:263–271

Clive J (2013) Global status of commercialized biotech/GM crops: 2013. ISAAA Brief No. 46. ISAAA: Ithaca. ISBN: 978-1-892456-55-9

Cotsaftis O, Sallaud C, Breitler JC, Meynard D, Greco R, Pereira A, Guiderdoni E (2002) Transposon-mediated generation of marker free rice plants containing a Bt endotoxin gene conferring insect resistance. Mol Breed 10:65–180

Curtin SJ, Voytas DF, Stupar RM (2012) Genome engineering of crops by designer nucleases. Plant Genome 5:42–50

Dan Y, Ow DW (eds) (2011) Plant transformation technology revolution in last three decades. Historical technology developments in plant transformation. Bentham Science Publisher 148 pp, eISBN: 978-1-60805-248-6

Doebley JF, Gaut BS, Smith BD (2006) The molecular genetics of crop domestication. Cell 127:1309–1321

EFSA (2004) Opinion of the scientific panel on genetically modified organisms on the use of antibiotic resistance genes as marker genes in genetically modified plants. EFSA J 48:1–18

EFSA (2012) Panel on genetically modified organisms (GMO) scientific opinion addressing the safety assessment of plants developed through cisgenesis and intragenesis. EFSA J 10:1–33

Elo A, Lemmetyinen J, Novak A, Keinonen K, Porali I, Hassinen M, Sopanen T (2007) BpMADS4 has a central role in the inflorescence initiation in silver Birch (*Betula pendula*, Roth). Physiol Plant 131:149–158

Elo A, Lemmetyinen J, Turunen ML, Tikka L, Sopanen T (2001) Three MADS-box genes similar to APETALA1 and FRUITFULL from silver birch (*Betula pendula*). Physiol Plant 112:95–103

Endo S, Sugita K, Sakai M, Hiroshi T, Ebinuma H (2002) Single-steptransformation for generating marker-free transgenic rice using *ipt*-type MAT vector system. Plant J 30:115–122

EUR 24473—A decade of EU-funded GMO research (2001–2010), 2010. 264 pp. Luxembourg: Publications Office of the European Union. ISBN: 978-92-79-16344-9. doi:10.2777/97784

Ewald D, Hu J, Yang M (2006) Transgenic forest trees in China. Tree transgenesis—recent developments. In: Fladung M, Ewald E (eds) Tree transgenesis—recent developments. Springer, Berlin and Heidelberg

Fillatti J, Sellmer J, McCown B, Haissih B, Comai L (1987) *Agrobacterium* mediated transformation and regeneration of *Populus*. Mol Gen Genet 206:192–199

Flachowsky H, Hanke MV, Peil A, Strauss SH, Fladung M (2009) A review on transgenic approaches to accelerate breeding of woody plants. Plant Breed 128:217–226

Flachowsky H, Peil A, Sopanen T, Elo A, Hanke V (2007) Overexpression of *BpMADS4* from silver birch (*Betula pendula*) in apple (*Malus* x *domestica*) induces early flowering. Plant Breed 126:137–145

Fladung M (2008) Frühe Blüte bei Forstgehölzen zur Beschleunigung der Züchtung. Vorträge für Pflanzenzüchtung 74:73–82

Fladung M (2011) Analysis of re-integrated Ac element positions in the genome of *Populus* provides a basis for Ac/Ds-transposon activation tagging in trees. Trees 25:551–557

Fladung M, Ahuja MR (1997) Excision of the maize transposable element Ac in periclinal chimeric leaves of *35S-Ac-rolC* transgenic aspen - *Populus*. Plant Mol Biol 33:1097–1103

Fladung M, Becker D (2010) Targeted integration and removal of transgenes in hybrid aspen (*Populus tremula* L. x *P. Tremuloides* Michx.) using site-specific recombination system. Plant Biol 12:334–340

Fladung M, Nowitzki O, Ziegenhagen B, Kumar S (2003) Vegetative and generative dispersal capacity of field released transgenic aspen trees. Trees Struct Funct 17:412–416

Fladung M, Nowitzki O, Kumar S, Hoenicka H (2005) The site-specific recombination systems Cre-lox and FLP-FRT are functionally active in poplar. Forest Genet 12:121–130

Fladung M, Polak O (2012) *Ac*/Ds-transposon activation tagging in poplar: a powerful tool for gene discovery. BMC Genom 13:61

Fladung M, Tusch A, Markussen T, Ziegenhagen B (2000) Analysis of morphological mutants in *Picea*. In: Espinel HS, Ritter E (eds) Proceedings of the international congress applications of biotechnology to forest genetics (Biofor 99), Vitoria-Gasteiz, 22–25 Sept 1999, Vitoria-Gasteiz, Spain

Fladung M, Schenk TMH, Polak O, Becker D (2010) Elimination of marker genes and targeted integration via FLP/FRT-recombination system from yeast in hybrid aspen (*Populus tremula* L. × *P. tremuloides* Michx.). Tree Genet Genomes 6:205–217

Gleba YY, Tusé D, Giritch A (2014) Plant viral vectors for delivery by *Agrobacterium*. Curr Top Microbiol Immunol 375:155–192. doi:10.1007/82_2013_352

Häggman H, Aronen T, Stomp AM (1996) Early-flowering Scots pines through tissue culture for accelerating tree breeding. Theor Appl Genet 93:840–848

Häggman H, Raybould A, Borem A, Fox T, Handley L, Hertzberg M, Lu M, Macdonald P, Oguchi T, Pasquali G, Pearson L, Peter G, Quemada H, Séguin A, Tattersall K, Ulian E, Walter C, McLean M (2013) Genetically engineered trees for plantation forests: key considerations for environmental risk assessment. Plant Biotechnol J 11:785–798

Häggman H, Sutela S, Walter C, Fladung M (2014) Biosafety considerations in the context of deployment of GE trees. In: Fenning T (ed) Challenges and opportunities for the world's forests in the 21st century in forestry sciences. Springer, Netherlands, pp 491–524

Hoenicka H, Fladung M (2006) Biosafety in *Populus* spp. and other forest trees: from non-native species to taxa derived from traditional breeding and genetic engineering. Trees 20:131–144

Hoenicka H, Nowitzki O, Debener T, Fladung M (2006) Faster evaluation of induced floral sterility in transgenic early flowering poplar. Silvae Genet 55:285–291

Hoenicka H, Nowitzki O, Hanelt D, Fladung M (2008) Heterologous overexpression of the birch *FRUITFULL*-like MADS-box gene *BpMADS4* prevents normal senescence and winter dormancy in *Populus tremula* L. Planta 227:1001–1011

Hoenicka H, Lautner S, Klingberg A, Koch G, El-Sherif F, Lehnhardt D, Zhang B, Burgert I, Odermatt J, Melzer S, Fromm J, Fladung M (2012a) Influence of over-expression of the *FLOWERING PROMOTING FACTOR 1* gene (*FPF1*) from *Arabidopsis* on wood formation in hybrid poplar (*Populus tremula* L. x *P. tremuloides* Michx.). Planta 235:359–373

Hoenicka H, Lehnhardt D, Polak O, Fladung M (2012b) Early flowering and genetic containment studies in transgenic poplar. iForest 5:138–146

Hsu CY, Liu Y, Luthe DS, Yuceer C (2006) Poplar FT2 shortens the juvenile phase and promotes seasonal flowering. Plant Cell 18:1846–1861

Hu J, Wang L, Yan D, Lu MZ (2014) Research and application of transgenic poplar in China. In: Fenning T (ed) Challenges and opportunities for the world's forests in the 21st century in forestry sciences. Springer, Netherlands, pp 567–584

IPCC (2007) Climate change 2007: impacts, adaptation and vulnerability. In: Parry ML, Canziani OF, Palutikof JP, van der Linden PJ, Hanson CE (eds) Contribution of working group II to the fourth assessment report of the intergovernmental panel on climate change. Cambridge University Press, Cambridge, 976 pp

Isik F (2014) Genomic selection in forest tree breeding: the concept and an outlook to the future. New Forests 45:379–401

Khush GS (2001) Green revolution: the way forward. Nat Rev Genet 2:815–882

Klümper W, Qaim M (2014) A meta-analysis of the impacts of genetically modified crops. PLoS ONE 9(11):e111629

Kumar S, Fladung M (2001) Gene stability in transgenic aspen (*Populus*) II. Molecular characterization of variable expression of transgene in wild and hybrid aspen. Planta 213:731–740

Kumar S, Fladung M (2003) Somatic mobility of the maize element *Ac* and its utility for gene tagging in aspen. Plant Mol Biol 51:643–650

Langner W (1957/58) Einführung in die Forstpflanzenzüchtung. 1. –16. Allgem. Forstz, 48 bis 16 = Arbeitsbericht der Bundesforschungsanst. Forst- Holzwirtsch. Nr. 22

Lemmetyinen J, Keinonen K, Sopanen T (2004) Prevention of flowering of a tree, silver birch. Mol Breed 13:243–249

Liu C, Thong Z, Yu H (2009) Coming into bloom: the specification of floral meristems. Development 136:3379–3391

Mauro ML, Trovato M, De Paolis A, Gallelli A, Costantino P, Altamura MM (1996) The plant oncogene rolD stimulates flowering in transgenic tobacco plants. 180:693–700

Meilan R (1997) Floral induction in woody angiosperms. New Forest 14:179–202

Meilan R, Brunner Skinner J, Strauss S (2001) Modification of flowering in transgenic trees. In: Komamine A, Morohoshi N (eds) Molecular breeding of woody plants. Elsevier Science BV, Amsterdam, pp 247–256

Meyer RS, Purugganan MD (2013) Evolution of crop species: genetics of domestication and diversification. Nat Rev Gen 14:840–852

Miki B, Huang B, Bird S, Kemble R, Simmonds D, Keller W (1989) A procedure for the microinjection of the plant cells and protoplasts. J Tissue Cult Meth 12:139–140

Mullin TJ, Lee SJ (2013) Best practice for tree breeding in Europe. Skogforsk, Stockholm

Nagatani N, Honda H, Shimada T, Kobayashi T (1997) DNA delivery into rice cells and transformation using silicon carbide whiskers. Biotechnol Tech 11:471–473

Neale DB, Kremer A (2011) Forest tree genomics: growing resources and applications. Nat Rev 12:111–122

Neale DB, Langley CH, Salzberg SL, Wegrzyn JL (2013) Open access to tree genomes: the path to a better forest. Genome Biol 14:120

Nellemann C, MacDevette M, Manders T, Eickhout B, Svihus B, Prins AG, Kaltenborn BP (eds) (2009) The environmental food crisis—the environment's role in averting future food crises. A UNEP rapid response assessment. United Nations Environment Programme, GRID-Arendal, http://www.grida.no/files/publications/FoodCrisis_lores.pdf

Nieminen K, Robischon M, Immanen J, Helariutta Y (2012) Towards optimizing wood development in bioenergy trees. New Phytol 194:46–53

Ow DW (2002) Recombinase-directed plant transformation for the post-genomic era. Plant Mol Biol 48:183–200
Pâques LE (2013) Forest tree breeding in Europe: current state-of-the-art and perspectives. Springer, Dordrecht
Plan D, Van den Eede G (2010) The EU Legislation on GMOs. JRC57223
Rottmann WH, Meilan R, Sheppard LA, Brunner AM, Skinner JS, Ma C, Cheng S, Jouanin L, Pilate G, Strauss SH (2000) Diverse effects of overexpression of LEAFY and PTLF, a poplar (*Populus*) homolog of LEAFY/FLORICAULA, in transgenic poplar and *Arabidopsis*. Plant J 22:235–245
Sander JD, Joung JK (2014) CRISPR-Cas systems for editing, regulating and targeting genomes. Nat Biotechnol 32:347–355
Sanford JC (1988) The biolistic process. Trends Biotechnol 6:299–302
Sanford JC (1990) Biolistic plant transformation. Physiol Plant 79:206–209
Sasaki S, Yamagishi N, Yoshikawa N (2011) Efficient virus-induced gene silencing in apple, pear and Japanese pear using *Apple latent spherical virus* vectors. Plant Methods 7:15. doi:10.1186/1746-4811-7-15
Schell J, Van Montagu M, De Beuckeleer M, De Block M, Depicker A, De Wilde M, Engler G, Genetello C, Hernalsteens JP, Holsters M, Seurinck J, Silva B, Van Vliet F, Villarroel R (1979) Interactions and DNA transfer between *Agrobacterium tumefaciens*, the Ti-plasmid and the plant host. Proc R Soc Lond B Biol Sci 204:251–266
Skinner JS, Meilan R, Ma C, Strauss S (2003) The *Populus PTD* promoter imparts floral-predominant expression and enables high levels of floral-organ ablation in *Populus*, *Nicotiana* and *Arabidopsis*. Mol Breed 12:119–132
Taylor G (2002) *Populus*: *Arabidopsis* for forestry. do we need a model tree? Ann Bot 90:681–689
Theissen G, Becker A, Di Rosa A, Kanno A, Kim JT, Münster T, Winter KU, Saedler H (2000) A short history of MADS-box genes in plants. Plant Mol Biol 42:115–149
Tränkner C, Lehmann S, Hoenicka H, Hanke MV, Fladung M, Lenhardt D, Dunemann F, Gau A, Schlangen K, Malnoy M, Flachowsky H (2011) Note added in proof to: Over-expression of an FT-homologous gene of apple induces early flowering in annual and perennial plants. Planta 233:217–218
Trick HN, Finer JJ (1997) SAAT: sonication associated *Agrobacterium-mediated* transformation. Transgenic Res 6:329–336
Trontin JF, Walter C, Klimaszewska K, Park Y-S, Lelu-Walter M-A (2007) Recent progress in genetic transformation of four *Pinus* spp. Transgenic Plant J 1:314–329
Venter M (2007) Synthetic promoters: genetic control through cis engineering. Trends Plant Sci 12:118–124
Vik NI, Gjerde H, Bakke K, Hvoslef-Eide AK (2001) Stable transformation of Poinsettia *via* DNA-electrophoresis. Acta Hort 560:101–103
von Lochow P (1929) Etwas über Forstpflanzenzüchtung. Der Züchter 1:73–79
Wada M, Cao QF, Kotoda N, Soejima JI, Masuda T (2002) Apple has two orthologues of FLORICAULA/LEAFY involved in flowering. Plant Mol Biol 49:567–577
Wei H, Meilan R, Brunner AM, Skinner JS, Ma C, Gandhi HT, Strauss SH (2007) Field trial detects incomplete *barstar* attenuation of vegetative cytotoxicity in *Populus* trees containing a poplar *LEAFY* promoter:*barnase* sterility transgene. Mol Breed 19:69–85
Weigel D, Nilsson O (1995) A developmental switch sufficient for flower initiation in diverse plants. Nature 377:95–500
White TL, Adams WT, Neale DB (2007) Forest genetics. CAB International, Cambridge
Wolf H (2003) Forstpflanzenzüchtung in Deutschland. Allg. Forstz 8/2006
Wood DW, Setubal JC, Kaul R, Monks DE et al (2001) The genome of the natural genetic engineer *Agrobacterium tumefaciens* C58. Science 294:2317–2323
Wullschleger SD, Jansson S, Taylor G (2002) Genomics and forest biology: *Populus* emerges as the perennial favorite. Plant Cell 14:2651–2655

Zambryski P, Joos H, Genetello C, Leemans J, Monatagu MV, Schell J (1983) Ti plasmid vector for introduction of DNA into plant cells without alteration of their normal regeneration capacity. EMBO J 2:2143–2150

Zhang H, Harry DE, Ma C, Yuceer C, Hsu CY, Vikram V, Shevchenko O, Etherington E, Strauss SH (2010) Precocious flowering in trees: the *FLOWERING LOCUS T* gene as a research and breeding tool in *Populus*. J Exp Bot 61:2549–2560

Zhang J, Nieminen K, Alonso Serra JS, Helariutta Y (2014) The formation of wood and its control. Curr Opin Plant Biol 17:56–63

Zhu X, Pattahil S, Mazumdar K, Brehm A, Hahn MG, Dinesh-Kumar SP, Joshi CP (2010) Virus-induced gene silencing offers a functional genomics platform for studying plant cell wall formation. Mol Plant 3:818–833

New Transformation Technologies for Trees

Nuria Alburquerque, Fabien Baldacci-Cresp, Marie Baucher, Josep M. Casacuberta, Cécile Collonnier, Mondher El Jaziri, Fabien Nogué and Lorenzo Burgos

Abstract Application of transgenic techniques to trees has emerged as a powerful tool for their genetic improvement. However, for some recalcitrant or transformation time-consuming species transgenic research should be strengthened and further efforts will be necessary to improve regeneration and transformation efficiencies. This review focuses on the recent advances and techniques for genetic manipulation that can be applied to obtain transgenic trees with enhanced biosafety. After selection of transformed plants, marker genes presence becomes useless and undesirable. For generation of marker-free transgenic plants or resolving complex

---

N. Alburquerque · L. Burgos (✉)
Fruit Biotechnology Group, Department of Plant Breeding, CEBAS-CSIC Campus Universitario de Espinardo, 30100 Murcia, Spain
e-mail: burgos@cebas.csic.es

N. Alburquerque
e-mail: nalbur@cebas.csic.es

F. Baldacci-Cresp · M. Baucher · M. El Jaziri
Laboratoire de Biotechnologie Végétale, Université libre de Bruxelles (ULB), 6041 Gosselies, Belgium
e-mail: fbaldacc@ulb.ac.be

M. Baucher
e-mail: mbaucher@ulb.ac.be

M. El Jaziri
e-mail: jaziri@ulb.ac.be

J.M. Casacuberta
Center for Research in Agricultural Genomics, CRAG (CSIC-IRTA-UAB-UB), Campus UAB, Cerdanyola Del Vallès, 08193 Barcelona, Spain
e-mail: Josep.casacuberta@cragenomica.es

C. Collonnier · F. Nogué
INRA AgroParisTech, IJPB, UMR 1318, INRA Centre de Versailles, route de Saint Cyr, 78026 Versailles Cedex, France
e-mail: Cecile.Collonnier@versailles.inra.fr

F. Nogué
e-mail: fabien.nogue@versailles.inra.fr

C. Vettori et al. (eds.), *Biosafety of Forest Transgenic Trees*,
Forestry Sciences 82, DOI 10.1007/978-94-017-7531-1_3

transgene integration structures, several methodologies based on site-specific recombination have been developed. Precise homologous recombination-mediated integration of a DNA sequence of interest at a particular site within a genome is the ultimate tool for genetic engineering. Genome editing allows a much more precise manipulation of tree genomes. Expression of multiple transgenes is often required for engineering metabolic pathways. Recent progress has made powerful techniques available and gene stacking can be achieved in trees by two methodologies: co-transformation and serial transformation. The application of technologies based on small RNAs allowing silencing target genes (RNA interference, artificial microRNAs, and artificial trans-acting siRNAs) is of particular interest to produce directed down-regulation of target genes in tree species. Mixing of genetic materials between species that cannot hybridize by natural means is one of the major criticisms to transgenic crops. Cisgenesis and intragenesis were developed as alternatives to transgenesis.

1 Introduction

Adventitious organogenesis and embryogenesis are the bottleneck for the development of efficient and genotype-independent transformation of trees. They are usually very sensitive processes, strongly affected by small changes. Meristem transformation may eliminate the need for regeneration during the production of transgenic trees, allowing genetic manipulation of different cultivars or even species. However, meristems are organized tissues which make almost impossible to obtain uniformly transformed plants from them. Nevertheless, meristem transformation has been used in seed propagated species to produce chimerical plants where the germ lines may be transformed and, hence, uniformly transformed plants would be produced in the segregating progeny (Gambley et al. 1993; Sautter et al. 1995). This approach is not possible with vegetatively propagated species, particularly woody plants with long generation cycles.

The future of genetic transformation as a tool for the breeding of trees requires the development of genotype-independent procedures based on the transformation of meristematic cells with high regeneration potential and/or the use of regeneration-promoting genes (Petri and Burgos 2005).

Unless new regeneration technologies or transformation systems that do not require regeneration are developed, the technologies described in this review will be difficult to apply to tree crops or their implementation will be too difficult to make the effort worth. However, much research is underway and technologies described here may be part of the breeding tools in tree research.

2 Marker Gene Elimination

2.1 Site-Specific Recombination (SSR)

The use of selectable genes for screening transformed cells is a traditional approach to produce transgenic plants (Hare and Chua 2002). Most commonly used selectable marker genes are those conferring resistance to antibiotics or herbicides; the products of these genes permit growth of transformed plants or cells on selective medium. Expression of other selectable genes in transgenic plants may give a metabolic advantage to transgenic plants, accelerating their growth and differentiation. Reporter genes, such as *gusA* (*uidA*) of β-glucuronidase or *gfp* of green fluorescent protein, permit a visual screening of transformed plant cells. However, after selection of transformed plants, the presence of marker genes in their genomes becomes useless and even highly undesirable. Controversial social apprehensions of genetically modified (GM) plants are their potential impacts on both food safety and environment. Selection of GM plants using antibiotic-resistant marker genes has raised concerns particularly regarding risks associated with horizontal gene transfer from plants to bacteria in soil, or from plant products consumed as food to intestinal microorganisms, which could possibly result in reduced effectiveness of antimicrobial therapy. Therefore, the use of resistant marker genes for selection has been the subject of many scientific papers and opinions. The major trend coming out of these papers is that there is no scientific evidence to date suggesting that either antibiotic-resistant genes currently used in GM plants have been harmful for human or animal health or have significantly contributed to the problem of clinical antibiotic resistance (Breyer et al. 2014). Nevertheless, the removal of these genes from plants is sought after not only to alleviate the above-mentioned concerns but also to address technical problems such as stacking genes of interest in vegetatively propagated species, where segregation of the marker gene is not possible by crossings, or to avoid problems of gene silencing associated with gene homology and also when cisgenic plants are the final objective. Several methodologies have been developed for the production of marker-free transgenic plants.

Site-specific recombinases are highly specialized enzymes that promote DNA rearrangements between specific target sites (Grindley et al. 2006). In nature, these enzymes control and coordinate a number of diverse eukaryotic and prokaryotic functions including the integration and excision of viral genomes, the activation of developmentally relevant genes, and the transposition of mobile genetic elements. Most known site-specific recombinases exhibit distinct and strict sequence specificities, an evolutionary result of the tightly regulated role that DNA re-organization plays in key biological pathways. For generation of marker-free plants, several systems of site-specific recombination, which were previously established in studies with bacteria and fungi, are used.

2.2 Types and Modes of Action

The recombinase super family can be split into two fundamental groups, the tyrosine and serine recombinases. This division is based on the active amino acid (Tyr or Ser) within the catalytic domain of the enzymes in each family. The first and best characterized group has members that include the *Cre-lox* (Dale and Ow 1991), *FLP-FRT* (Kilby et al. 1995), and *R-RS* (Onouchi et al. 1991) systems where *Cre*, *FLP*, and *R* are bidirectional tyrosine recombinases and *lox*, *FRT*, and *RS* are the respective DNA recognition sites. These systems are the most commonly used in plant biotechnology (Puchta 2003). Because of the identical nature of the recognition sites the recombination reaction is fully reversible, although intra-molecular recombination (excision) is highly favored over inter-molecular reactions (integration).

The unidirectional tyrosine subfamily has nonidentical recognition sites typically known as *attB* (attachment site bacteria) and *attP* (attachment site phage) and performs irreversible recombination: HK022 (Gottfried et al. 2005).

The serine recombinase family also has two different members with the division being based on size of the enzyme. The small serine subfamily contains small serine recombinases (b, cd, CinH, and ParA), with their respective DNA recognition sites (six, res, RS2, and MRS) (Wang et al. 2011). While recombination mediated by these small serine recombinases uses identical recognition sites, only intra-molecular excision events are observed. An excision event mediated by the small serine recombinases is considered irreversible. The large serine subfamily is represented by phiC31, TP901-1, R4, and Bxb1 (Wang et al. 2011). These enzymes act on two recognition sites that differ in sequence, typically known as recognition sites *attB* and *attP*, to yield hybrid product sites known as *attL* and *attR*. Excision, inversion, or integration reactions can occur, but because the recognition site sequences of *attB* and *attP* are changed to *attL* and *attR*, the reverse reaction cannot occur.

Site-specific recombination was among the first methods applied to create transgenic plants without retention of a selectable marker transgene (Dale and Ow 1991). Removal of the selectable marker also allows reuse of the same selection regime for subsequent rounds of gene transfer.

The most used strategy consists on cloning the trait gene(s), the marker gene, the recombinase, and its recognition sites into a single construct with the recombinase gene under the control of a constitutive or an inducible promoter. The selection gene is placed in a cassette flanked by directly oriented recognition sites, while the trait gene is inserted outside of the region flanked by recognition sites. When recombinase expression is under a constitutive promoter, the excision of the cassette occurs spontaneously. The use of a MAT (multi-auto-transformation) vector, pEXMGFP1, in the genetic transformation of Carrizo citrange and Pineapple sweet orange was studied by Ballester et al. (2007). This vector contains the *R* recombinase gene under CaMV 35S constitutive promoter and the *ipt* as selectable reporter gene. Results indicated that the IPT phenotype was clearly distinguishable in sweet orange but not in citrange. After site-specific recombination and excision of the DNA fragment between RS sites, marker gene was successfully removed from 65 % of the sweet orange transformants, giving rise to fully normal *ipt*-marker-free transgenic plants. However, spontaneous excision of marker genes gave way to many cases of chimerism and incorrect excision. Transformation with the MAT vector pEXMGFP1 was also attempted in the apricot cultivar 'Helena' (López-Noguera et al. 2009) demonstrating that MAT vectors may be used to eliminate marker genes from transformed apricot plants. In addition, this work evidenced that the regeneration-promoting genes, such as *ipt*, may improve transformation of difficult-to-transform species.

If recombinase gene is driven by an inducible promoter after transformation, the putative transgenic plants are induced to initiate the expression of a recombinase. The heat shock promoter HSP81-1 controlled the expression of the *Cre* recombinase and marker-free transgenic plants were generated in maize (Zhang et al. 2003), tobacco (Liu et al. 2005), and potato (Cuellar et al. 2006). The *flp* recombinase gene driven by the soybean Gmhsp17.5-E heat shock-induced promoter was used to establish a transformation protocol for the apple cultivar 'PinS' based on site-specific elimination of the *nptII* marker gene after selection of transgenic apple plants (Herzog et al. 2012).

Likewise, chemical-regulated promoters have been used in site-specific DNA removal systems. Excision of both the *R* and the *ipt* genes was regulated by placing *R* recombinase under control of the maize GST-II-27 promoter, which is induced by the herbicide antidote 'Safener,' using a MAT vector system in *Populus* (Matsunaga et al. 2002). Ballester et al. (2008) reported the production of marker-free transgenic citrus plants with this MAT system, and perfect excision was predominant in Carrizo citrange and sweet orange transformants.

In the binary plasmid pX6-GFP (Zuo et al. 2001), the expression of the *Cre* recombinase is controlled by the ß-estradiol-inducible XVE transactivator. Upon induction by the estrogen, sequences encoding the selectable marker (*nptII* gene), *Cre*, and XVE sandwiched by two LoxP sites are excised from the construct, leading to activation of the downstream green fluorescent protein (*gfp*) reporter gene, which represents the possible gene of interest in this experimental construct. This system was originally used in *Arabidopsis* (Zuo et al. 2001) and after was also utilized for generating marker-free plants in food crops like rice (Sreekala et al. 2005) and tomato (Zhang et al. 2006). Recently, this chemical-inducible *Cre-LoxP* system has been used for marker gene elimination in apricot cv. 'Helena' (Petri et al. 2012). The DNA site-specific recombination in different transgenic lines was precise and tightly controlled by the inducer ß-estradiol. Expression of the *gfp* gene was only detected when 3 μM ß-estradiol was added to the medium at an average frequency of 11.3 %, based on GFP expression.

Other chemically inducible recombinase system reported by Schaart et al. (2004) was applied to obtain marker-free strawberry plants. After *Agrobacterium*-mediated transformation regeneration took place on a kanamycin selective medium. The recombinase *R* was then chemically activated by addition of dexamethasone to the medium and, after excision, recombinants were selected on a negative selection medium containing 5-fluorocytosine. Apple lines of the cultivar 'Gala' carrying the scab resistance gene HcrVf2 were produced adopting this method (Vanblaere et al. 2011).

If chemicals or heat shock treatments are required for recombinase activation, marker gene deletion may be limited to certain plant species and/or may present complications for the transformation process due to premature recombinase expression (Li et al. 2007).

An alternative approach is the use of a developmentally inducible promoter to activate recombinase expression only within specific organs or tissues during development. Some germline-specific promoters have been used for recombinase expression in soybean (Li et al. 2007) and in model plants (Kopertekh et al. 2010). Chong-Pérez et al. (2013) developed a system to excise the selectable marker and the *cre* recombinase genes from transgenic banana cv. 'Grande Naine' using the embryo-specific REG-2 promoter, which confers *Cre* functionality in the late stage of development of somatic embryos. This system allowed obtaining completely marker-free banana plants with an efficiency of 41.7 %.

Another strategy is to provide transient expression of the recombinase transforming cells directly with a recombinase expression cassette (Srivastava and Ow 2001). The recombinase is transiently expressed in cells and should not stably be integrated into the genome of the host cell. However, 40 % of host cells underwent recombinase-mediated excision when genomic integration of the recombinase gene occurred (Srivastava and Ow 2001). There are two main expression vectors specifically designed for transient C*re* expression: one utilizes *A. tumefaciens* transformation proteins (Kopertekh and Schiemann 2005) and the other is a *cre*-virus vector (Jia et al. 2006). This approach is mainly applicable to vegetatively propagated species.

2.3 *Resolution of Transgene Concatomers*

Recombinases' ability has been used to resolve complex transgene integration structures to single-copy units. The technique was originally demonstrated in wheat by Srivastava et al. (1999). Four multi-copy transgenic lines were resolved to single copy by Cre-mediated excision. At least one recognition site within the transgene structure is required. The complexed DNA or tandem arrays are removed by recombinase-mediated excision due to the presence of the multiple recognition sites in direct orientation. Excision will continue until a single recognition site is left. Fragmented T-DNA could be located outside the outer most recognition site but this possibility should be detectable with appropriate molecular characterization. Single-copy transgene structures are generally the most desirable due to their consistent expression pattern, stability within the genome, heritability, low incidence of silencing, and simplicity of structural characterization (Day et al. 2000). This procedure has the advantage of reducing the total number of transgenic plants required in order to find a properly expressing single-copy line.

2.4 SSR Combined with Conditional Genes

There is a demand for new markers for both research and commercial crop productions. Alternatives are being pursued and a new marker gene, *dao1*, encoding D-amino acid oxidase (DAAO, EC 1.4.3.3) has the potential to provide one interesting option. It can be used for either positive or negative selection, depending on the substrate. Selection is based on differences in the toxicity of different D-amino acids and their metabolites to plants.

The low capacity for D-amino acid metabolism in plants has major consequences for the way plants respond to D-amino acids. On the other hand, some D-amino acids, like D-valine and D-isoleucine, have minor effects on plant growth. DAAO catalyzes the oxidative deamination of a range of D-amino acids. Thus, D-alanine and D-serine are toxic to plants, but are metabolized by DAAO into nontoxic products, whereas D-isoleucine and D-valine have low toxicity, but are metabolized by DAAO into the toxic keto acids 3-methyl-2-oxopentanoate and 3-methyl-2-oxobutanoate, respectively. Hence, both positive and negative selections are possible with the same marker gene. This marker gene has been successfully introduced in *Arabidopsis thaliana*, allowing selection immediately after germination of seeds (Erikson et al. 2004).

Integrating existing techniques such as inducible site-specific recombinase systems (Hare and Chua 2002) with the *dao1* gene may have the advantage that both its insertion and loss can be screened, by positive and negative selections, respectively.

An innovative construct (pX6-DAO1), combining the chemical-inducible Cre-LoxP system and the conditional selectable marker gene *dao1*, was designed to obtain marker-free transgenic tobacco plants (García-Almodovar et al. 2014).

3 Targeted Genome Modification Using Site-Directed Nucleases

Conventional plant breeding, which since the 1950s includes mutagenesis techniques, has been tremendously successful in securing plant production and increasing food quality for human population over history. However, it has some limitations, in particular to create new variability. Indeed, conventional mutagenesis requires extensive screens on large populations, and its random nature limits its use to obtain gene knockouts. In the last 15 years, these techniques have been complemented by transgenesis, which has proven to be a very useful tool for plant breeding by offering the possibility to introduce or modify defined characters in a single step. However, transgenesis also has its limitations, which include the random insertion of the transgene in the genome and the time-consuming and expensive safety analyses that are mandatory prior to commercialization. Insertion of DNA sequences at a defined genomic locus by homologous recombination, called gene targeting (GT), remains hardly feasible in higher plants where the efficiency of homologous recombination is very low (0.01–0.1 % targeted events per transformed plant (Hanin and Paszkowski 2003). One way to enhance HR-dependent gene targeting is to induce genomic DNA double-strand breaks (DSBs) at the target locus (Wehrkamp-Richter et al. 2009).

In the last few years, different site-directed nucleases (SDNs) allowing the introduction DSBs at precise locations in the genome have been developed. They provide a very innovative way of engineering the genome by gene-targeted modifications allowing gene disruption, gene correction, and gene addition (Fig. 1).

These different applications are usually referred to as SDN-1 (targeted random mutation), SDN-2 (targeted gene modification), and SDN-3 (targeted transgene insertion) (Podevin et al. 2013). NHEJ-mediated (Nonhomologous end joining) repair of nuclease-induced DSBs leads to the introduction of small insertions or deletions at the targeted site, resulting in the knockout of gene function via frameshift mutations. By co-delivering a SDN with a donor plasmid bearing locus-specific homology sequences, a donor DNA can be efficiently integrated at the target locus by homologous recombination (HR). This method can be used to insert single or multiple transgenes into endogenous loci or to replace endogenous loci by superior variants (knock-in). The use of several SDNs targeting different loci at the same time can also induce deletions of large chromosomal segments (allowing deletion of tandem repeated genes), large chromosomal inversions (Lee et al. 2012), and translocations (Brunet et al. 2009).

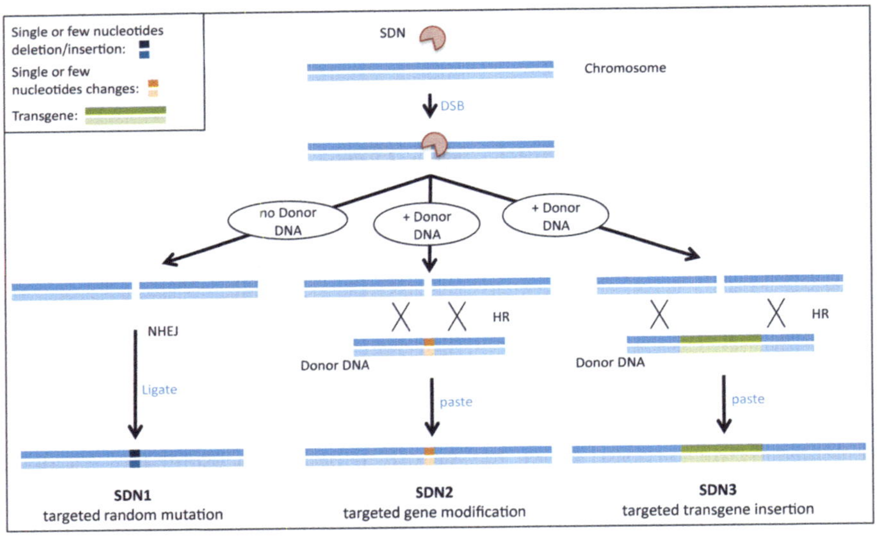

Fig. 1 Different site-directed nuclease (SDN) techniques (SDN-1, 2, and 3). Induced double-strand breaks by the nuclease. The repair can take place via nonhomologous end joining (NHEJ) or homologous recombination (HR). SDN-1 can result in targeted random mutation by NHEJ. In SDN-2, a homologous donor DNA is used to induce targeted gene modification by HR. In SDN-3 a transgene is integrated in the plant genome via HR

3.1 Description of SDNs

Four classes of SDNs have been developed for precise modifications of genomes: meganucleases (Stoddard 2011), zinc finger nucleases (ZFNs) (Carroll 2011), TALE nucleases (Bogdanove and Voytas 2011), and more recently the CRISPR/Cas system (Jinek et al. 2012).

3.1.1 Meganucleases

Meganucleases are natural rare-cutting endodeoxyribonucleases called homing endonucleases. They cleave long recognition sites (from 12 to 40 bp, typically 20–30 bp) which make them highly specific (Stoddard 2011). They are encoded by mobile introns and inteins (sequences coding for protein segments edited after translation) from a variety of organisms such as bacteria, yeast, algae, and some plant organelles. These nucleases are classified into five families of which the LAGLIDADG family is the largest and best characterized one. Within this family, the most widely used proteins are I-SceI (discovered in the mitochondria of baker's yeast Saccharomyces cerevisiae), I-CreI (from the chloroplasts of the green alga Chlamydomonas reinhardtii), and I-DmoI (from the archaebacterium Desulfurococcus mobilis) (Hafez and Hausner 2012).

The DNA-binding domain of the meganucleases, which encompasses the catalytic domain, is composed of two components located on either side of the DNA cleavage catalytic site (Stoddard 2011). For this reason, meganucleases are either homodimers (like I-CreI) or monomers with an internal symmetry (like I-SceI).

Because the natural variability of the meganucleases is not sufficient to cleave at any given site in a genome, different methodologies have been developed for the production of re-engineered endonucleases that are capable of recognizing DNA sequences found at loci of interest (Pâques and Duchateau 2007). However, as the DNA-binding domain overlaps with the catalytic domain, the modification of the DNA-binding specificity without altering the catalytic activity of the protein is challenging.

3.1.2 ZFNs

ZFNs are artificial endonucleases consisting of a synthetic Cys2-His2 DNA-binding zinc finger domain fused to a nuclease which is generally the non-specific catalytic domain of the type IIS restriction enzyme FokI from Flavobacterium okeanokoites (Carroll 2011). Zinc fingers are protein motifs stabilized by a Zn molecule. The C2–H2 zinc finger domain is among the most common types of DNA-binding motifs found in eukaryotes and they are present in many transcription factors. Each zinc finger recognizes specifically 3 bp. By combining several zinc fingers (3–6) with known recognition sites, one can create binding domains specific of a given sequence in the genome (modular assembly) (Maeder et al. 2008).

The FokI enzyme must dimerize in order to cleave DNA; therefore, a pair of ZFNs is required to target non-palindromic DNA sites. The two ZFNs hybridize on both sides of the target on opposite DNA strands. Their respective FokI monomers dimerize and cleave the DNA between the two binding domains in a 5–6 bp spacer.

FokI is the most widely used nuclease but other catalytic domains are available for genome editing. Some of them, like the ones derived from the GIY-YIG homing endonucleases, are monomeric. They require the design of only one ZFN to target a given sequence but their use still remains anecdotic (Kleinstiver et al. 2012).

Several engineering approaches have been developed to create custom ZFNs by modular assembly of characterized ZF modules (Maeder et al. 2008; Urnov et al. 2010; Tovkach et al. 2009). ZFNs are delicate to design since their affinity to the target does not depend only on the association of fingers matching triplets of nucleotides. Incompatibilities at the finger–finger inter-face, called context-dependent effects, often modify the DNA-binding specificity (Ramirez et al. 2008). Significant progress has been achieved, thanks to new methodologies, like the Context-dependent Assembly (CoDA), taking into account context-dependent effects between adjacent fingers (Sander et al. 2011).

3.1.3 TALENs

TALENs are chimeric nucleases whose DNA-binding domain is derived from transcription activator-like effectors (TALE), a family of proteins used in the infection process of plant pathogens from the Xanthomonas genus. The TAL effectors are transcription factors injected by the bacteria into the host plant cells via a type III secretion system. They activate the expression of specific plant genes involved in pathogenesis. About 100 TAL effectors have been identified so far, the first one being the avirulence factor AvrBs3 (Bogdanove et al. 2010; Boch and Bonas 2010).

TALEs are composed of a series of tandem repeats (12–27) (17.5 for ArBs3). Each repeat contains 33–35 amino acids (+1 truncated repeat of 20 amino acids) all identical except for two, in positions 12 and 13, called the repeat variable diresidue (RVD). Each repeat is specific of one DNA base pair, the specificity resulting from the two RVDs. The TALE binds specifically to a given DNA region whose sequence is determined by the RVDs of the successive repeats. Although some RVDs can be specific of more than one base pair, a simple code of recognition with RVD modules having sufficient affinity for each of the four base pairs could be identified. Engineering of specific DNA-binding domains is done by combining repeats containing RVDs corresponding to the target sequence (Moscou and Bogdanove 2009; Boch et al. 2009). TALE repeats seem to be less subjected to context effects than zinc fingers (Christian et al. 2010). This, and the simplicity of the recognition code, makes the customization of specific DNA-binding domains by modular assembly simpler for TALEs than for ZFNs.

TALENs consist of an engineered TALE fused to the catalytic domain of FokI. As for ZFNs, dimerization of the FokI domain is required and a pair of TALENs has to be designed to cleave the desired target sequence. Each half target site is 14–20 bp long, with a spacer of 12–19 bp. To help with the design of TALENs, different softwares are accessible online like the TAL Effector Nucleotide Targeter 2.0 tool (https://tale-nt.cac.cornell.edu/). Different protocols have been developed for the assembly of TALE modules, such as the GoldenGate® (Cermak et al. 2011), the FLASH™ (Reyon et al. 2012), and other variant technologies of modular cloning (Li et al. 2011).

In addition to the creation of DSBs, TALEs can also be used to regulate the expression of genes by being fused to transcriptional activator or repressor factors instead of a nuclease domain (Zhang et al. 2013b) or for epigenome editing by being fused to epigenetic modifiers (De Lange et al. 2014).

3.1.4 CRISPR/Cas System

The CRISPR/Cas system (clustered regularly interspaced short palindromic repeats / CRISPR-associated protein) is a type II prokaryotic adaptive immune system that a number of bacteria use to defend themselves from invading viral and plasmid DNAs (Sorek et al. 2013). In the cell, the invasive DNA is cut into 20 bp fragments that are integrated as spacers between the short palindromic repeats of the CRISPR locus of the bacterial chromosome. If the same invasion occurs again, the CRISPR region is transcribed and cut into small interference CRISPR RNAs (crRNAs) that are processed and fused to trans-activating RNAs (tracrRNAs). Each crRNA/tracrRNA complex associates to a Cas protein that is guided to the foreign DNA homologous to the 20 bp crRNA. The Cas protein contains two independent nuclease domains that can cleave each strand of DNA: the HNH nuclease domain cleaves the complementary DNA strand, whereas the RuvC-like domain cleaves the noncomplementary strand. As a result, a blunt cut is introduced in the target DNA just upstream of a protospacer adjacent motif (PAM) composed of a few nucleotides. This system was adapted for genome editing by associating to the Cas a chimerical single-guide RNA (sgRNA) containing a 20 bp sequence homologous to the target, fused to a synthetic tracRNA (Jinek et al. 2012). The most widely used Cas protein is Cas9 derived from Streptococcus pyogenes. Its PAM sequence is NGG, which enables gene targeting on any gene. Cas9 was codon-optimized for use on eukaryotic organisms (Mali et al. 2013).

The Cas9 gene and the sgRNA are transferred together into the host cell for a functional RNA-guided gene disruption. Simultaneous transformation with one Cas9 gene and multiple sgRNA genes is possible to target multiple genes (Wang et al. 2013) and even to set up loss-of-function genetic screens (Wang et al. 2014). The system can also be used to modify the expression (repression or activation) of a target gene using a mutated Cas9 and dCAS9, fused to effector domains with distinct regulatory functions (Gilbert et al. 2013).

3.1.5 Nickase Versions of the SDNs

In higher plants and animals, SDN-induced DSBs are repaired via NHEJ mechanisms more than by HR. To favor the HR pathway and help gene targeting, meganucleases (Katz et al. 2014; Davis and Maizels 2011; McConnell Smith et al. 2009), ZFNs (Sanders et al. 2009; Ramirez et al. 2012), and the CRISPR/Cas9 systems (Cong et al. 2013) can be turned into nickases that make only single-strand breaks (SSBs), by inactivating one of the nuclease catalytic domains. However, the drawback to the use of nickases is that their absolute frequency of cleavage is significantly lower than the SDNs making DSBs.

3.1.6 Off-Target Activity of the SDNs

Sequence specificity of SDNs is not absolute and cleavage can occur at sites similar to but different from the target site (Boissel et al. 2014; Cong et al. 2013; Fu et al. 2013; Hsu et al. 2013; Mussolino and Cathomen 2011; Mussolino et al. 2011; Pattanayak et al. 2011). Off-target activity can vary from one SDN type to another and can also vary from one target site to another (Pauwels et al. 2014). Whereas for applications to model organism specificity of the targetable nucleases is not a big issue, it is paramount for applications to human gene therapy or to modified plants for food and feed uses, although in the latter, backcrossing and selection of the desired modifications should help eliminate the undesired off-target mutations. For this reason different approaches have been followed in order to increase SDNs' specificity.

DNA cleavage specificity of ZFNs has been increased by modifying the cleavage domain to require the formation of Fok1 heterodimers (Doyon et al. 2011); these modified Fok1 can be used also as the cleavage domain of TALENs. More specifically, for TALENs, broad specificity profiling resulted in engineered nucleases with decreased off-target cleavage (Guilinger et al. 2014a) and tools for the genome-wide prediction of TALEN off targets have been developed facilitating the choice of TALENs in order to minimize the risk of off-target mutagenesis (Grau et al. 2013). In the CRISPR-CAS system, truncated gRNAs, with shorter regions of target complementarity <20 nucleotides in length, can decrease undesired mutagenesis at some off-target sites by 5000-fold or more without sacrificing on-target genome-editing efficiencies (Fu et al. 2014). Recently, it has been shown that the association of two modified CAS9 monomers (fusion of catalytically inactive Cas9 and FokI nuclease) permits a 140-fold increase in specificity compared to a wild-type CAS9 (Tsai et al. 2014; Guilinger et al. 2014b). The field of SDNs improvement is very active and new generations of SDNs with higher specificity can be expected in the next years.

3.2 *Use of SDNs in Plants*

The ability of SDNs to produce DSBs and generate targeted mutagenesis and gene targeting in plants has been tested either through transient expression assays such as protoplasts transformation, calli bombardment, and agroinfiltration, or through the production of stable transformants by Agrobacterium-mediated transformation. Error-prone NHEJ-mediated repair of the SDNs-induced DSBs enabled the production of plants with targeted mutations. HR-mediated repair of the SDNs-induced DSBs triggered either sequences recombination allowing reconstitution of genes at the target locus (single-strand annealing assays), or integration of cassettes presenting homology with the target.

When stably transformed, mutated plants can either be heterozygous/monoallelic (one mutated allele at the target locus), biallelic (two different mutated alleles at the target locus), homozygous (two identical mutated alleles at the target locus), or chimeric. When regenerated from tissue culture, stably transformed plants will be homogenous for the targeted mutation(s) only if they are regenerated from a unique meristematic cell and if the targeted mutation(s) were completed in this first cell. By floral dip, depending on the level and timing of expression, the editing can occur during the fertilization stage but can also continue after the division of the fertilized egg generating multiple mutated alleles. Stably transformed T1 plants are then chimeric and only some of the T2 plants bearing germline-transmitted mutations will be homogenous. In this case, screening of heritable mutations is done in T2 or later generations. The use of germline-specific promoters to drive nucleases expression could help reducing the number of somatic mutations in the future (Feng et al. 2014).

3.2.1 NHEJ-Mediated Repair of SDNs-Induced DSBs in Plants

Natural and engineered versions of meganucleases have successfully been used in plants to induce targeted mutations via error-prone NHEJ-mediated repair at synthetic and endogenous sites. Tobacco plants containing a marker gene carrying a synthetic I-SceI site between the promoter and the coding region were transformed via Agrobacterium infiltration with I-SceI. DSB at the target site triggered loss of function of the marker (Salomon and Puchta 1998). Other transgenic tobacco plants bearing synthetic I-SceI or I-CeuI sites were transformed with I-SceI or I-CeuI and a donor T-DNA with no homology to the target but bearing a selection marker containing an I-SceI or a I-CeuI site. Indel mutations and NHEJ-dependant integration of the intact or digested donor T-DNA were detected at the DSB site (Chilton and Que 2003; Tzfira et al. 2003). In maize, I-Sce1 was used to target a synthetic I-SceI site in embryos (Yang et al. 2009). Short deletions were observed in about 1 % of analyzed F1 plants and were transmitted to the progeny. Re-engineered I-CreI nucleases were used to target the endogenous liguleless1 (LG1) gene promoter (Gao et al. 2010) and the endogenous MS26 fertility gene

(Djukanovic et al. 2013) by Agrobacterium-mediated transformation of maize immature embryos. Small deletions (2–220 bp) and/or insertions were detected at the target site in 3 and 5.8 % of the T0 plants, respectively.

ZFNs have successfully been applied to trigger NHEJ-induced targeted mutagenesis in Arabidopsis (Qi et al. 2013; Lloyd et al. 2005; de Pater et al. 2009; Even-Faitelson et al. 2011; Osakabe et al. 2010; Zhang et al. 2010), tobacco (Maeder et al. 2008; Tovkach et al. 2009; Townsend et al. 2009; Marton et al. 2010), petunia (Marton et al. 2010), and soybean (Curtin et al. 2011). ZFNs were delivered for stable integration or transient expression either by floral dipping (Qi et al. 2013; Lloyd et al. 2005; de Pater et al. 2009; Even-Faitelson et al. 2011; Osakabe et al. 2010; Zhang et al. 2010), crossing between plants (Petolino et al. 2010), protoplasts electroporation (Maeder et al. 2008; Townsend et al. 2009), hairy-root transformation (Curtin et al. 2011), or infection with a tobacco rattle RNA virus (TRV) (Marton et al. 2010). The promoters controlling the ZFN genes were either constitutive, inducible (estrogen, heat shock) (Zhang et al. 2010; Osakabe et al. 2010), or tissue-specific (like the egg apparatus-specific enhancer (EASE)-driving protein expression in the egg cell to ensure mutations in the germline) (Even-Faitelson et al. 2011). The efficiency of ZFNs was demonstrated by inducing mutations resulting in frame shifts or deletions restoring or inactivating protein expression in reporter genes such as the GUS gene (Tovkach et al. 2009; Petolino et al. 2010; Even-Faitelson et al. 2011). ZFNs were also used to induce mutations in endogenous genes like the ABI4 gene in Arabidopsis which led to the production of ABA and glucose insensitive mutants (Osakabe et al. 2010) and the acetolactate synthase genes (ALS, SuRA, and SuRB) in tobacco resulting in mutants resistant to the imidazolinone and sulphonylurea herbicides (Maeder et al. 2008; Townsend et al. 2009). ZFN-based mutagenesis also provided an efficient way of making mutations in duplicate genes in Arabidopsis (Qi et al. 2013) and soybean (Curtin et al. 2011) inducing independent mutations in tandemly arrayed genes but also deletions of chromosomal fragments ranging from 4.5 to 55.3 kb (efficiency ~1 %) reaching even 9 Mb (efficiency ~0.1 %). Globally, ZFNs-induced mutations were deletions of 1–200 bp and insertions of 1–14 bp that were, when tested, efficiently transmitted to the progeny. Considering all species, the frequency of mutations at the target sites ranged from 1 to 35 % (with ~20 % biallelic mutations).

NHEJ-mediated targeted mutagenesis induced by TALENs has been described in both dicotyledon and monocotyledon plant species including Arabidopsis (Christian et al. 2013; Cermak et al. 2011), tobacco (Zhang et al. 2013a; Mahfouz et al. 2011), rice (Chen et al. 2014; Shan et al. 2013a; Li et al. 2012), maize (Liang et al. 2014), Brachypodium (Shan et al. 2013a), and barley (Wendt et al. 2013). Transient or stable expressions of the TALENs were obtained via Agrobacterium leaf infiltration (Mahfouz et al. 2011), Agrobacterium transformation of embryonic cells (Chen et al. 2014; Shan et al. 2013a; Li et al. 2012), Agrobacterium transformation of immature embryos (Wendt et al. 2013), protoplasts PEG transformation (Liang et al. 2014; Zhang et al. 2013a; Cermak et al. 2011), and floral dip (Christian et al. 2013). A pair of TALENs was designed to induce NHEJ-mediated mutagenesis in an exogenous GUS reporter gene activated by the frameshift resulting from the targeted mutation in tobacco (Mahfouz et al. 2011). Otherwise, TALENs were mainly used to target endogenous genes such as the ADH1, TT4, MAPKKK1, DSK2B, and NATA2 genes in Arabidopsis (Christian et al. 2013; Cermak et al. 2011); the BRI1, DEP1, BADH2, CKX, SD1, and SWEET14 genes in rice (Chen et al. 2014; Shan et al. 2013a; Li et al. 2012); the PDS, IPK1A, IPK, and MRP4 genes in maize (Liang et al. 2014); the ABA1, CKX2, SMC6, SPL, SBP, COI1, RHT, and HTA1 genes in Brachypodium (Shan et al. 2013a); and the PAPhy_a gene in barley (Wendt et al. 2013). Only the targeting of the sucrose-efflux transporter gene OsSWEET14 had an agronomic application: indeed, mutating the effector-binding element (EBE) located in the promoter region of this gene, which is the natural binding site of the TAL effector AvrXa7 of Xanthomonas, led to an increased resistance to bacterial blight in rice (Li et al. 2012). Additionally, TALENs enabled the targeting of tandemly duplicated genes, triggering either mutations at the target sites, like for the duplicate acetolactate synthase (ALS) genes (SurA and SurB) in tobacco (Zhang et al. 2013a), or large deletions, like in the GLL22 gene cluster where TALENs targeting the same sequence in two tandemly duplicated genes induced a 4.4 kb deletion in Arabidopsis (Christian et al. 2013). Globally, TALENs-induced mutations were mainly deletions of 1–55 bp, with a few insertions ranging from 3 to 9 bp and some nucleotide substitutions that were, when tested, transmitted to the progeny. Mosaicism was observed in T0 plants (Wendt et al. 2013). All species considered, the frequency of mutations at the target sites ranged from 2 to 100 % in individual transgenic plant lines expressing the TALENs (with 3.4–22.2 % biallelic mutations).

Up to now, the CRISPR/Cas9 system has been applied for the induction of NHEJ-mediated targeted mutagenesis in Arabidopsis, tobacco, sorghum, rice, wheat, and maize (Feng et al. 2013, 2014; Liang et al. 2014; Xie and Yang 2013; Shan et al. 2013b; Nekrasov et al. 2013; Miao et al. 2013; Li et al. 2013; Jiang et al. 2013; Belhaj et al. 2013). Mutations were generated in endogenous loci including the BRI1, JAZ1, GAI, CHLI, TT4, PDS3, and FLS2 genes in Arabidopsis (Li et al. 2013; Feng et al. 2013); the promoter of the bacterial blight susceptibility genes SWEET14 and SWEET11, as well as the PDS, BADH2, Os02g23823, MPK2, CAO1, LAZY1, ROC5, SPP, and YSA genes in rice (Feng et al. 2013; Miao et al. 2013; Shan et al. 2013b; Jiang et al. 2013); the PDS, IPK1A, IPK, and MRP4 genes in maize (Liang et al. 2014); the PDS in N. benthamiana (Li et al. 2013; Nekrasov et al. 2013); and the TaMLO gene in wheat (Shan et al. 2013b). Deletions of 1–20 bp (with some rare cases of deletions >100 bp) and/or insertions of 1–3 bp with rare single-nucleotide substitutions were detected in 2–91 % of the transformants, with as much as 13–50 % biallelic mutations. One sgRNA targeting a sequence common to 2 genes from the RACK1 family in Arabidopsis triggered mutations in both target genes with a similar mutagenesis frequency (2.5–2.7 %) (Li et al. 2013).

In the NbPDS and AtPDS3 genes, larger deletions (up to 76 and 48 bp, respectively) were produced using two sgRNAs targeting two sites (distant of 50 and 24 bp, respectively) in the same target gene (Li et al. 2013; Belhaj et al. 2013). No mutation was observed in Arabidopsis when using the ratio gRNA:Cas9 = 1:19 (instead of the ratio 1:1) (Li et al. 2013). Many experiments generated whole plants carrying mutations at the targeted loci, but very few studied their transmission to the progeny. In one report on Arabidopsis, the proportion of plants bearing mutations was 71.2 % at T1, 58.3 % at T2, and 79.4 % at T3 generations, and a Mendelian segregation of monoallelic and biallelic mutations was observed in the T2 and T3 plants (Feng et al. 2014). The efficiency of CRISPR/Cas9 system and TALENs was compared by transient expression on two endogenous genes in rice: the frequency of mutations for the first gene (Os02g23823) was lower with CRISPR/Cas9 (26.0 %) than with TALENs (36.5 %), whereas those of the second gene (OsBADH2) was higher with Crispr (26.5 %) than with TALENs (8.0 %) (Shan et al. 2013a, b).

3.2.2 HR-Mediated Repair of Nuclease-Induced DSB in Plants

HR-mediated repair of meganucleases-induced DSBs has been initially demonstrated in plants through reporter gene-based assays. Nicotiana plumbaginifolia protoplasts co-transfected with a plasmid carrying a synthetic I-Scel gene and GUS recombination substrates carrying an I-Scel-site showed extrachromosomal recombination (Puchta et al. 1993, 1996). Nicotiana tabacum transformants bearing a codA gene flanked by two overlapping halves of a GUS gene separated by two I-SceI sites were transformed with I-SceI by Agrobacterium infiltration of seedlings. In the resulting plantlets, the elimination of the codA gene and the restoration of a functional GUS gene by HR (and not by NHEJ that would have led to a non-functional GUS gene) showed that if genomic DSBs are induced in close proximity to homologous repeats, they can be repaired in up to one-third of cases by HR in tobacco (Siebert and Puchta 2002). HR-mediated repair of meganucleases-induced DSBs can also enable gene targeting of cassettes bearing homology to the target. I-SceI was for example used to introduce a 35S promoter in a Bar cassette allowing the expression of this reporter gene in maize. DNA integration occurred very accurately at the DSB site and in 30–40 % of the recovered events, no random insertions were observed whether the DNA was delivered by Agrobacterium or particle bombardment (D'Halluin et al. 2008). Meganucleases-induced gene targeting was also successfully performed for stacking in cotton using a I-CreI re-engineered for specific cleavage of an endogenous target sequence adjacent to a transgenic insect control locus. By bombardment of embryogenic calli, two herbicide tolerance genes (hppd, epsps) were inserted next to a pre-existing Bt locus. Targeted insertion events were recovered at a frequency of about 2 % of the independently transformed embryogenic callus lines. All trait genes were inherited as a single genetic unit (D'Halluin and Ruiter 2013).

With ZFNs, HR-mediated repair of DSBs was demonstrated with and without donor DNA bearing or not homology to synthetic or endogenous targets. Without donor DNA, intra-chromosomal reconstitution of a disabled reporter gene was obtained in tobacco with ZFNs inducing a DSB between two redundant fragments of a GFP gene (Cai et al. 2009). A donor DNA without homology to the target was used to trigger the ZFN-induced insertion of an herbicide tolerance gene at the endogenous locus IPK1 in maize (Shukla et al. 2009). A donor DNA with homology to the target was used to enable ZFNs-induced targeted insertion at artificial loci in Arabidopsis (de Pater et al. 2009), tobacco (Wright et al. 2005; Cai et al. 2009), and maize (Ainley et al. 2013), and at endogenous loci in Arabidopsis (PPO and ADH1) (de Pater et al. 2009; Qi et al. 2013) and in tobacco (endo-chitinase) (Cai et al. 2009). In tobacco, when using a donor DNA to restore a defective GUS reporter gene, HR occurred in more than 10 % of the transformed protoplasts and approximately 20 % of the GUS reporter genes were repaired only by HR, whereas 80 % had associated DNA insertions or deletions consistent with repair by both HR and NHEJ (Wright et al. 2005). In Arabidopsis, ZFNs allowed gene targeting with a frequency of 1–3.1 ‰ (de Pater et al. 2009; Qi et al. 2013) in wild-type plants, and 3–16 times more in mutant backgrounds impacted in NHEJ-dependant DNA repair pathways (Qi et al. 2013). ZFNs were successfully used for gene stacking in maize (Ainley et al. 2013): immature embryos already containing the herbicide resistance PAT transgene flanked with a TLP (Trait Landing Pad) were bombarded with a donor DNA containing a second herbicide resistance gene, aad1, flanked by sequences homologous to the TLP. Five percents of the embryo-derived transgenic events integrated the aad1 transgene precisely at the TLP and both herbicide resistance traits cosegregated in subsequent generations.

HR-mediated repair of DSBs induced by TALENs was reported in tobacco and rice through single-strand annealing assays in which TALENs' cleavage enables the reconstitution of reporter genes such as YFP and LacZ (Zhang et al. 2013a; Li et al. 2014a). HR-mediated gene targeting using TALENs and a donor DNA has only been reported in tobacco. Calli derived from protoplasts transformed with TALENs targeting the ALS genes and a 322 bp donor molecule differing by 6 bp from the ALS coding sequences showed targeted gene replacement with a frequency of 4 % (Zhang et al. 2013a).

HR-mediated repair of DSBs induced by the CRISPR/Cas9 system was demonstrated through single-strand annealing assays in which cleavage enables the reconstitution of artificial reporter genes, such as YFP in Arabidopsis (Feng et al. 2013), GFP in Arabidopsis and tobacco (Jiang et al. 2013), DsRED2 in sorghum (Jiang et al. 2013), and GUS in rice and Arabidopsis (Miao et al. 2013; Feng et al. 2014). HR-mediated gene targeting of a donor DNA using the CRISPR/Cas9 system has only been reported by transient expression in N. benthamiana protoplasts. A DNA donor containing a 533 bp left homology arm and a 114 bp right homology arm to the NbPDS locus was incorporated in the target locus with a frequency of 10.7 %, and simultaneously, NHEJ-mediated targeted mutagenesis at the NbPDS locus was detected with a frequency of 14.2 % (Li et al. 2013).

In addition to the activities listed above, SDNs are presently being used by plant breeding companies to accelerate the introduction of new traits in crops such as maize, oilseed rape, or tomato (Lusser et al. 2012).

3.3 SDNs and the Regulatory Framework

SDNs are promising tools for plant breeding. Research laboratories are improving their design and methods for delivery and plant breeding companies have started to use them to develop a new generation of plants with improved traits. However, uncertainty remains over the legal status of plants generated with this technology. Indeed, in the last few years, there has been an open debate at the international level on whether or not these plants should be regulated similarly to GMPs. This is a highly relevant decision, as it will heavily impact on the costs and the public acceptance of these products, in particular in the EU.

The use of SDNs to introduce random mutations at a specific site (SDN-1) results in changes of one of few nucleotides, which are undistinguishable from those obtained by conventional mutagenesis. For this reason, and in the absence of any foreign DNA in the final product, several authors have proposed to consider these plants as non-GM (Hartung and Schiemann 2013; Lusser and Davies 2013; Pauwels et al. 2014; Podevin et al. 2013). On the contrary, as SDN-3 uses SDNs for targeted introduction of transgenes, there is little doubt that the plants obtained should be considered as GMPs. In addition to the public debate in scientific journals, several national and international risk assessment and regulatory bodies have published statements and opinion papers. In the USA, the Department of Agriculture has informed crop companies that plants obtained by SDN-1, as well as those where transgenes coding for SDNs have been removed, will not be regulated, while those obtained by SDN-2 and SDN-3 will be analyzed case-by-case (Waltz 2012). In the EU National Competent Authorities such as the German ZKBS and COGEM from the Netherlands (Podevin et al. 2013) and ACRE from the UK (https://www.gov.uk/government/publications/genetically-modified-organisms-new-plant-growing-methods) have published statements in favor or a non-regulated status of SDN-1 products when no transgene is present, while, in line with the Opinion of the European Food Safety Authority, plants obtained by SDN-3 techniques should be considered as GMPs although the targeted nature of the technique may allow for an alleviated risk assessment in some cases (EFSA Panel on GMO 2012). This is also in line with the opinion expressed by the European Academies Advisory Council (EASAC) "Planting the future: opportunities and challenges for using crop genetic improvement technologies for sustainable agriculture" (EASAC policy report 21 June 2013, ISBN: 978-3-8047-3181-3), which also expresses the urgency for this legal framework clarification that "would give strong, immediate support to the competitiveness of the EU plant breeding sector which, thus far, has been responsible for a significant proportion of the worldwide research on New Breeding Techniques." In summary, there is a clear consent on the importance of clarifying the legal status of these techniques and, while the consensus has not been reached yet, there is a general agreement in that the different uses of SDNs should be regulated differently, and that while SDN-3 will probably be considered a particular type of GMP, most SDN-1 approaches should be treated similar to conventional mutagenesis.

3.4 SDN Strategy in Trees

The targeted introduction of mutations or modifications utilizing SDNs and particularly transcription activator-like effector nucleases (TALENs) and the CRISPR/Cas system is currently revolutionizing reverse genetic approaches in numerous model organisms. The usefulness of these strategies for genome modification has been demonstrated for the model plant Arabidopsis but also for crops, like wheat, maize, rice, barley, sorghum, or soybean, and is used now in breeding programs (Podevin et al. 2013). At the moment there is no published examples of the use of SDN strategies in trees, although there are many traits interesting for the fruit tree and/or paper/wood industry that could potentially be modified by SDNs, including virus or insects resistances, herbicide tolerance, reduced antinutrients/allergenics, improved nutrients/vitamins, slowed spoilage/senescence, improved biomass conversion, sterility (GMO containment), or male sterility (hybrids production). However, it must be noticed that the biology of most of the trees used as crops (i.e., very long life cycle compared to other crops and obligate vegetative propagation) in most cases imposes some additional constraints to the routine use of these new techniques. First, if classical transformation strategies are used, like Agrobacterium-mediated transformation or direct gene transfer using the gene gun, the difficulty or impossibility to go through a seed step will increase the risk of chimerism of the modification (i.e., targeted mutation or modification of a gene or targeted integration of a transgene). This will not necessarily impede the obtainment of the desired trait but could pose a problem for the definition of the new variety. Second, when stable transformation with the transgene bearing the expression unit of the SDN is needed, the impossibility to go through a seed step will also prevent the segregation of this transgene in the final product. However, it must be noticed that removal of transgenes has been demonstrated in trees using site-specific recombination systems (Fladung and Becker 2010) and could be an alternative for the SDN strategies. One solution to circumvent these limitations would be the transient transformation of protoplasts. Protoplasts transformation is used in routine for assessing SDN nuclease activity (Shan et al. 2013b; Liang et al. 2014; Jiang et al. 2013; Cermak et al. 2011; Chen et al. 2014) and has been used successfully for inheritable modification of genes in crops like tobacco (Zhang et al. 2013a). Transient transformation and regeneration from protoplasts is not routine but feasible in trees (Le Roux and Staden 1991; Park and Son 1992; Guo et al. 2012; Kanwar et al. 2003) and the next step now will be to test whether the existing protocols can be easily transferred to different tree species for gene modification via SDN strategies. The use of protoplast transient transformation in tree would eliminate the risk of chimerism and in the case of the SDN1 and SDN2 strategies could facilitate the classification of the modified plants as non-GMO as no stable integration of the transgene is needed. Therefore, although there are still some technical difficulties that need to be solved,

SDNs may become a particularly promising alternative to other techniques for introducing new variability, like tilling for example, that require large mutant collections and the elimination of undesired mutations by backcrossing, which are particularly difficult for long life cycle organisms like trees.

4 Gene Stacking

Engineering metabolic pathways in plants often requires simultaneous expression of several genes. Expression of multiple transgenes has been challenging with traditional transgenic approaches but recent progress has made powerful techniques available.

The traditional approach of introducing more than one transgene involves the stacking of expression cassettes in the transformation vector. Each cassette consists of its own promoter, the coding region of the transgene-of-interest and a terminator. The cassettes are usually arranged in tandem and in the same transcriptional orientation, which may trigger transgene silencing (Hamilton and Baulcombe 1999; Stam et al. 1997; Waterhouse et al. 1998). This transgene stacking approach is only suitable for a small number of genes, because the need to build large vectors poses a natural limit to the number of transgenes that can be introduced simultaneously. Three different strategies are used to overcome this limitation: (1) co-transformation with two or more transformation vectors, (2) serial transformation in which sets of transgenes are introduced by successive transformation experiments, or (3) combination of sets of transgenes initially inserted into different plants by sexual crosses.

In trees, which are vegetatively propagated and with generation times of several years, the last methodology is not useful and stacking has to be achieved by one of the first two.

4.1 Co-transformation

Co-transformation can be achieved with direct DNA transfer methods (e.g., particle gun-mediated transformation, DNA microinjection, or polyethylene glycol-assisted protoplast transformation) or with *Agrobacterium*-mediated transformation. In this last case, two plasmids in one or two bacterial strains, or two different T-DNAs in just one vector and one strain (Rukavtsova et al. 2013).

Combinatorial transformation (Zhu et al. 2008) is a special case of co-transformation. It takes advantage of the regular integration of multiple copies of plasmid DNA into the nuclear genome, and the usual co-integration of these copies into a single chromosomal locus (Agrawal et al. 2005). In combinatorial transformation all plasmids required for the introduction of the desired trait (e.g., a novel metabolic pathway requiring the concerted expression of many enzymes) are mixed together with an additional plasmid harboring a selectable marker gene. The mix is loaded onto the gun particles and, due to the very high number of plasmid molecules coating each particle, bombardment typically introduces copies of many or all of the vectors present in the mix. Following antibiotic selection of transgenic lines, plants are regenerated and can be analyzed by phenotypic screening for the accumulation of the desired metabolite (Zhu et al. 2008). There is no theoretical limit to the number of transgene cassettes that can be co-transformed. A disadvantage of the method is the lack of control over the structure of the transgenic locus.

4.2 *Serial Transformation*

Serial transformation is performed by stacking additional transgenes in an already transgenic plant line. It is more time consuming than co-transformation since needs to be done in successive transformation experiments and has the additional drawback of requiring new selectable marker gene for each transformation round. Alternatively, selectable marker genes can be removed by site-specific recombination (See “Marker-gene elimination” in this chapter) and the same gene can be used in successive transformation rounds but this represents an additional step in the process.

5 Potential for Genetic Manipulation of Forest Trees Through Short Non-translated RNA-Based Technologies

Elucidation of gene function remains a major challenge in forest tree biotechnology mainly because large mutant collections are not yet available. Since the discovery of the activity of short non-translated RNA molecules in the genetic fine-tuning regulation mechanisms in eukaryotes, new technologies using small RNA (sRNA) have been developed to suppress/downregulate gene expression. In addition, since RNA transgenes are dominant suppressors, these methods are of substantial interest in forest trees breeding programs for which long intervals are required to obtain homozygotes through sexual reproduction.

5.1 *Small RNAs (SRNAs)*

sRNAs are ~18–25-nucleotide-long molecules produced from double-stranded RNA (dsRNA) precursors by the action of Dicer-like (DCL) RNase-III classes of enzymes (reviewed by Bologna and Voinnet 2014). Amongst the large number of different sRNA categories being still elucidated (Ghosh and Mallick 2012), the two main classes are the short interfering RNAs (siRNAs) and the microRNAs (miRNAs), which differ in the biosynthesis of their precursors. The precursors of siRNAs are fully complementary dsRNAs arising either from viral replication, transposon, or transgene, and result from the activity of an RNA-dependent RNA polymerase (RDR). The precursors of miRNAs are nuclear gene-derived transcript produced by RNA polymerase II which form imperfect dsRNAs (pri-miRNAs) (Voinnet 2009). Pri-miRNAs are cleaved by DCL producing successively a miRNA hairpin (hp) precursor (pre-miRNA) and a mature miRNA/miRNA* duplex (Jones-Rhoades et al. 2006).

One strand of the siRNA or miRNA duplex is recruited by an Argonaute (AGO) nuclease family protein, the catalytic core of protein RNA-induced silencing complex (RISC), which will then recognize mRNAs complementary to the mature sRNA and either lead to translational repression or to the slicing of the mRNA in the region corresponding to the middle of the sRNA. These silencing mechanisms are mainly at the origin of post-transcriptional gene silencing (mRNA cleavage/destabilization or translational inhibition). However, siRNAs are also involved in transcriptional silencing by inducing methylation of DNA homologous sequences (Lippman and Martienssen 2004).

Several technologies based on sRNAs allowing the silencing of target genes have been developed for plants, including RNA interference (RNAi), artificial miRNAs (amiRNAs), and artificial trans-acting siRNAs (atasiRNAs) (Fig. 2). Since there are no available genome-wide mutant collections for forest trees, the application of these methods is of particular interest to produce directed down-regulation of target genes.

5.2 RNAi

Reverse genetics through RNAi approaches, based on the production of siRNAs targeting candidate genes, have been largely used for plant biotechnology and have enhanced the ability to manipulate gene expression. The RNAi methods that have been developed include the virus-induced gene silencing (VIGS) and the hpRNAi. VIGS consists in the use of an RNA virus, found as dsRNA during its replication and processed by DCL into siRNAs, in which the sequence of a target gene is added and is then introduced in plants (Waterhouse and Helliwell 2003). hpRNAi constructs contain two DNA sequences (100–800 bp long) placed in inverted orientation and connected by a linker (Chuang and Meyerowitz 2000; Wesley et al. 2001). hpRNAi constructs produce dsRNAs which are processed by DCL, leading to the production of siRNAs which will drive silencing of the target gene (Fig. 2a). However, since a heterogeneous population of siRNAs is produced, it may affect other genes than those targeted. In addition, secondary siRNAs may be coming from the target mRNA sequence and affect non-desired genes (Molesini et al. 2012).

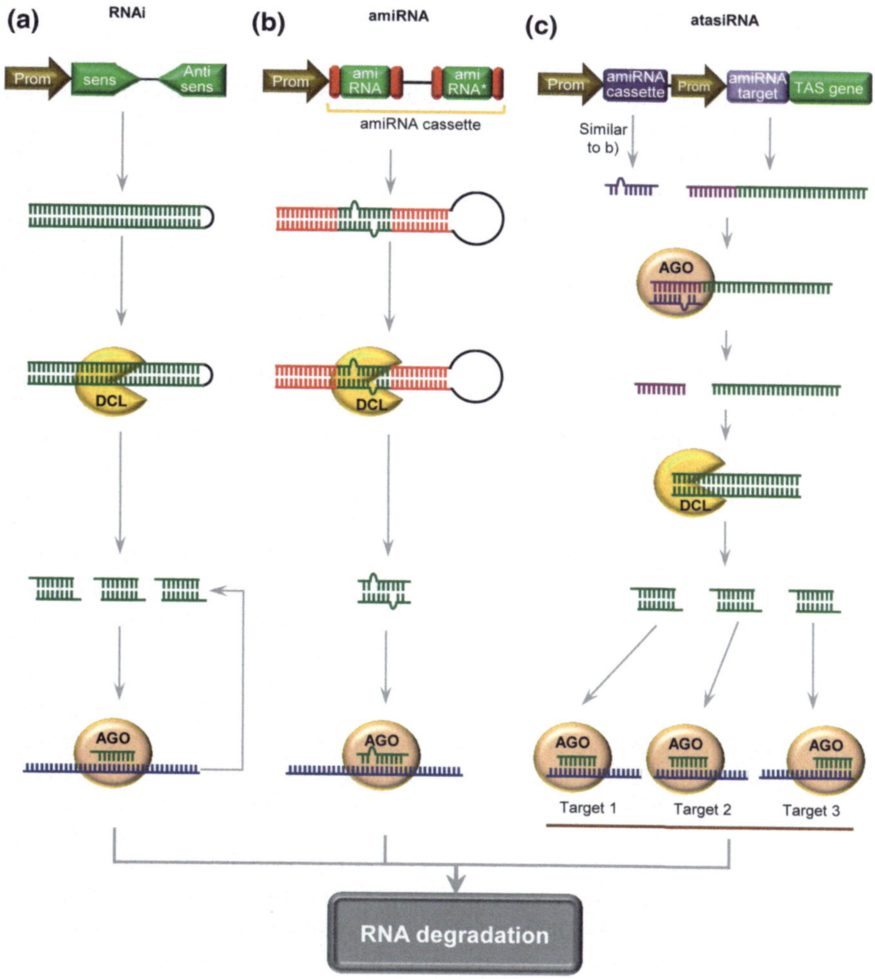

Fig. 2 Schematic illustration of different sRNA-based technologies. **a** RNAi. Hairpin (hp)RNAi constructs contain two DNA sequences (100–800 bp) placed in inverted orientation and connected by a linker. The transcript deriving from hpRNAi constructs will produce a dsRNA that will be cleaved by DCL into siRNA duplexes. These siRNAs, recruited by AGO within the RISC complex, will be processed as single-strand siRNA and will guide the recognition of homologous mRNA sequences that will be sliced. Following the degradation of the mRNA, secondary siRNAs can be produced by the activity of RDR and be different than those generated by the hpRNAi constructs. **b** amiRNA. To produce an amiRNA targeting a particular gene, sequences corresponding to amiRNA and amiRNA* are cloned in a natural pre-miRNA backbone in place of endogenous miRNA and miRNA* sequences. This precursor will be processed by DCL into an amiRNA/amiRNA* duplex and mediate target mRNA recognition through AGO recruitment. **c** atasiRNA. Constructs for the atasiRNA technology contain cassettes for producing both a pre-miRNA (such as the Arabidopsis miR173, miR390, or an amiRNA) and a TAS transcript, which have a binding site for that (a)miRNA and a sequence corresponding to target gene(s). The processed (a)miRNA (see **b**) will bind to the TAS transcript. After conversion into dsRNA by RDR, DCL4 will slice the transcript into phased 21 nt siRNAs. The siRNAs produced have the capability to target other genes in trans

5.3 Artificial MicroRNAs (amiRNAs)

The modification of a miRNA sequence within its precursor was shown to not alter the biogenesis of this miRNA if the hp-loop structure is not affected (Zeng et al. 2002). The amiRNA precursors are thus modified endogenous pre-miRNAs in which the miRNA and miRNA* are replaced with sequences homologous to any target gene (Ossowski et al. 2008). The amiRNA will be processed as a native miRNA and will direct AGO to downregulate the target gene(s) (Fig. 2b). The potential of this technique is high as it may lead to the degradation of single target but also of multiple targets if the miRNA is designed against conserved sequences within a target gene family (Alvarez et al. 2006).

The design of amiRNAs involves the identification of an amiRNA sequence complementary to the target gene, with optimal specificity and efficiency. Specificity for particular or multiple genes can be checked by scanning the genome sequence of the studied plant species and unwanted potential targets can be avoided from the design (Schwab et al. 2006). The efficiency can be optimized by selecting amiRNA with a low hybridization energy (<−30 kcal/mol) to the desired target. Web application resources such as the Web MicroRNA Designer (WMD) (wmd3.weigelworld.org) have been developed to support the design of plant amiRNAs (Schwab et al. 2006). WMD3 allows the identification of amiRNAs for several forest tree species including, *Eucalyptus globulus*, *E. gunnii*, *Fraxinus excelsior*, *Ginkgo biloba*, *Juglans hindsii* × *J. regia*, *Liriodendrum tulipifera*, *Picea abies*, *Pinus* sp., *Quercus* sp., *Populus* sp., and *Pseudotsuga menziesii*. The designed amiRNA/amiRNA* sequences are then incorporated in place of the naturally occurring miRNA/miRNA* through guided mutagenesis, overlapping PCR or by cloning of double-strand DNA oligonucleotides (Schwab et al. 2006; Carbonell et al. 2014). Different pre-miRNAs coming from various plants species have been used (reviewed by Pérez-Quintero and López 2010; Li et al. 2014b) and it has been demonstrated that precursors from one species can be effective in other ones (Alvarez et al. 2006). After integration in plant transformation vector and transgenesis, the mRNA cleavage at the amiRNA site can be demonstrated by RACE-PCR. amiRNA-based technology is considered to be more controlled than hpRNAi since the pre-miRNA will produce theoretically only one effective sRNA with a known sequence.

Recently, Hauser et al. (2013) reported the generation of a genome-wide scale library of 22,000 amiRNAs targeting over 18,000 Arabidopsis genes. To evaluate the efficiency of amiRNA in vivo, Li et al. (2014b) reported and epitope-tagged protein (etp)-based amiRNA method, which consists to co-express target genes encoding etp and amiRNA candidates in plant protoplasts and to measure the accumulation of the tagged proteins. The development of such tools is potentially of great interest for the characterization of genes with unknown function and would be highly valuable if developed for forest tree species.

Several reports demonstrating the specificity and the efficiency of amiRNA-mediated gene silencing have been done mainly for eudicots, monocots, mosses, and algae model species (reviewed by Pérez-Quintero and López 2010; Sablok et al. 2011; Liu and Chen 2010). Regarding forest tree species, amiRNAs have been reported to be effective in *Populus*. For instance, Du et al. (2009) integrated, into the Arabidopsis pre-miR164b, an amiRNA sequence targeting the class-I KNOX homeobox *ARBORKNOX2* (*ARK2*) gene in *P. tremula* × *P. alba*. They obtained transgenic poplars with ~10–20 % residual *ARK2* expression and evidenced a role for this gene in the regulation of secondary growth, affecting the differentiation of cambial daughter cells and lignified cell types. Shi et al. (2010) generated transgenic *P. trichocarpa* expressing an amiRNA targeting two subsets of genes encoding phenylalanine ammonia lyase (PAL), using the *Ptr-miR408* precursor. These subset transcripts, 80 % identical at and around the binding region of the amiRNA, were found to be specifically down-regulated in respective transgenic plants, with an efficiency up to 60 % for the first subset (including *PAL2*, *4*, and *5*) and up to 23 % for the second subset (including *PAL1* and *3*). A significant up-regulation of *PAL* transcripts of the second subset was measured in transgenic plants down-regulated for the first subset, evidencing a differential regulation amongst the different members of the gene family. No phenotype was described.

5.4 Artificial Trans-Acting SiRNAs (AtasiRNAs)

Trans-acting siRNAs (tasiRNAs) are a type of siRNA involved in post-transcriptional gene silencing whose processing and activity involve both siRNA and miRNAs machineries (reviewed by Bologna and Voinnet 2014). tasiRNAs are initiated from cleavage of tasiRNA-producing (*TAS*) loci by a specific miRNA, such as the miR173 (Arabidopsis), miR828, and miR390 (in both Arabidopsis and poplar) (Klevebring et al. 2009; Zhang 2014). After miRNA cleavage, the mRNA strand is made dsRNA by RDR and processed by DCL into several phased 21 nt siRNAs, which will target genes different to that from which they originate (*in trans*). Until now, 8 *TAS* loci (*TAS1a-c*, *TAS2*, *TAS3a-c*, and *TAS4*) have been identified in plants (reviewed by Bologna and Voinnet 2014). In Arabidopsis, tasiRNAs mediate the regulation of large gene families such as the pentatricopeptide repeat proteins (*TAS1* and *TAS2*), several auxin response transcription factors (*TAS3*), and MYB transcription factors (*TAS4*) (reviewed by Yoshikawa 2013). Since the presence of only one target site for the miRNA is necessary for tasiRNA production (Montgomery et al. 2008), successful silencing of target genes has been obtained either by replacing the siRNA region in the *TAS* gene (Gutiérrez-Nava et al. 2008) or by a fusion of gene fragments to an upstream miR173 target site (De Felippes et al. 2012). By co-expression of the miRNA, these last authors showed that this technology can be applied in other species such as *Nicotiana benthamania*. Recently, Carbonell et al. (2014) provided a protocol to perform high throughput generation of atasiRNA constructs. This mixed technology, integrating both amiRNA and siRNA strategies, is very powerful as a single *TAS* locus can be engineered to generate tasiRNAs targeting either different parts within a particular gene or different genes within a genome (Fig. 2c). The atasiRNA technology may be applied to specific traits of forest trees such as lignin biosynthesis pathway where one or simultaneous branches can be targeted.

5.5 *Stability of Down-Regulation*

Considering the perennial growth of trees, the stability of new traits during their long life span is a main concern. Although reports on efficiency and stability of gene suppression in forest trees are still limited, several studies aiming at evaluating such stability have been performed. For instance, Pilate et al. (2002) demonstrated that the down-regulation of cinnamyl alcohol dehydrogenase (CAD) or caffeate/5-hydroxy-ferulate *O*-methyltransferase (COMT) encoding genes in transgenic poplar through antisense gene suppression was stable over four years. Li et al. (2008) investigated the stability of the suppression by RNAi of the herbicide resistance gene *BAR*, encoding phosphinothricin acetyltransferase, under the control of the Arabidopsis *atS1A ribulose-1,5-biphosphate carboxylase small subunit* (*rbcS*) promoter. Four transgenic lines, harboring a stable expression of *BAR* for several years in the field, were again transformed with four different hpRNAi constructs targeting either the promoter (to induce transcriptional gene silencing) or the coding sequence (to induce post-transcriptional gene silencing). As shown by these authors for plants co-transformed for *BAR* and hpRNAi against *BAR* and grown over 2 years in the field, stable and efficient *BAR* transgene suppression was found when using hpRNAi against the coding sequence (more than 90 % suppression for 80 % of the transgenic lines) than against the promoter sequence (90 % suppression for 6 % of the transgenic lines). The low efficiency of the RNAi construct designed against the promoter was suggested to be due to inherent properties of the invert repeat tested or to the promoter itself, as the intensity of suppression may depend on the number and the arrangement of potential methylation sites within the promoter sequence (Li et al. 2008). In *Chlamydomonas* amiRNA activity has remained over 6 months (Zhao et al. 2008) and over 500 generations (Molnar et al. 2009) without losing the expression of amiRNAs and phenotypes.

6 Cisgenesis and Intragenesis

Mixing of genetic materials between species that cannot hybridize by natural means is one of the major criticisms to transgenic crops. Two transformation concepts cisgenesis and intragenesis were developed as alternatives to transgenesis. Both concepts imply that plants must only be transformed with genetic material derived from the species itself or from closely related species capable of sexual hybridization.

Intragenesis differs from cisgenesis by allowing use of new gene combinations created by in vitro rearrangements of functional genetic elements.

Currently, intragenic/cisgenic plants are regulated as transgenic plants worldwide. However, as the gene pool exploited by intragenesis and cisgenesis is identical to the gene pool available for conventional breeding, less comprehensive regulatory measures are expected.

Most of the present-day commercial apple cultivars are susceptible to scab caused by *Venturia inaequalis*. The most commonly used resistance in conventional breeding of apple is the Vf locus from the wild apple Malus floribunda 821 (Szankowski et al. 2009). Although this wild apple resistance has been transferred into different apple varieties through classical breeding, the procedure is extremely long and is associated with undesirable linkage drag. Recently, a cisgenic apple with resistance to scab was developed through the transfer of the HcrVf2 genomic clone including its own promoter and terminator into apple cv Gala (Vanblaere et al. 2011). The same gene was used to produce resistant intragenic apples (Joshi et al. 2011), containing the promoter and terminator of the small subunit of the apple rubisco gene.

Pathogenesis-related (PR) proteins from grapevine are being used to produce cisgenic grape plants after elimination of the marker genes (Dutt et al. 2008). Dhekney et al. (2011) have used a PR protein named VVTL-1 (*Vitis vinifera* thaumatin-like protein) to produce transgenic plants demonstrating that the protein constitutively expressed inhibits spore germination and hypha growth and the transformants show delay in powdery mildew disease development and decreased severity of black rot.

7 Conclusions

Public concern with the use of transgenic plants has speed up the research to produce plants without marker genes. Therefore, the number of tree species where the elimination of marker genes has been demonstrated has largely increased in the last years. The design of vectors where the recombinase is controlled by an inducible promoter and placed, together with marker genes, within its recognition sites has allowed the application of this technology to trees where segregation of the recombinase in the progeny is not an option. Additionally, this technology can be used to stack genes when complex metabolic routes are to be modified using subsequent transformation rounds but the same marker gene and therefore the same selection procedure. In fact, this is a methodology becoming routinely used for the production of trees with less commercialisation impediments.

The use of SDNs for targeted genome modification is revolutionizing reverse genetic approaches in numerous model organisms, and is also starting to be used in crops. The very long life cycles of trees, compared to other crops and their obligate vegetative propagation, make the use of these techniques in trees more challenging. However, there are different strategies that may allow circumvent these problems.

The use of protoplast transient transformation, and regeneration from modified protoplast, could be one of such strategies. Therefore, although there are still some technical difficulties that need to be solved, SDNs may become a particularly promising alternative to other techniques for introducing new variability, like tilling for example, that require large mutant collections and the elimination of undesired mutations by backcrossing, which are particularly difficult for long life cycle organisms like trees.

Down-regulation technologies based on amiRNA and atasiRNA are specific and effective post-transcriptional gene silencing approaches that can applied to decipher the huge genomic information and to improve plant molecular breeding. Although very few reports on these new tools exist for forest trees species, they pave the way for future perspectives of tree genetic engineering, with more controlled and more stable modification strategies.

Hopefully, SDNs as well as cisgenesis and transgenesis will ease the regulatory process and make some of these trees, which are important improvements in the breeding of these difficult species, publicly available in a next future.

References

Agrawal PK, Kohli A, Twyman RM, Christou P (2005) Transformation of plants with multiple cassettes generates simple transgene integration patterns and high expression levels. Mol Breed 16:247–260

Ainley WM, Sastry-Dent L, Welter ME, Murray MG, Zeitler B, Amora R, Corbin DR, Miles RR, Arnold NL, Strange TL, Simpson MA, Cao Z, Carroll C, Pawelczak KS, Blue R, West K, Rowland LM, Perkins D, Samuel P, Dewes CM, Shen L, Sriram S, Evans SL, Rebar EJ, Zhang L, Gregory PD, Urnov FD, Webb SR, Petolino JF (2013) Trait stacking via targeted genome editing. Plant Biotechnol J 11:1126–1134

Alvarez JP, Pekker I, Goldshmidt A, Blum E, Amsellem Z, Eshed Y (2006) Endogenous and synthetic microRNAs stimulate simultaneous, efficient, and localized regulation of multiple targets in diverse species. Plant Cell 18:1134–1151

Ballester A, Cervera M, Peña L (2007) Efficient production of transgenic *Citrus* plants using isopentenyl transferase positive selection and removal of the marker gene by site-specific recombination. Plant Cell Rep 26:39–45

Ballester A, Cervera M, Peña L (2008) Evaluation of selection strategies alternative to nptII in genetic transformation of citrus. Plant Cell Rep 27:1005–1015

Belhaj K, Chaparro-Garcia A, Kamoun S, Nekrasov V (2013) Plant genome editing made easy: targeted mutagenesis in model and crop plants using the CRISPR/Cas system. Plant Methods 9:39

Boch J, Bonas U (2010) Xanthomonas AvrBs3 family-type III effectors: discovery and function. Annu Rev Phytopathol 48:419–436

Boch J, Scholze H, Schornack S, Landgraf A, Hahn S, Kay S, Lahaye T, Nickstadt A, Bonas U (2009) Breaking the code of DNA binding specificity of TAL-type III effectors. Science 326:1509–1512

Bogdanove AJ, Voytas DF (2011) TAL effectors: customizable proteins for DNA targeting. Science 333:1843–1846

Bogdanove AJ, Schornack S, Lahaye T (2010) TAL effectors: finding plant genes for disease and defense. Curr Opin Plant Biol 13:394–401

Boissel S, Jarjour J, Astrakhan A, Adey A, Gouble A, Duchateau P, Shendure J, Stoddard BL, Certo MT, Baker D, Scharenberg AM (2014) megaTALs: a rare-cleaving nuclease architecture for therapeutic genome engineering. Nucleic Acids Res 42:2591–2601

Bologna NG, Voinnet O (2014) The diversity, biogenesis, and activities of endogenous silencing small RNAs in *Arabidopsis*. Annu Rev Plant Biol 65:28.1–28.31

Breyer D, Kopertekh L, Dirk R (2014) Alternatives to antibiotic resistance marker genes for *in vitro* selection of genetically modified plants. Scientific developments, current use, operational access and biosafety considerations. Critical Rev Plant Sci 33(4):286–330

Brunet E, Simsek D, Tomishima M, DeKelver R, Choi VM, Gregory P, Urnov F, Weinstock DM, Jasin M (2009) Chromosomal translocations induced at specified loci in human stem cells. Proc Natl Acad Sci USA 106:10620–12625

Cai CQ, Doyon Y, Ainley WM, Miller JC, Dekelver RC, Moehle EA, Rock JM, Lee YL, Garrison R, Schulenberg L, Blue R, Worden A, Baker L, Faraji F, Zhang L, Holmes MC, Rebar EJ, Collingwood TN, Rubin-Wilson B, Gregory PD, Urnov FD, Petolino JF (2009) Targeted transgene integration in plant cells using designed zinc finger nucleases. Plant Mol Biol 69:699–709

Carbonell A, Takeda A, Fahlgren N, Johnson SC, Cuperus JT, Carrington JC (2014) New generation of artificial microRNA and synthetic trans-acting small interfering RNA vectors for efficient gene silencing in Arabidopsis. Plant Physiol. doi:10.1104/pp.113.234989

Carroll D (2011) Genome engineering with zinc-finger nucleases. Genetics 188:773–782

Cermak T, Doyle EL, Christian M, Wang L, Zhang Y, Schmidt C, Baller JA, Somia NV, Bogdanove AJ, Voytas DF (2011) Efficient design and assembly of custom TALEN and other TAL effector-based constructs for DNA targeting. Nucleic Acids Res 39:e82

Chen K, Shan Q, Gao C (2014) An efficient TALEN mutagenesis system in rice. Methods. doi:10.1016/j.ymeth.2014.02.013

Chilton MDM, Que Q (2003) Targeted integration of T-DNA into the tobacco genome at double-stranded breaks: new insights on the mechanism of T-DNA integration. Plant Physiol 133:956–965

Chong-Pérez B, Reyes M, Rojas L, Ocaña B, Ramos A, Kosky RG, Angenon G (2013) Excision of a selectable marker gene in transgenic banana using a Cre/lox system controlled by an embryo specific promoter. Plant Mol Biol 83:143–152

Christian M, Cermak T, Doyle EL, Schmidt C, Zhang F, Hummel A, Bogdanove AJ, Voytas DF (2010) Targeting DNA double-strand breaks with TAL effector nucleases. Genetics 186:757–761

Christian M, Qi Y, Zhang Y, Voytas DF (2013) Targeted mutagenesis of Arabidopsis thaliana using engineered TAL effector nucleases. G3 3:1697–1705

Chuang CF, Meyerowitz EM (2000) Specific and heritable genetic interference by double-stranded RNA in *Arabidopsis thaliana*. Proc Natl Acad Sci USA 97:4985–4990

Cong L, Ran FA, Cox D, Lin S, Barretto R, Habib N, Hsu PD, Wu X, Jiang W, Marraffini LA, Zhang F (2013) Multiplex genome engineering using CRISPR/Cas systems. Science 339:819–823

Cuellar W, Gaudin A, Solórzano D, Casas A, Ñopo L, Chudalayandi P, Medrano G, Kreuze J, Ghislain M (2006) Self-excision of the antibiotic resistance gene *nptII* using a heat inducible *Cre-loxP* system from transgenic potato. Plant Mol Biol 62:71–83

Curtin SJ, Zhang F, Sander JD, Haun WJ, Starker C, Baltes NJ, Reyon D, Dahlborg EJ, Goodwin MJ, Coffman AP, Dobbs D, Joung JK, Voytas DF, Supar RM (2011) Targeted mutagenesis of duplicated genes in soybean with zinc-finger nucleases. Plant Physiol 156:466–473

D'Halluin K, Ruiter R (2013) Directed genome engineering for genome optimization. Int J Dev Biol 57:621–627

D'Halluin K, Vanderstraeten C, Stals E, Cornelissen M, Ruiter R (2008) Homologous recombination: a basis for targeted genome optimization in crop species such as maize. Plant Biotechnol J 6:93–102

Dale EC, Ow DW (1991) Gene-transfer with subsequent removal of the selection gene from the host genome. P Natl Acad Sci USA 88:10558–10562
Davis L, Maizels N (2011) DNA nicks promote efficient and safe targeted gene correction. PLoS ONE 6:e23981
Day CD, Lee E, Kobayashi J, Holappa LD, Albert H, Ow DW (2000) Transgene integration into the same chromosome location can produce alleles that express at a predictable level, or alleles that are differentially silenced. Gene Dev 14:2869–2880
De Felippes FF, Wang JW, Weigel D (2012) MIGS: miRNA-induced gene silencing. Plant J 70:541–547
De Lange O, Binder A, Lahaye T (2014) From dead leaf, to new life: TAL effectors as tools for synthetic biology. Plant J 78(5):753–771
De Pater S, Neuteboom LW, Pinas JE, Hooykaas PJJ, van der Zaal BJ (2009) ZFN-induced mutagenesis and gene-targeting in Arabidopsis through Agrobacterium-mediated floral dip transformation. Plant Biotechnol J 7:821–835
Dhekney SA, Li ZT, Gray DJ (2011) Grapevines engineered to express cisgenic Vitis vinifera thaumatin-like protein exhibit fungal disease resistance. In Vitro Cell Dev Plant 47:458–466
Djukanovic V, Smith J, Lowe K, Yang M, Gao H, Jones S, Nicholson MG, West A, Lape J, Bidney D, Falco SC, Jantz D, Lyznik LA (2013) Male-sterile maize plants produced by targeted mutagenesis of the cytochrome P450-like gene (MS26) using a re-designed I-CreI homing endonuclease. Plant J 76:888–899
Doyon Y, Vo TD, Mendel MC, Greenberg SG, Wang J, Xia DF, Miller JC, Urnov FD, Gregory PD, Holmes MC (2011) Enhancing zinc-finger-nuclease activity with improved obligate heterodimeric architectures. Nat Methods 8:74–79
Du J, Mansfield SD, Groover AT (2009) The *Populus* homeobox gene *ARBORKNOX2* regulates cell differentiation during secondary growth. Plant J 60:1000–1014
Dutt M, Li ZT, Dhekney SA, Gray DJ (2008) A co-transformation system to produce transgenic grapevines free of marker genes. Plant Sci 175:423–430
EFSA Panel on GMO (2012) Scientific opinion addressing the safety assessment of plants developed using Zinc Finger Nuclease 3 and other Site-Directed Nucleases with similar function. EFSA J 10:2943
Erikson O, Hertzberg M, Näsholm T (2004) A conditional marker gene allowing both positive and negative selection in plants. Nat Biotechnol 22:455–458
Even-Faitelson L, Samach A, Melamed-Bessudo C, Avivi-Ragolsky N, Levy A (2011) Localized egg-cell expression of effector proteins for targeted modification of the Arabidopsis genome. Plant J 68:929–937
Feng Z, Mao Y, Xu N, Zhang B, Wei P, Yang DL, Wang Z, Zhang Z, Zheng R, Yang L, Zeng L, Liu X, Zhu JK (2014) Multigeneration analysis reveals the inheritance, specificity, and patterns of CRISPR/Cas-induced gene modifications in Arabidopsis. Proc Natl Acad Sci USA 111:4632–4637
Feng Z, Zhang B, Ding W, Liu X, Yang DL, Wei P, Cao F, Zhu S, Zhang F, Mao Y, Zhu JK (2013) Efficient genome editing in plants using a CRISPR/Cas system. Cell Res 23:1229–1232
Fladung M, Becker D (2010) Targeted integration and removal of transgenes in hybrid aspen (Populus tremula L. x P. tremuloides Michx.) using site-specific recombination systems. Plant Biol 12:334–340
Fu Y, Foden JA, Khayter C, Maeder ML, Reyon D, Joung JK, Sander JD (2013) High-frequency off-target mutagenesis induced by CRISPR-Cas nucleases in human cells. Nat Biotechnol 31:822–826
Fu Y, Sander JD, Reyon D, Cascio VM, Joung JK (2014) Improving CRISPR-Cas nuclease specificity using truncated guide RNAs. Nat Biotechnol 32:279–284
Gao H, Smith J, Yang M, Jones S, Djukanovic V, Nicholson MG, West A, Bidney D, Falco SC, Jantz D, Lyznik LA (2010) Heritable targeted mutagenesis in maize using a designed endonuclease. Plant J 61:176–187
Gambley RL, Ford R, Smith GR (1993) Microprojectile transformation of sugarcane meristems and regeneration of shoots expressing B-Glucuronidase. Plant Cell Rep 12:343–346

García-Almodovar RC, Petri C, Padilla IMG, Burgos L (2014) Combination of site-specific recombination and a conditional selective marker gene allows for the production of marker-free tobacco plants. Plant Cell Tiss Organ Cult 116:205–215

Ghosh Z, Mallick B (2012) Renaissance of the regulatory RNAs. In: Mallick B (ed) Regulatory RNAs. Springer, Berlin. doi:10.1007/978-3-642-22517-8_1

Gilbert LA, Larson MH, Morsut L, Liu Z, Brar GA, Torres SE, Stern-Ginossar N, Brandman O, Whitehead EH, Doudna JA, Lim WA, Weissman JS, Qi LS (2013) CRISPR-mediated modular RNA-guided regulation of transcription in eukaryotes. Cell 154:442–451

Gottfried P, Lotan O, Kolot M, Maslenin L, Bendov R, Gorovits R, Yesodi V, Yagil E, Rosner A (2005) Site-specific recombination in Arabidopsis plants promoted by the Integrase protein of coliphage HK022. Plant Mol Biol 57:435–444

Grau J, Boch J, Posch S (2013) TALENoffer: genome-wide TALEN off-target prediction. Bioinformatics 29:2931–2932

Grindley NDF, Whiteson KL, Rice PA (2006) Mechanisms of site-specific recombination. Annu Rev Biochem 75:567–605

Guilinger JP, Pattanayak V, Reyon D, Tsai SQ, Sander JD, Joung JK, Liu DR (2014a) Broad specificity profiling of TALENs results in engineered nucleases with improved DNA-cleavage specificity. Nat Methods 11:429–435

Guilinger JP, Thompson DB, Liu DR (2014b) Fusion of catalytically inactive Cas9 to FokI nuclease improves the specificity of genome modification. Nat Biotechnol 32:577–582

Guo J, Morrell-Falvey JL, Labbé JL, Muchero W, Kalluri UC, Tuskan GA, Chen JG (2012) Highly efficient isolation of Populus mesophyll protoplasts and its application in transient expression assays. PLoS ONE 7:e44908

Gutiérrez-Nava ML, Aukerman MJ, Sakai H, Tingey S, Williams RW (2008) Artificial trans-acting siRNAs confer consistent and effective gene silencing. Plant Physiol 147:543–551

Hafez M, Hausner G (2012) Homing endonucleases: DNA scissors on a mission. Genome 55:553–569

Hamilton AJ, Baulcombe DC (1999) A species of small antisense RNA in posttranscriptional gene silencing in plants. Science 286:950–952

Han KH, Wilbert S, Perkins EJ, Gordon MP (1992) Development of transgenic trees for use in removal and detoxification of chemical wastes. In: Proceedings of IUFRO working party S2.04.06 workshop "Molecular Biology of Forest Trees", Carcans-Maubuisson, France

Hanin M, Paszkowski J (2003) Plant genome modification by homologous recombination. Curr Opin Plant Biol 6:157–162

Hare PD, Chua NH (2002) Excision of selectable marker genes from transgenic plants. Nat Biotechnol 20:575–580

Hartung F, Schiemann J (2013) Precise plant breeding using new genome editing techniques: opportunities, safety and regulation in the EU. Plant J 78:742–752

Hauser F, Chen W, Deinlein U, Chang K, Ossowski S, Fitz J, Hannon GJ, Schroeder JI (2013) A genomic-scale artificial microRNA library as a tool to investigate the functionally redundant gene space in *Arabidopsis*. Plant Cell 25:2848–2863

Herzog K, Flachowsky H, Deising HB, Hanke MV (2012) Heat-shock-mediated elimination of the nptII marker gene in transgenic apple (Malus × domestica Borkh). Gene 498:41–49

Hsu PD, Scott DA, Weinstein JA, Ran FA, Konermann S, Agarwala V, Li Y, Fine EJ, Wu X, Shalem O, Cradick TJ, Marraffini LA, Bao G, Zhang F (2013) DNA targeting specificity of RNA-guided Cas9 nucleases. Nat Biotechnol 31:827–832

Jia H, Pang Y, Chen X, Fang R (2006) Removal of the selectable marker gene from transgenic tobacco plants by expression of Cre recombinase from a tobacco mosaic virus vector through agroinfection. Transgenic Res 15:375–384

Jiang W, Zhou H, Bi H, Fromm M, Yang B, Weeks DP (2013) Demonstration of CRISPR/Cas9/sgRNA-mediated targeted gene modification in Arabidopsis, tobacco, sorghum and rice. Nucleic Acids Res 41:e188

Jinek M, Chylinski K, Fonfara I, Hauer M, Doudna JA, Charpentier E (2012) A programmable dual-RNA-guided DNA endonuclease in adaptive bacterial immunity. Science 337:816–821

Jones-Rhoades MW, Bartel DP, Bartel B (2006) MicroRNAs and their regulatory roles in plants. Annu Rev Plant Biol 57:19–53

Joshi SG, Schaart JG, Groenwold R, Jacobsen E, Schouten HJ, Krens FA (2011) Functional analysis and expression profiling of HcrVf1 and HcrVf2 for development of scab resistant cisgenic and intragenic apples. Plant Mol Biol 75:579–591

Kanwar K, Bhardwaj A, Agarwal S, Sharma DR (2003) Genetic transformation of Robinia pseudoacacia by Agrobacterium tumefaciens. Indian J Exp Biol 41:149–153

Katz SS, Gimble FS, Storici F (2014) To nick or not to nick: comparison of I-SceI single- and double-strand break-induced recombination in yeast and human cells. PLoS ONE 9:e88840

Kilby NJ, Davies GJ, Snaith MR, Murray JAH (1995) FLP recombinase in transgenic plants: constitutive activity in stably transformed tobacco and generation of marked cell clones in Arabidopsis. Plant J 8:637–652

Kleinstiver BP, Wolfs JM, Kolaczyk T, Roberts AK, Hu SX, Edgell DR (2012) Monomeric site-specific nucleases for genome editing. Proc Natl Acad Sci USA 109:8061–8066

Klevebring D, Street NR, Fahlgren N, Kasschau KD, Carrington JC, Lundeberg J, Jansson S (2009) Genome-wide profiling of populus small RNAs. BMC Genom 10:620

Kopertekh L, Schiemann J (2005) Agroinfiltration as a tool for transient expression of cre recombinase in vivo. Transgenic Res 14:793–798

Kopertekh L, Schulze K, Frolov A, Strack D, Broer I, Schiemann J (2010) Cre-mediated seed-specific transgene excision in tobacco. Plant Mol Biol 72:597–605

Le Roux JJ, Staden JV (1991) Micropropagation and tissue culture of Eucalyptus-a review. Tree Physiol 9:435–477

Lee HJ, Kweon J, Kim E, Kim S, Kim J (2012) Targeted chromosomal duplications and inversions in the human genome using zinc finger nucleases. Genome Res 22:539–548

Li ZS, Xing AQ, Moon BP, Burgoyne SA, Guida AD, Liang HL, Lee C, Caster CS, Barton JE, Klein TM, Falco SC (2007) A Cre/loxP-mediated self-activating gene excision system to produce marker gene free transgenic soybean plants. Plant Mol Biol 65:329–341

Li J, Brunner AM, Shevchenco O, Meilan R, Ma C, Skinner JS, Strauss SH (2008) Efficient and stable transgene suppression via RNAi in field-grown poplars. Transgenic Res 17:679–694

Li T, Huang S, Zhao X, Wright DA, Carpenter S, Spalding MH, Weeks DP, Yang B (2011) Modularly assembled designer TAL effector nucleases for targeted gene knockout and gene replacement in eukaryotes. Nucleic Acids Res 39:6315–6325

Li T, Liu B, Spalding MH, Weeks DP, Yang B (2012) High-efficiency TALEN-based gene editing produces disease-resistant rice. Nat Biotechnol 30:390–392

Li JF, Norville JE, Aach J, McCormack M, Zhang D, Bush J, Church GM, Sheen J (2013) Multiplex and homologous recombination-mediated genome editing in Arabidopsis and Nicotiana benthamiana using guide RNA and Cas9. Nat Biotechnol 31:688–691

Li JF, Zhang D, Sheen J (2014a) Epitope-tagged protein-based artificial miRNA screens for optimized gene silencing in plants. Nat Protoc 9:939–949

Li T, Liu B, Chen CY, Yang B (2014b) TALEN utilization in rice genome modifications. Methods 69:9–16

Liang Z, Zhang K, Chen K, Gao C (2014) Targeted mutagenesis in Zea mays using TALENs and the CRISPR/Cas system. J Genet Genomics 41:63–68

Lippman Z, Martienssen R (2004) The role of RNA interference in heterochromatic silencing. Nature 431:364–370

Liu Q, Chen YQ (2010) A new mechanism in plant engineering: The potential roles of microRNAs in molecular breeding for crop improvement. Biotechnol Adv 28:301–307

Liu HK, Yang C, Wei ZM (2005) Heat shock-regulated site-specific excision of extraneous DNA in transgenic plants. Plant Sci 168:997–1003

Lloyd A, Plaisier CL, Carroll D, Drews GN (2005) Targeted mutagenesis using zinc-finger nucleases in Arabidopsis. Proc Natl Acad Sci USA 102:2232–2237

López-Noguera S, Petri C, Burgos L (2009) Combining a regeneration-promoting gene and site-specific recombination allows a more efficient apricot transformation and the elimination of marker genes. Plant Cell Rep 28:1781–1790

Lusser M, Davies HV (2013) Comparative regulatory approaches for groups of new plant breeding techniques. Nat Biotechnol 30:437–446
Lusser M, Parisi C, Plan D, Rodríguez-Cerezo E (2012) Deployment of new biotechnologies in plant breeding. Nat Biotechnol 30:231–239
Maeder ML, Thibodeau-Beganny S, Osiak A, Wright DA, Anthony RM, Eichtinger M, Jiang T, Foley JE, Winfrey RJ, Townsend JA, Unger-Wallace E, Sander JD, Müller-Lerch F, Fu F, Pearlberg J, Göbel C, Dassie JP, Pruett-Miller SM, Porteus MH, Sgroi DC, Iafrate AJ, Dobbs D, McCray PB Jr, Cathomen T, Voytas DF, Joung JK (2008) Rapid "open-source" engineering of customized zinc-finger nucleases for highly efficient gene modification. Mol Cell 31:294–301
Mahfouz MM, Li L, Shamimuzzaman M, Wibowo A, Fang X, Zhu JK (2011) De novo-engineered transcription activator-like effector (TALE) hybrid nuclease with novel DNA binding specificity creates double-strand breaks. Proc Natl Acad Sci USA 108:2623–2628
Mali P, Esvelt KM, Church GM (2013) Cas9 as a versatile tool for engineering biology. Nat Methods 10:957–963
Marton I, Zuker A, Shklarman E, Zeevi V, Tovkach A, Roffe S, Ovadis M, Tzfira T, Vainstein A (2010) Nontransgenic genome modification in plant cells. Plant Physiol 154:1079–1087
Matsunaga E, Sugita K, Ebinuma H (2002) Asexual production of selectable marker-free transgenic woody plants, vegetatively propagated species. Mol Breeding 10:95–106
McConnell Smith A, Takeuchi R, Pellenz S, Davis L, Maizels N, Monnat RJ, Stoddard BL (2009) Generation of a nicking enzyme that stimulates site-specific gene conversion from the I-AniI LAGLIDADG homing endonuclease. Proc Natl Acad Sci USA 106:5099–5104
Miao J, Guo D, Zhang J, Huang Q, Qin G, Zhang X, Wan J, Gu H, Qu LJ (2013) Targeted mutagenesis in rice using CRISPR-Cas system. Cell Res 23:1233–1236
Molesini B, Pii Y, Pandolfini T (2012) Fruit improvement using intragenesis and artificial microRNA. Trends Biotechnol 30:80–87
Molnar A, Bassett A, Thuenemann E, Schwach F, Karkare S, Ossowski S, Weigel D, Baulcombe D (2009) Highly specific gene silencing by artificial microRNAs in the unicellular alga *Chlamydomonas reinhardtii*. Plant J 58:165–174
Montgomery TA, Yoo SJ, Fahlgren N, Gilbert SD, Howell MD, Sullivan CM, Alexander A, Nguyen G, Allen E, Ahn JH, Carrington JC (2008) AGO1-miR173 complex initiates phased siRNA formation in plants. Proc Natl Acad Sci USA 105:20055–20062
Moscou MJ, Bogdanove AJ (2009) A simple cipher governs DNA recognition by TAL effectors. Science 326:1501
Mussolino C, Cathomen T (2011) On target? Tracing zinc-finger-nuclease specificity. Nat Methods 8:725–726
Mussolino C, Morbitzer R, Lütge F, Dannemann N, Lahaye T, Cathomen T (2011) A novel TALE nuclease scaffold enables high genome editing activity in combination with low toxicity. Nucleic Acids Res 39:9283–9293
Nekrasov V, Staskawicz B, Weigel D, Jones JDG, Kamoun S (2013) Targeted mutagenesis in the model plant Nicotiana benthamiana using Cas9 RNA-guided endonuclease. Nat Biotechnol 31:691–693
Onouchi H, Yokoi K, Machida C, Matsuzaki H, Oshima Y, Matsuoka K, Nakamura K, Machida Y (1991) Operation of an efficient site-specific recombination system of Zygosaccharomyces rouxii in tobacco cells. Nucleic Acids Res 19:6373–6378
Osakabe K, Osakabe Y, Toki S (2010) Site-directed mutagenesis in Arabidopsis using custom-designed zinc finger nucleases. Proc Natl Acad Sci USA 107:12034–12039
Ossowski S, Schwab R, Weigel D (2008) Gene silencing in plants using artificial microRNAs and other small RNAs. Plant J 53:674–690
Pâques F, Duchateau P (2007) Meganucleases and DNA double-strand break-induced recombination: perspectives for gene therapy. Curr Gene Ther 7:49–66
Park YG, Son SH (1992) In vitro shoot regeneration from leaf mesophyll protoplasts of hybrid poplar (Populus nigra × P. maximowiczii). Plant Cell Rep 11:2–6

Pattanayak V, Ramirez CL, Joung JK, Liu DR (2011) Revealing off-target cleavage specificities of zinc-finger nucleases by in vitro selection. Nat Methods 8:765–770

Pauwels K, Podevin N, Breyer D, Carroll D, Herman P (2014) Engineering nucleases for gene targeting: safety and regulatory considerations. Nat Biotechnol 31:18–27

Pérez-Quintero A, López C (2010) Artificial microRNAs and their applications in plant molecular biology. Agron Colomb 28:373–381

Petolino JF, Worden A, Curlee K, Connell J, Strange Moynahan TL, Larsen C, Russell S (2010) Zinc finger nuclease-mediated transgene deletion. Plant Mol Biol 73:617–628

Petri C, Burgos L (2005) Transformation of fruit trees. Useful breeding tool or continued future prospect? Transgenic Res 14:15–26

Petri C, López-Noguera S, Wang H, García-Almodovar RC, Alburquerque N, Burgos L (2012) A chemical-inducible *Cre-LoxP* system allows for elimination of selection marker genes in transgenic apricot. Plant Cell Tiss Org 110:337–346

Pilate G, Guiney E, Holt K, Petit-Conil M, Lapierre C, Leplé JC, Pollet B, Mila I, Webster EA, Marstorp HG, Hopkins DW, Jouanin L, Boerjan W, Schuch W, Cornu D, Halpin C (2002) Field and pulping performances of transgenic trees with altered lignification. Nat Biotechnol 20:607–612

Podevin N, Davies HV, Hartung F, Nogué F, Casacuberta JM (2013) Site-directed nucleases: a paradigm shift in predictable, knowledge-based plant breeding. Trends Biotechnol 31:375–383

Puchta H, Dujon B, Hohn B (1993) Homologous recombination in plant cells is enhanced by in vivo induction of double strand breaks into DNA by a site-specific endonuclease. Nucleic Acids Res 21:5034–5040

Puchta H, Dujon B, Hohn B (1996) Two different but related mechanisms are used in plants for the repair of genomic double-strand breaks by homologous recombination. Proc Natl Acad Sci USA 93:5055–5060

Puchta H (2003) Marker-free transgenic plants. Plant Cell Tiss Org 74:123–134

Qi Y, Li X, Zhang Y, Starker CG, Baltes NJ, Zhang F, Sander JD, Reyon D, Joung JK, Voytas DF (2013) Targeted deletion and inversion of tandemly arrayed genes in Arabidopsis thaliana using zinc finger nucleases. G3 3:1707–1715

Ramirez CL, Foley JE, Wright DA, Müller-Lerch F, Rahman SH, Cornu TI, Winfrey RJ, Sander JD, Fu F, Townsend JA, Cathomen T, Voytas DF, Joung K (2008) Unexpected failure rates for modular assembly of engineered zinc fingers. Nat Methods 5:374–375

Ramirez CL, Certo MT, Mussolino C, Goodwin MJ, Cradick TJ, McCaffrey AP, Cathomen T, Scharenberg AM, Joung JK (2012) Engineered zinc finger nickases induce homology-directed repair with reduced mutagenic effects. Nucleic Acids Res 40:5560–5568

Reyon D, Tsai SQ, Khayter C, Foden JA, Sander JD, Joung JK (2012) FLASH assembly of TALENs for high-throughput genome editing. Nat Biotechnol 30:460–465

Rukavtsova EB, Lebedeva AA, Zakharchenko NS, Buryanov Y (2013) The ways to produce biologically safe marker-free transgenic plants. Russ J Plant Physiol 60:14–26

Sablok G, Pérez-Quintero AL, Hassan M, Tatarinova T, López C (2011) Artificial microRNAs (amiRNAs) engineering—on how microRNA-based silencing methods have affected current plant silencing research. Biochem Biophys Res Com 406:315–319

Salomon S, Puchta H (1998) Capture of genomic and T-DNA sequences during double-strand break repair in somatic plant cells. EMBO J 17:6086–6095

Sander JD, Maeder ML, Joung JK (2011) Engineering designer nucleases with customized cleavage specificities In: Current protocols in molecular biology. John Wiley and Sons, Inc., New York, pp 12.13.1–12.13.16

Sanders KL, Catto LE, Bellamy SRW, Halford SE (2009) Targeting individual subunits of the FokI restriction endonuclease to specific DNA strands. Nucleic Acids Res 37:2105–2115

Sautter C, Leduc N, Bilang R, Iglesias VA, Gisel A, Wen X, Potrykus I (1995) Shoot apical meristems as a target for gene-transfer by microballistics. Euphytica 85:45–51

Schaart JG, Krens FA, Pelgrom KTB, Mendes O, Rouwendal GJA (2004) Effective production of marker-free transgenic strawberry plants using inducible site-specific recombination and a bifunctional selectable marker gene. Plant Biotechnol J 2:233–240

Schwab R, Ossowski S, Riester M, Warthmann N, Weigel D (2006) Highly specific gene silencing by artificial microRNAs in *Arabidopsis*. Plant Cell 18:1121–1133

Shan Q, Wang Y, Chen K, Liang Z, Li J, Zhang Y, Zhang K, Liu J, Voytas DF, Zheng X, Zhang Y, Gao C (2013a) Rapid and efficient gene modification in rice and Brachypodium using TALENs. Mol Plant 6:1365–1368

Shan Q, Wang Y, Li J, Zhang Y, Chen K, Liang Z, Zhang K, Liu J, Xi JJ, Qiu JL, Gao C (2013b) Targeted genome modification of crop plants using a CRISPR-Cas system. Nat Biotechnol 31:686–688

Shi R, Yang C, Lu S, Sederoff R, Chiang V (2010) Specific down-regulation of *PAL* genes by artificial microRNAs in *Populus trichocarpa*. Planta 232:1281–1288

Shukla VK et al (2009) Precise genome modification in the crop species Zea mays using zinc-finger nucleases. Nature 459:437–441

Siebert R, Puchta H (2002) Efficient repair of genomic double-strand breaks by homologous recombination between directly repeated sequences in the plant genome. Plant Cell 14: 1121–1131

Sorek R, Lawrence CM, Wiedenheft B (2013) CRISPR-mediated adaptive immune systems in bacteria and archaea. Ann Rev Biochem 82:237–266

Sreekala C, Wu L, Gu K, Wang D, Tian D, Yin Z (2005) Excision of a selectable marker in transgenic rice (*Oryza sativa* L.) using a chemically regulated *Cre/loxP* system. Plant Cell Rep 24:86–94

Srivastava V, Ow DW (2001) Single-copy primary transformants of maize obtained trhough the co-introduction of a recombinase-expressing construct. Plant Mol Biol 46:561–566

Srivastava V, Anderson OD, Ow DW (1999) Single-copy transgenic wheat generated through the resolution of complex integration patterns. Proc Natl Acad Sci USA 96:11117–11121

Stam M, Mol JNM, Kooter JM (1997) The silence of genes in transgenic plants. Ann Bot-London 79:3–12

Stoddard BL (2011) Homing endonucleases: from microbial genetic invaders to reagents for targeted DNA modification. Structure 19:7–15

Szankowski I, Waidmann S, Degenhardt J, Patocchi A, Paris R, Silfverberg-Dilworth E, Broggini G, Gessler C (2009) Highly scab-resistant transgenic apple lines achieved by introgression of HcrVf2 controlled by different native promoter lengths. Tree Genet Genomes 5:349–358

Tovkach A, Zeevi V, Tzfira T (2009) A toolbox and procedural notes for characterizing novel zinc finger nucleases for genome editing in plant cells. Plant J 57:747–757

Townsend JA, Wright DA, Winfrey RJ, Fu F, Maeder ML, Joung JK, Voytas DF (2009) High-frequency modification of plant genes using engineered zinc-finger nucleases. Nature 459:442–445

Tsai SQ, Wyvekens N, Khayter C, Foden JA, Thapar V, Reyon D, Goodwin MJ, Aryee MJ, Joung JK (2014) Dimeric CRISPR RNA-guided FokI nucleases for highly specific genome editing. Nat Biotech 32:569–576

Tzfira T, Frankman LR, Vaidya M, Citovsky V (2003) Site-specific integration of Agrobacterium tumefaciens T-DNA via double-stranded intermediates. Plant Physiol 133:1011–1023

Urnov FD, Rebar EJ, Holmes MC, Zhang HS, Gregory PD (2010) Genome editing with engineered zinc finger nucleases. Nat Rev Genet 11:636–646

Vanblaere T, Szankowski I, Schaart J, Schouten H, Flachowsky H, Broggini GAL, Gessler C (2011) The development of a cisgenic apple plant. J Biotechnol 154:304–311

Voinnet O (2009) Origin, biogenesis and activity of plant microRNAs. Cell 136:669–687

Waltz E (2012) Tiptoeing around transgenics. Nat Biotechnol 30:215–217

Wang H, Yang H, Shivalila CS, Dawlaty MM, Cheng AW, Zhang F, Jaenisch R (2013) One-step generation of mice carrying mutations in multiple genes by CRISPR/Cas-mediated genome engineering. Cell 153:910–918

Wang T, Wei JJ, Sabatini DM, Lander ES (2014) Genetic screens in human cells using the CRISPR-Cas9 system. Science 343:80–84

Wang Y, Yau YY, Perkins-Balding D, Thomson JG (2011) Recombinase technology: applications and possibilities. Plant Cell Rep 30:267–285

Waterhouse PM, Graham MW, Wang MB (1998) Virus resistance and gene silencing in plants can be induced by simultaneous expression of sense and antisense RNA. Proc Natl Acad Sci USA 95:13959–13964

Waterhouse PM, Helliwell CA (2003) Exploring plant genomes by RNA-induced gene silencing. Nat Genet 4:29–38

Wehrkamp-Richter S, Degroote F, Laffaire JB, Paul W, Perez P, Picard G (2009) Characterisation of a new reporter system allowing high throughput in planta screening for recombination events before and after controlled DNA double strand break induction. Plant Physiol Biochem 47:248–255

Wendt T, Holm PB, Starker CG, Christian M, Voytas DF, Brinch-Pedersen H, Holme IB (2013) TAL effector nucleases induce mutations at a pre-selected location in the genome of primary barley transformants. Plant Mol Biol 83:279–285

Wesley SV, Helliwell CA, Smith NA, Wang M, Rouse DT, Liu Q, Gooding PS, Singh SP, Abbott D, Stoutjesdijk PA, Robinson SP, Gleave AP, Green AG, Waterhouse PM (2001) Construct design for efficient, effective and high-throughput gene silencing in plants. Plant J 27:581–590

Wright DA, Townsend JA, Winfrey RJ, Irwin PA, Rajagopal J, Lonosky PM, Hall BD, Jondle MD, Voytas DF (2005) High-frequency homologous recombination in plants mediated by zinc-finger nucleases. Plant J 44:693–705

Xie K, Yang Y (2013) RNA-guided genome editing in plants using a CRISPR-Cas system. Mol Plant 6:1975–1983

Yang M, Djukanovic V, Stagg J, Lenderts B, Bidney D, Falco SC, Lyznik LA (2009) Targeted mutagenesis in the progeny of maize transgenic plants. Plant Mol Biol 70:669–679

Yoshikawa M (2013) Biogenesis of *trans*-acting siRNAs, endogenous secondary siRNAs in plants. Genes Genet Syst 88:77–84

Zeng Y, Wagner EJ, Cullen BR (2002) Both natural and designed micro RNAs can inhibit the expression of cognate mRNAs when expressed in human cells. Mol Cell 9:1327–1333

Zhang F, Maeder ML, Unger-Wallace E, Hoshaw JP, Reyon D, Christian M, Li X, Pierick CJ, Dobbs D, Peterson T, Joung JK, Voytas DF (2010) High frequency targeted mutagenesis in Arabidopsis thaliana using zinc finger nucleases. Proc Natl Acad Sci USA 107:12028–12033

Zhang Y, Zhang F, Li X, Baller JA, Qi Y, Starker CG, Bogdanove AJ, Voytas DF (2013a) Transcription activator-like effector nucleases enable efficient plant genome engineering. Plant Physiol 161:20–27

Zhang Z, Xiang D, Heriyanto F, Gao Y, Qian Z, Wu WS (2013b) Dissecting the Roles of miR-302/367 cluster in cellular reprogramming using TALE-based repressor and TALEN. Stem Cell Rep 1:218–225

Zhang ZJ (2014) Artificial trans-acting small interfering RNA: a tool for plant biology study and crop improvements. Planta 239:1139–1146

Zhang W, Subbarao S, Addae P, Shen A, Armstrong C, Peschke V, Gilbertson L (2003) *Cre/lox*-mediated marker gene excision in transgenic maize (*Zea mays* L.) plants. Theor Appl Genet 107:1157–1168

Zhang Y, Li H, Ouyang B, Lu Y, Ye Z (2006) Chemical-induced autoexcision of selectable markers in elite tomato plants transformed with a gene conferring resistance to lepidopteran insects. Biotechnol Lett 28:1247–1253

Zhao T, Wang W, Bai X, Qi Y (2008) Gene silencing by artificial microRNAs in *Chlamydomonas*. Plant J 58:157–164

Zhu C, Naqvi S, Breitenbach J, Sandmann G, Christou P, Capell T (2008) Combinatorial genetic transformation generates a library of metabolic phenotypes for the carotenoid pathway in maize. Proc Natl Acad Sci USA 105:18232–18237

Zuo J, Niu QW, Moller SG, Chua NH (2001) Chemical-regulated, site-specific DNA excision in transgenic plants. Nat Biotechnol 19:157–161

Lessons from 25 Years of GM Tree Field Trials in Europe and Prospects for the Future

G. Pilate, I. Allona, W. Boerjan, A. Déjardin, M. Fladung, F. Gallardo, H. Häggman, S. Jansson, R. Van Acker and C. Halpin

Abstract It is common agronomic practice to perform a formal evaluation of the behaviour of new varieties under natural field conditions. Accordingly, shortly after the optimization of genetic engineering techniques on trees, a number of field trials were set up to assess GM trees modified for different genes. Here, we review the work that has been done in this arena in Europe over the last 25 years, and summarize what we learned from these experiments. GM tree field trials remain the exception rather than the rule in Europe. Several trials have been destroyed by anti-GMO activists and it is becoming increasingly difficult to obtain authorization for a GM tree field trial. These increasing constraints on GM tree trials within

G. Pilate (✉) · A. Déjardin
UR0588, Amélioration Génétique et Physiologie Forestières,
Institut National de La Recherche Agronomique (INRA), 45075 Orléans, France
e-mail: gilles.pilate@orleans.inra.fr

I. Allona
Centre for Plant Biotechnology and Genomics (UPM-INIA), 28223 Madrid, Spain

W. Boerjan · R. Van Acker
Department of Plant Systems Biology, VIB, Technologiepark 927, B-9052 Ghent, Belgium

W. Boerjan · R. Van Acker
Belgium Department of Plant Biotechnology and Bioinformatics, Ghent University, Technologiepark 927, B-9052 Ghent, Belgium

M. Fladung
Thünen Institute of Forest Genetics (TI-FG), 22927 Grosshansdorf, Germany

F. Gallardo
Dpto Biología Molecular y Bioquímica, Universidad de Málaga, 29071 Málaga, Spain

H. Häggman
Genetics and Physiology Unit, Faculty of Science, University of Oulu, 90014 Oulu, Finland

S. Jansson
Department of Plant Physiology, Umea Plant Science Center, Umea University, 901 87 Umea, Sweden

C. Halpin
Division of Plant Sciences, College of Life Sciences, University of Dundee at the JHI, Invergowrie, Dundee DD2 5DA, Scotland, UK

C. Vettori et al. (eds.), *Biosafety of Forest Transgenic Trees*,
Forestry Sciences 82, DOI 10.1007/978-94-017-7531-1_4

Europe are both surprising and counter-productive as we learned a lot from the past 25 years of experiments and the results were promisingly positive: (1) Phenotypic effects resulting from transgene expression in GM trees grown in the field appears to be stable, albeit variable; (2) most field studies have validated earlier observations made under greenhouse conditions, although in some cases the modification of target traits was less obvious in fluctuating field environments, and in a few cases had severe growth and developmental penalties; (3) non-target effects were consistently within the range of natural variation. Overall, the European GM tree field trials failed to exemplify any significant tangible risks. Based on this evidence, it seems appropriate that Europe should now move forward beyond small confined trials to larger scale experiments better fitted to a broader context of evaluation and environmental assessment.

1 Introduction

Forests are crucial natural resources for the earth ecosystem, as they provide a diversity of services including recycling carbon dioxide into oxygen, acting as a carbon sink, participating in climate regulation, protecting soil, and water and serving as a genetic reserve for biodiversity. For humankind, forests also serve as recreational areas, as a source of lumber, pulp, and paper, as a source of energy for cooking and heating, and now as a feedstock for bioenergy, biofuels and biomaterials. To respond to the growing demand for wood and alleviate pressure on natural forests, intensively managed forest plantations have been developed to increase productivity and shorten rotation time. Among other opportunities, GM technology is one potentially useful tool to create improved tree genotypes that enhance the efficiency and value of such plantations and of downstream wood uses.

GM technology was first developed more than 30 year ago, in the early 1980s: the first articles on GM tobacco transformed through *Agrobacterium tumefaciens*-mediated DNA transfer were published in 1983 (Barton et al. 1983; Herrera-Estrella et al. 1983). This transformation method was very rapidly transferred to other dicot plant species and in 1987 the first GM poplar tree was produced (Fillatti et al. 1987). Conifers, like monocots, proved to be more recalcitrant to *A. tumefaciens* infection and the first report of the production of a GM conifer by particle acceleration was not published until 1993 (Ellis et al. 1993). Taking advantage of the power of somatic embryogenesis in conifers, an efficient *A. tumefaciens* transformation procedure was eventually developed, first for larch (Levée et al. 1997) and later on for other conifer species. Today, GM materials can be produced for a wide array of tree species, even though this success is still often limited to a small number of genotypes within a given species.

Tree GM technology was first used in basic research studies to gain a better knowledge of biological mechanisms specific to these large, long-lived, perennial species. The evaluation of GM trees was initially mostly restricted to greenhouses.

Table 1 List of GM tree field trials planted in Europe from 1988-present and described in this chapter

Trial name and clones used	Promoter/transgene	Years	Area	Publications
BELGIUM				
Herbicide tolerance				
clone 064 (*P. deltoides* × *P. trichocarpa*) and clone 357 (*P. alba* × *P. tremula*)	*bar*	1988–1990 and 1991–1999	nd	De Block (1990)
Lignin modification				
717-1B4	35S::*Sccr*; 35S::AS*ccr*	2009–2017	331 m^2	Van Acker et al. (2014)
717-1B4	35S::RNAi*cad*	2014–2022	1311 m^2	Unpublished
FINLAND				
Pollination trials				
P. sylvestris, *P. abies*	35S::*uid*A	1996–2002		Aronen et al. (2003), Clapham et al. (2003)
Growth				
B. pendula	Nos::*npt*II; *lac*Z	1990–1991		Unpublished
B. pendula	35S::*Rbc*S; 35S::*Nr*; 35S::*npt*II	2000–2005 destruction		Unpublished
Tree interactions with fungi and insects				
B. pendula	35S::*chitinase* IV; 35S::*Bp4CL1*	2000–2003		Pasonen et al. (2005), Pasonen et al. (2009)
Sterility				
B. pendula	pMADS::*barnase*	2005–2012		Unpublished
T89, 51 (hybrid aspen)	pLMX5::*ERF*; *SLAC*1	2013–2018		Unpublished
FRANCE				
Herbicide tolerance				
Huneghem, Beaupré (*P. deltoides* × *P. trichocarpa*)	*bar*	1994–2002	8000 m^2	Unpublished results

(continued)

Table 1 (continued)

Trial name and clones used	Promoter/transgene	Years	Area	Publications
Stability and expression of foreign genes				
717-1B4 (*P. tremula* × *P. alba* [hybrid aspen])	35S::*uid*A	1991–1999	<100 m^2	Hawkins et al. (2003)
717-1B4	35S::*uid*A; pcad::*uid*A	1996–2003	1400 m^2	Hawkins et al. (2003)
Insect tolerance				
353-38 (*P. tremula* × *P. tremuloides* [hybrid aspen])	35S::*OCI*	1995–1999	945 m^2	Unpublished
Lignin modification				
717-1B4	35S::AS*cad*; 35S::AS*comt*	1995–2007	440 m^2	Lapierre et al. (1999), Pilate et al. (2002)
717-1B4	35S::AS*ccr*; 35S::AS*ccoamt*	1999–2007	925 m^2	Leplé et al. (2007)
717-1B4	35S::AS*cad*; 35S::AS*comt*; 35S::AS*ccr*; 35S::AS*ccoamt*	2007–2013	1365 m^2	Van Acker et al. (2014)
GERMANY				
Stability and expression of foreign genes				
Esch5 (*P. tremula* × *P. tremuloides* [hybrid aspen]), Brauna11, W52 (*P. tremula* [aspen])	35S::*rol*C; rbcS::*rolC*	1996–2001	1500 m^2	Fladung et al. (2004), Fladung and Hoenicka (2012)
Horizontal gene transfer				
T89, Esch5 (*P. tremula* × *P. tremuloides* [hybrid aspen])	GPD::*bar*	2000–2003	1000 m^2	Nehls et al. (2006)
Phytoremediation				
717-1B4	*Gsh*I	2002–2004	2500 m^2	Peuke and Rennenberg (2006)
717-1B4	*Gsh*I	2003–2005	2 × 2500 m^2	

(continued)

Table 1 (continued)

Trial name and clones used	Promoter/transgene	Years	Area	Publications
SPAIN				
Increased yield				
717-1B4	*gs*	nd	nd	Jing et al. (2004), El-Khatib et al. (2004), Pascual et al. (2008)
717-1B4	*CsRAV*1	2012-ongoing	204 m^2	Moreno-Cortés et al. (unpublished)
SWEDEN				
Stability and expression of foreign genes				
T89	13 different	1994	500 m^2	
Modification of photoprotection				
T89	PsbS RNAi	2004, 2007	2 m^2	Unpublished
Increased yield				
T89	16 different	2010-ongoing	1000 m^2	Unpublished
T89	7 different	2011-ongoing	500 m^2	Unpublished
T89	14 different	2012-ongoing	1000 m^2	Unpublished
Altered wood properties				
T89	7 different	2011-ongoing	500 m^2	Unpublished
T89	21 different	2014-ongoing	500 m^2	Unpublished
Altered phenology				
T89	27 different	2014-ongoing	1000 m^2	Unpublished
UNITED KINGDOM				
Lignin modification				
717-1B4	AS*cad*; AS*comt*	1996-destruction 1999		Pilate et al. (2002), Halpin et al. (2007)

Note Some other field trials with GM trees have been planted in Europe during this period, but data on these assays are not available

But field trials rapidly became necessary especially to take into account the effects of environmental variations on plant phenotype. This was particularly relevant for GM trees, as trees grown in the greenhouse quickly became cramped, with severely limited space for trunk, canopy and root development. Indeed, studies carried out in the greenhouse are necessarily restricted to very young trees (less than two-year-old) still at a juvenile stage. With the prospect of applied research to deploy economically interesting GM tree genotypes, it rapidly became necessary to evaluate GM trees in field tests, in just the same way that new non-GM genotypes bred through classical controlled crosses would be trialled.

Safety and containment procedures for such trials had to take several issues specific to trees into account (e.g. long lifetime, phase changes from juvenile to mature, bud dormancy, long-distance seed dispersal) in order to limit potential risks and accidental escape. The purpose of a rather high proportion of the first field trials with GM trees was precisely to assess what these potential risks might be. The first full field trials with GM trees were installed France in 1991 and a number of others were established in subsequent years throughout the nineties, in France, England, Belgium, Germany, Spain, Finland and Sweden (Table 1). At this time, there was already an administrative framework and a scientific evaluation procedure to control the establishment, maintenance and the final destruction of every field trial. With time, the conditions for obtaining authorization became more and more cumbersome, the risk of vandalism of trials increased, and the funding for studies involving GM plants became rarer. As a consequence, the years 2000–2010 were characterized by a decrease in the number of GM tree field trials in Europe, while the number was increasing in other parts of the world.

What did we learn from these experiments? This is the topic of this chapter which will review the different field tests performed in seven European countries over the last 25 years (in Belgium, Finland, France, Germany, Spain, Sweden and United Kingdom). Most European GM tree field trials for which published data is available are reported on in this paper.

2 Field Tests in Belgium (4 Field Trials with GM Poplar)

A first field trial with transgenic poplars was established in 1988 in Geraardsbergen, Belgium, with trees engineered for herbicide tolerance (De Block 1990). Three years later, in 1991, a second field trial with the same herbicide tolerant poplar lines was established (Report WIV-ISP 2010). Between then and 2002, more than 120 GM crop field trials including three on GM trees (apple, poplar) were conducted in Belgium, the majority by companies such as Plant Genetic Systems NV, Aventis Cropscience NV and Monsanto Europe. After 2002, the flourishing field trial culture in Belgium came to an end. The combination of uncertainty about what the implementation of the 2001/18 EU Directive would bring, and a number of field trial destructions, made companies decide to stop doing field trials (Custers 2009). This situation lasted until November 2007, when VIB requested permission for a field trial with poplars (*Populus tremula* × *P. alba*) genetically modified in lignin biosynthesis with the purpose of using them for bioethanol production. Despite positive opinions from the Biosafety Advisory Council and the Flemish minister for environment, the federal ministers declined the field trial request, resulting in a political conflict (The Poplar Files, www.vib.be). Ultimately, VIB was forced to go to the Council of State to have the negative decision suspended. In December 2008, the Council of State ruled in favour of VIB and finally, in February 2009, VIB received permission to initiate a field trial with GM poplars for a period of 8 years. This permission can be considered a milestone in Belgian field trial history (Custers 2009).

The field trial in question (Fig. 1) was planted in May 2009 and contained two transgenic lines (FS3 and FS40) and a control wild type line (all in *P. tremula* × *P. alba*). The field itself consisted of six randomized blocks for each line, each block containing 20 clonally propagated trees. The transgenic trees were down-regulated in the gene encoding cinnamoyl-coenzymeA reductase (CCR), an enzyme involved in the biosynthesis of lignin. These GM poplars were made in the framework of the EU-funded project TIMBER. Downregulation of CCR typically results in reduced lignin content and an orange-wine red coloration of the xylem upon debarking (Leplé et al. 2007). As lignin is considered as the most important limiting factor in the conversion of plant biomass into fermentable sugars

Fig. 1 Belgium CCR field trial. **a** 2 weeks after planting in May 2009. **b** 20 August 2009: after 3 months of growth. **c** March 2013: harvest

(Chen and Dixon 2007; Van Acker et al. 2013), the aim of this field trial was to investigate whether wood from field-grown lignin-deficient poplars was easier to process into bioethanol, compared to wild type poplars. The study demonstrated a large variation in the level of downregulation in individual stems, reflected by a variation in the intensity and pattern of red xylem coloration (from fully white stems, through patchy red coloured stems, to completely red stems). This variability in red coloration allowed the researchers to demonstrate a correlation between the level of CCR downregulation and the efficiency of wood processing into fermentable sugars (saccharification) and ethanol. However, strong downregulation of CCR also affected biomass yield, which outweighed the improved bioethanol production in most lines, a problem that probably could be solved by targeting the downregulation of lignin genes specifically to the fibres (Van Acker et al. 2014).

After the difficulties in obtaining regulatory permission for the establishment of a field trial with the CCR down-regulated poplars, a new era for field trial experiments in Belgium had started. In 2011, a two-year field trial with GM potato lines with improved resistance to *Phytophthora infestans* was established, while in 2012, permission was granted for a field trial with GM maize to demonstrate altered growth characteristics. In 2013, VIB obtained permission, for a period of 8 years, for the fourth GM poplar field trial in Belgian history (GMO register European Commission).

In this most recent poplar field trial, *P. tremula* × *P. alba* is down-regulated in another gene involved in lignin biosynthesis, *cad*,encoding cinnamyl alcohol dehydrogenase. These GM poplars were made by INRA-AGPF and the field trial has been established in Belgium under the framework of the EU-funded project MultiBioPro. The CAD enzyme catalyses the last step in the biosynthesis of the building blocks of lignin. It was demonstrated previously that CAD-deficient poplar wood, both greenhouse- and field-grown, was easier to delignify in simulated Kraft pulping experiments (Baucher et al. 1996; Lapierre et al. 1999; Pilate et al. 2002). The aim of the current field trial is to investigate the influence of CAD downregulation on the efficiency of wood processing into fermentable sugars and bioethanol when these trees are grown in a short rotation coppice culture, similar to the CCR field trial. The CAD field trial consists of three transgenic poplar lines in which CAD is down-regulated using an RNA interference strategy. The field itself is divided in six randomized blocks per line with 40 clonal replicates in each block. This on-going field trial was planted in May 2014 and a first coppice and harvest was performed in February 2015.

3 Field Tests in Finland (8 Field Trials, 6 with GM Trees)

3.1 Pollination Experiments with Transgenic Pollen (2 Field Trials)

Two approved field experiments in Finland are notable as they were not conventional GM tree field trials but were set up in order to evaluate pollen transformation as an alternative gene transfer method for Scots pine (*Pinus sylvestris* L.) and Norway spruce (*Picea abies* L. Karst.). The marker genes GUS (β-glucuronidase, *uid*A) and *npt*II (neomycin phosphotransferase) were transferred biolistically to pollen grains (Häggman et al. 1997; Aronen et al. 1998). To enable the seeds to develop under natural conditions, the pollen was transferred to female cones on trees growing in the outdoor seed orchards at the Research Park of the Finnish Forest Research Institute (Metla), Punkaharju. However, all cones and female catkins were covered by specific protection bags for containment of transgenic material (Fig. 2a, b). It was obvious that the bags affected cone development negatively, particularly in the case of Scots pine. The seed production of Scots pine and Norway spruce varied yearly. Based on the GUS assay, 601 of the Scots pine seedlings germinated in the greenhouse were further tested by PCR. One of the seedlings turned out to be transgenic (Aronen et al. 2003; Clapham et al. 2003). None of the Norway spruce seedlings were confirmed to be transgenic.

3.2 *Silver Birch (5 Field Trials)*

Altogether, five genetically modified silver birch (*Betula pendula* Roth) field trials were established in Finland between the years 1990 and 2013 in three different locations. The first short trial (1990–1991) was established by the company Kemira Oy at Vihti, Kotkaniemi test farm. The purpose of this small-scale trial (60 birches) was to study the effect of the marker genes *npt*II and *lac*Z (beta-galactosidase, EC 3.2.1.23) on birch growth. Even though the results were positive, the field trial was ended in 1991 partly due to the company decision to focus on fertilizer production and also due to corporate acquisition (pers. comm. Prof. Matti Sarvas, Dr. Pauli Seppälä and Dr. Kari Jokinen).

Another early research-oriented GM birch field trial in Eastern Finland at the Research Park of the Finnish Forest Research Institute (Metla), Punkaharju, was originally planned to run from 2000–2005 (Fig. 2d). The purpose of the trial was to study the effect of genetic modification on the key enzymes of carbon and nitrogen metabolism, the Rubisco small subunit gene (*Rbc*S, Kontunen-Soppela et al. 2010) and the nitrate reductase gene (*Nr*), and the marker gene *npt*II (Ryynänen et al. 2002). Unfortunately, the trial was destroyed, the field arrangement was lost and the trial had to be ended.

At the University of Helsinki field experimental site of Viikki, a silver birch field trial was established from 2000–2003 to study biotic interactions between trees, fungi (Pasonen et al. 2005, 2009), and insects, and to evaluate growth characteristics and adaptive traits (Pasonen et al. 2008) and palatability for mountain hare and roe deer (Vihervuori et al. 2012). The policy-relevance of the GM field trial and its public context were also evaluated (Valve et al. 2010). The trial consisted in total of 1050 seedlings from 15 randomly selected chitinase-expressing (Lohtander et al. 2008) and four lignin-modified (Seppänen et al. 2007) transgenic lines along with the corresponding non-transgenic control clone birch lines and eight other

Fig. 2 Finnish GM forest tree field trials. **a** *P. abies* pollinated by GM pollen in 1996 and covered by protective bags (*photo* Teijo Nikkanen), **b** *P. abies* female cone treated with Triton prior to isolation (*photo* Jouko Lehto), **c** Chitinase *B. pendula* at the Viikki field trial (*photo* Ari Pappinen), **d** RbcS *B. pendula* at the Punkaharju field trial in year 2004 (*photo* Jouko Lehto), **e** The effect of enhanced cytokinin and ethylene signalling on wood formation in *P. tremula* × *tremuloides* (*photo* Tuomas Puukko)

non-transgenic "wild" lines (Fig. 2c). Chitinase expressing lines with high or intermediate accumulation of the chitinase transcript were more resistant to birch rust than those showing a low accumulation, but in general GM birch plants were less resistant to leaf spot disease than corresponding control lines (Pasonen et al. 2004). The level of transgene expression did not affect growth or leaf phenology, but instead it influenced parameters related to the stress status of a tree. Some transgenic lines showed deviant performance compared to the control plants but these differences were suggested to be due to position effects of the transgene, not due to the level of transgene expression. Tree genotype was a more important factor determining the structure of fungal communities than the transgenic status of the plants (Pasonen et al. 2009). Thus, the conceivable changes in the litter quality, possibly derived from pleiotropic effects due to gene modification, were either absent or too weak to affect the decomposability of the litter in the soil (Vauramo et al. 2006). One transgenic birch line was significantly less palatable to hares than the wild-type control. The results of this study indicated that plant genotype may be related to the palatability of plant material when transgenic and wild-type woody plant material is used as winter food for hares (Vihervuori et al. 2012). However, the results did not reveal changes in the palatability of the studied transgenic lines that could be related to the functioning of the transgenes.

In Joensuu, at the University of Joensuu (at present University of Eastern Finland) within the area of the botanical garden Botania, a silver birch field trial was established in 2005–2008, and continued until 2012. The purpose of the release was to study environmental effects associated with the prevention of flowering by genetic modification (Lännenpää et al. 2005; Lemmetyinen et al. 2008). Potential unforeseen adverse effects were also a special object of interest in the experiment, and they were studied by examining e.g. changes in growth habit and branching, chemical composition of plants, and plant resistance to herbivores and pathogens. Flowering was prevented by introducing a barnase gene regulated by the promoter (BpMADS5pro) of the inflorescence-specific gene (*BpFULL*1) from silver birch. The marker gene was *npt*II and the transgenic material was produced by *Agrobacterium*-mediated transformation. The experiment included four genetically modified lines of birch clone BPM5. The results indicated that the non-flowering phenotype was stable: none of the individual plants of the transgenic lines had inflorescences during the study period (2005–2012), although the early flowering non-transgenic control line flowered every season. Compared to the wild type, the transgenic lines had slightly slower height growth, increased branching, and smaller leaf size. The increased branchiness was positively correlated with the abundance of spiders on saplings, which may affect the number of herbivorous insects. However, there were no statistically significant differences in visual damage estimates of the lines. The transgenic lines did not differ from the wild type either in chemical composition or in insect feeding assays, but rather, the results varied depending on the transgenic line and insect species.

3.3 *Hybrid Aspen (1 Field Trial)*

At the University of Helsinki field experimental site of Viikki, a field trial with genetically modified hybrid aspen (*P. tremula* × *P. tremuloides*) clones T89 and 51 was established in 2013 and is scheduled to run until 2018 (Fig. 2e). The general aim of the release is to conduct basic research on hybrid aspen and understand the role and function of the plant hormones ethylene and cytokinin in wood formation. The second major aim is to understand the function of stomatal regulation in hybrid aspen under natural growth conditions. The results are also expected to provide important knowledge for traditional tree breeding.

The transgenes introduced overexpress ethylene response factor (*ERF*) genes or a gene enhancing cytokinin signalling under a constitutive LMX5 promoter, which directs the expression to wood forming tissues. Overexpression of these genes modifies wood properties and enhances growth. In another set of genetically modified hybrid aspens the function of stomata has been modified by silencing the hybrid aspen orthologues of *Arabidopsis thaliana* slow anion channel-associated 1 (*SLAC*1) and *SLAC*1 homolog 3 (*SLAH*3) using transgenes under the control of a guard-cell-specific promoter. Also, the *A. thaliana SLAC*1 gene was expressed in hybrid aspen since poplar *AtSLAC*1 orthologs might have a different function. The hygromycin phosphotransferase (*hpt*) marker gene, conferring hygromycin B-resistance, and/or the neomycin phosphotransferase (*npt*II) marker gene, conferring kanamycin-resistance, were used to select the genetically modified hybrid aspens. Genetic transformation was performed by *Agrobacterium*-mediated transformation.

4 Field Tests in France (7 Field Trials with GM Poplar)

Over the last 25 years, seven different field trials with GM poplar trees have been set up in France, mainly by INRA with one by AFOCEL (now part of FCBA), to explore the behaviour of GM trees in natural conditions.

The first French GM tree field trial was set up in 1991 by the INRA-AGPF (Forest Tree Breeding Research Unit) in their tree nursery near Orléans and lasted for 8 years until 1999 (Fig. 3a). The aim was to assess the stability of transgene expression in a model poplar tree (INRA clone 717-1B4, a *P. tremula* × *P. alba* hybrid) over several years growth and development. Toward this end, expression of a reporter gene driven by a constitutive promoter was measured in many samples (buds, leaves, stems and roots) collected at different seasons and over the years. This first experiment was completed by a second field trial, involving a higher number of transgenic lines in order to evaluate the stability in trees of tissue-specific expression of a transgene over time and space (Pilate et al. 1997; Hawkins et al. 2003). Overall, all these experiments led to the conclusion that transgene expression was stable although highly variable.

Fig. 3 French field trials as regular poplar plantation. **a** First European small-scale field trial with GM trees set up in 1991. **b** Rows of 5-year-old lignin-modified aspen trees (photo taken February 2004), **c** Catkin removal is easy as flowering takes place before leaf emergence, **d** Final harvest of 10 year-old lignin-modified aspen trees (photo taken February 2007)

The field trial established in April 1994 by AFOCEL evaluated GM poplar cv Hunnegem expressing the *bar* gene which confers resistance to ammonium gluphosinate (a herbicide also known as phosphinotricin or Basta). The transgenic plants were produced in collaboration with M. De Block (PGS, Belgium; De Block 1990). The objective of the trial was to assess the durability of herbicide resistance during a regular short rotation coppice cycle and assess whether initial direct herbicide treatment on the GM trees would insure maximized tree growth in the first year and potentially save on additional treatments in following years (the currently recommended cultural practice). Prior to field installation, the transgenic poplar lines had been evaluated for 2 years in confined greenhouse conditions in order to check for any obvious effects of the genetic modification on tree growth and development. The field trial was set up on an area of 0.8 ha, with 4 blocks of 20 plants per line (3 different transgenic lines and 2 WT lines, Hunnegem and Beaupré) and per treatment (3 herbicide treatments). The trial was monitored by the Regional Plant Protection Services.

The cv Hunnegem is a male poplar clone (*P. trichocarpa* × *P. deltoides*) that flowers in the field after 8–10 years. The trees were grown as SRC in an agricultural area, with 3 seven-year rotations scheduled, and the first coppicing occurred in 2001. However, the trial was prematurely terminated in 2005. A "safety" zone was regularly maintained around the plantation with a 2 m deep ditch to prevent any risk of sucker spreading beyond the trial site. Each year, herbicide spray tests consistently revealed complete Basta resistance in the transgenic lines, while soil treatment in and around the plantation never revealed any evidence of the emergence of Basta resistant suckers. The observations and measurements were performed at least three times per year during the whole duration of the trial. After 7 years, no flowering was observed and vegetative growth was similar in GM and wild-type trees. The field trial was prematurely terminated in 2005, in part because of the general bad shape of the trees due to severe rust attacks in 2000 and the following years, as Hunnegem trees, whether GM or not, are very sensitive to this disease. For 3 years following the termination of the trail, the plot was carefully monitored and regularly treated with Basta in order to identify potential regrowth of transformed material, but none occurred.

The other five French GM tree field trials were again located at the INRA tree nursery near Orléans and assessed the effects of several different genetic modifications on tree growth and development in near natural conditions in the model INRA clone 717-1B4 (*P. tremula* × *P. alba* hybrid). Unlike the first 1991 trial using marker genes, these later trials involved trees modified for agronomical or technological traits where the objective was to verify that valuable phenotypes observed in the greenhouse were also present under plantation conditions. All field trials were set up and monitored in accordance with the French regulations on GM dissemination and after getting the suitable authorizations (Fig. 4b). Generally, these authorizations were granted provided suitable containment was ensured to prevent any reproductive or vegetative dissemination, irrespective of the transgene present (Fig. 3c).

A trial between 1995 and 1999 evaluated GM poplars transformed with a cysteine proteinase inhibitor gene. The poplars had been shown to be tolerant in laboratory feeding tests to *Chrysomelae tremula*, a coleopteran phytophagous beetle that can cause important damage to poplar nurseries and short rotation coppices (Leplé et al. 1995). In the controlled conditions of a greenhouse, the expression of the cysteine proteinase inhibitor gene translated into an increased mortality of *C. tremulae* fed on transgenic foliage. However, this effect was probably too weak to induce any notable effects on *Coleoptera* feeding on field-grown transgenic trees.

Three field trials were carried out to evaluate GM trees modified for lignin quality or quantity (Figs. 3b, d, and 4a, e). Lignin is a major cell wall component in xylem and fibre cells that provides the mechanical support necessary for large trees to support their own weight, and the impermeability needed for efficient sap conduction between roots and leaves. Variations in lignin content or composition can influence wood technological properties either chemically (for pulp, paper, or

Fig. 4 French field trial as SRC. **a** Plantation of SRC lignin-modified aspens. **b** Field trial inspection by Plant Protection Service (photo taken spring 2008), **c** Sampling of soil cores/roots for assessment of environmental impact of GM modification on soil fungal communities by L. Danielsen and colleagues (photo taken October 2009), **d** Visit of the French GM field trial during a workshop of the COST FP0905 action in September 2011, **e** SRC of lignin modified GM aspens in February 2012 just before harvest

bioethanol production) or physically (mechanically) for solid wood use, and could also potentially lead to major disturbance in tree growth and development, both effects requiring field test evaluation.

These evaluations have been performed on trees down-regulated for any one of 4 key genes from the monolignol biosynthetic pathway, each of them leading to specific modifications in lignin polymer structure and in turn in wood properties. Notably, wood from one transgenic line down-regulated for CAD, the last enzyme from the monolignol pathway, appeared easier to delignify during the pulping process making it possible to use less harsh chemicals to produce, with the same

Fig. 4 (continued)

yield, a paper of better quality (Pilate et al. 2002). In addition, the genetic modification did not seem to affect tree growth and development. Other lignin-modified field-grown GM trees were down-regulated in caffeic acid *O*-methyl transferase (COMT), caffeoyl-coenzyme A *O*-methyl transferase (CCoAOMT) or CCR. Several CCR-down-regulated transgenic lines were also superior to wood from control trees for pulping (Leplé et al. 2007) as well as for biofuel production (Van Acker et al. 2014). However, in these CCR-suppressed GM trees, the lignin modifications led to a decreased growth rate most likely due to problems with vessel conductance and sap transport. Other analyses performed on these field trials included global assessment of the effects of genetic modifications on wood decomposition (Pilate et al. 2002) as well as the effects of lignin modifications on soil and root fungal diversity (Fig. 4c) (Danielsen et al. 2012, 2013).

At the moment, field trials with GM crops and trees have been interrupted in France. This is the case even though 25 years of safe field evaluations of transgenic trees have yielded a lot of new knowledge on the behaviour of GM trees, have evaluated their potential effects on the environment, have identified promising genotypes for more applied studies, and have suggested new routes for producing trees with improved wood for industrial applications such as biofuel production. In addition, during these 25 years of experiments, a lot of expertise was gained in the management of GM tree trials, and how to control vegetative and sexual dissemination. Finally, GM field trials offered a useful platform for communicating directly with the public and the media about experiments with GM trees (e.g. Fig. 4d).

5 Field Tests in Germany (4 Field Trials with GM Poplar)

In Germany, the first release experiment with GM trees was established in 1996 by the Federal Research Centre of Forestry and Forest Products, Institute of Forest Genetics and Forest Tree Breeding, Grosshansdorf. Permission for this field trial was given for five years, ending in autumn 2001. In total, eight transgenic aspen clones comprising two different gene constructs (35S::*rol*C in aspen lines Esch5 [*P. tremula* × *P. tremuloides*], Brauna11, W52 [both *P. tremula*]; and rbcS::*rol*C in aspen line Esch5) and the three control lines were planted out on a field of approximately 1500 m^2 (Fladung et al. 2004; Fig. 5).

The aim of this experiment was to investigate the stability and expression of foreign genes in transgenic trees on a long-term basis under field environmental conditions. The transgenic plants were measured with respect to growth (height, stem diameter, leaf size) and physiological parameters (dormancy, flushing, flower formation, hormone levels, metabolites) (Fladung et al. 2004). Phenotypic instabilities detected in individual plants as reversions to wild type from the dwarfy phenotype of the 35S::*rol*C transgenic trees were also investigated in the lab using molecular methods (Fladung et al. 1996; Fladung 1999; Kumar and Fladung 2000a, 2001; Fladung and Kumar 2002). Based on these results, a quick and easy molecular test was developed to assess the possible stability/instability of a foreign gene in transgenic trees at the genomic level (Kumar and Fladung 2000b).

The mycorrhizal status of the roots was analyzed in the transgenic and non-transgenic trees. No differences were found in the mycorrhizal status between control and transgenic trees with one exception in the transgenic clone Esch5:35S::*rol*C#5 (Kaldorf et al. 2002), however, the ectomycorrhizal spatial distribution was not the same in the different aspen clones (Kaldorf et al. 2004a). The mycorrhizal symbiosis is believed to provide the highest probability for horizontal gene transfer due to its special anatomical features forming the plant/fungus interface. Therefore,

Fig. 5 German field trial as SRC. **a** RolC-transgenic aspen few days after planting. **b** Field trial in spring 1998. **c** Field trial in December 2000. **d** Harvest and sampling of trees in October 2001. **e** Removal of rootstocks from the soil. **f** Burning of transgenic material after clearing the field trial

to unravel possible horizontal gene transfer events, small cuttings of transgenic aspen containing the rbcS::*rol*C gene construct were grown in vitro together with the ectomycorrhizal ascomycete *Phialocephala fortinii*, which was isolated from mycorrhized roots obtained from the field-grown aspen (Fladung et al. 2004). After 12 weeks of co-cultivation, genomic DNA was isolated only from hyphae of the *P. fortinii* mycorrhizas. No *rol*C signal could be detected in any of the analyzed *P. fortinii* colonies (Kaldorf et al. 2004b).

The susceptibility of the leaves to *Melampsora* and *Venturia* was analyzed according to a special evaluation technique. The 35S::*rol*C transgenic trees had a higher degree of susceptibility than the control plants (Fladung et al. 2004). Potential risks of transgene escape via vegetative dispersal were also evaluated. In the 4th and the 5th year after establishment of the field trial, an increasing number of root suckers could be observed developing from roots of transgenic and non-transgenic trees (Fladung et al. 2003). The phenotype of all suckers was wild type or rbcS::*rol*C. It could clearly be demonstrated that vegetative propagation via root suckering is easily controllable by best practice tree cultivation.

The second field release experiment was mainly designed to quantitate possible horizontal gene transfer from transgenic trees to mycorrhiza fungi under optimized conditions. The field experiment was again initiated by the Federal Research Centre of Forestry and Forest Products, Institute of Forest Genetics and Forest Tree Breeding, Grosshansdorf, in close cooperation with the University of Tuebingen. This experiment was conducted in a closed 1000 m^2 large field station and approved for 3 years from 2000 to 2003. Transgenic aspen trees were generated containing the *Streptococcus hygroscopicus* bar gene conferring herbicide (BASTA) resistance under the control of a fungal GPD promoter (Nehls et al. 2006). Roots containing ectomycorrhiza were sampled twice a year (May and September in 2001 and 2002, respectively). Mycorrhizas were isolated from soil cores under a dissection microscope and morphotyped according to Agerer (1987–1993). In total, 120,000 fungal samples were isolated and transferred to BASTA-containing selection medium.

Out of these, 40,000 isolates were revealed to be BASTA-tolerant and could be divided into two groups. The first group (majority 95 %) contained fast-growing fungi that colonized a Petri dish within a few days and were classified as saprophytic fungi (mainly *Aspergillus*) (Nehls et al. 2006). The second, smaller group (5 %) grew more slowly and morphological and molecular analyses revealed them also to be ascomycetes, representing endophytic but not ectomycorrhizal fungi (Nehls et al. 2006). PCR amplification assays were initiated to search for the presence of the bar gene in all fungal isolates. No bar gene signal could be obtained from any of the slow growing BASTA-tolerant isolates, suggesting that the fungal isolates were naturally herbicide resistant (Nehls et al. 2006).

Two further field trials carried out from the University of Freiburg dealt with the topic of phytoremediation. The experiments took place in Helbra (Saxony-Anhalt, from 2002–2004), and in Helbra and Mansfeld (Saxony-Anhalt 2003–2005) (Peuke and Rennenberg 2006). The transgenic poplars (*P. tremula* × *P. alba*) overexpressed the γ-glutamylcysteine synthetase (*gsh*I or γ-ECS) gene from *E. coli* to enhance glutathione (GSH) levels in the cytosol leading to an elevated capacity for phytochelatins (PC) synthesis. PCs are naturally occurring peptides with good binding affinities for a wide range of heavy metals (Kang et al. 2007). The expression of the *gsh*I gene in transgenic poplars led to a two- to four-fold enhanced GSH concentration in the leaves. In greenhouse experiments, the *gsh*I transgenic poplars revealed a high capacity for heavy metals detoxification of soils.

The capacity of GshI overexpressing poplar for phytoremediation was evaluated in the two field trials at different levels of copper contamination and under different climatic conditions (Peuke and Rennenberg 2006). The transgenic poplars were genetically stable when grown under field conditions. No horizontal gene transfer to rhizosphere microorganisms or to mycorrhiza was detected and no general impact on the environment could be observed (Peuke and Rennenberg 2006). *gsh*I transgenic poplars possess a higher capacity for accumulation of heavy metals when grown on highly contaminated soils. However, on control sites or sites with low contamination of heavy metals, no differences in phytoremediation capacity could be observed between *gsh*I transgenic and control poplars (Peuke and Rennenberg 2005, 2006).

6 Field Tests in Spain (2 Field Trials with GM Poplar)

The number of field trials with transgenic trees in Spain is very limited. Fifteen records for authorization of field trials are available at the Spanish Commission on Biosafety (http://www.magrama.gob.es), but published data are only available from a few studies including assays of transgenic poplar for increased yield using the hybrid *P. tremula* × *P. alba* 717-1B4 clone, a classical clone in forest tree research used in many studies.

The work of Jing et al. (2004) describes a field trial of transgenic poplar overexpressing a glutamine synthetase gene. This research was carried out by the University of Málaga with the aim of evaluating the stability of transgene expression and to confirm previous results indicating an effect on plant growth rate. Glutamine synthetase is involved in nitrogen assimilation and amino acid biosynthesis, representing a key enzyme in the biosynthesis of nitrogen compounds (Gallardo et al. 2003). Transgenic poplars overexpressing glutamine synthetase were characterized by higher protein, chlorophyll and growth in chamber or greenhouse studies (Gallardo et al. 1999; Fu et al. 2003). The transgenic trees also exhibited increased photosynthetic and photorespiratory capacities, improving their tolerance to drought stress compared with control plants (El-Khatib et al. 2004). The higher production of the glutamine synthetase enzyme also conferred a higher tolerance to phosphinotricin, a common herbicide used in controlling weeds in plantations (Pascual et al. 2008). The results from a 3-year field trial confirmed the previous results obtained in greenhouse studies and the transgenic poplars reached average heights 41 % taller than control plants after the third year of growth (Jing et al. 2004). This study was carried out in the province of Granada (Spain) with plants from eight independent transgenic lines. No flowering was observed during this study, and no apparent effect on nitrate content in soil was observed besides the differences in growth between transgenic and control plants. In addition, no differences in lignin content between transgenic and control plants were observed in 1-year and 5-year-old plants (Jing et al. 2004), but a recent study indicates differences in wood chemistry with respect to control plants in samples from three different lines in a 3-year field study in Virginia (USA; Coleman et al. 2012). Together, the findings suggest a higher capacity of the transgenic trees for both primary N assimilation and re-assimilation of ammonium released in different metabolic procedures, and the potential for improving growth and biomass production of productive clones or other tree species.

(a)
(b)

◀ **Fig. 6** Spanish field trial. Field trial established in Madrid (Spain) in 2012 to test biomass yield under SRC management of genetically modified hybrid poplars (*P. tremula* × *P. alba* INRA clone 717-1B4) developing a high degree of sylleptic branching. The out-door study included the wild type genotype as control, two events over-expressing the RAV1 (Related to ABI3/VP1 1) transcription factor from chestnut, and two events showing a reduced expression of the poplar endogenous genes PtaRAV1 and PtaRAV2. Thirty individuals per genotype were planted into three blocks of 10 plants each. The experimental plot area was 204 m^2, and the chosen plantation density was 10,000 trees/ha. It consisted of 12 × 17 rows with a spacing of 2 × 0.5 m, surrounded by a border row of *P.* × *euramericana* clone I-214 individuals. **a** July 2012 **b** January 2015

The Technical University of Madrid is also carrying out a field trial to evaluate the biomass yield of two events of GM poplars showing a high degree of sylleptic branchiness after over-expression of the chestnut transcription factor *Castanea sativa* RELATED TO ABI3/VP1 (CsRAV1) (Moreno-Cortés et al. 2012). For this purpose, the plantation was established at high density (10,000 trees/ha) and managed under a short rotation coppice system, with irrigation and weed control (Fig. 6). Several growth and biomass productivity determinants were measured before coppicing in 2013, two growing seasons after the establishment of the plantation. One of the *CsRAV1* over-expressing events showed a ca. 17 % larger basimetric area and dried aerial biomass than non-modified poplars. Histochemical and composition analysis of this modified wood did not reveal any change compared to the non-modified poplars. Calorific value determination also showed that modified and non-modified woods were comparable (Moreno-Cortés et al., unpublished results). This specific genetic modification is potentially applicable to any woody species, taking profit from the adaptive characteristics of each species to a particular habitat.

7 Field Tests in Sweden (8 Field Trials with GM Hybrid Aspen)

The first field trial with GM hybrid aspens in Sweden was established in 1994 in Sävar outside Umeå by researchers at the Swedish University of Agricultural Sciences. This, and subsequent *Populus* field trials in Sweden, have been performed with the hybrid aspen (*P. tremula* × *P. tremuloides*) clone T89, and in comparison with field trials in other European countries the Swedish trials have in general been larger in terms of the number of different genes modified and studied in the same trial. In the first trial, either one of five genes from *A. tumefaciens* affecting auxin or cytokinin metabolism, two genes from *Agrobacterium rhizogenes* affecting root formation, *iaalys* from *Pseudomonas savastenoi*, three genes originating from *E. coli* (*uid*A, *hpt* and *npt*II), one gene from winter flounder coding for an antifreeze protein (AFP) and finally the luciferase from American firefly, were introduced into the plants, The purpose was to characterize how growth in plants was affected by modified hormone metabolism, as well as more general studies on long-term stability of transgene expression. The AFP protein was included to investigate if chilling tolerance could be modified. The trial led to a fierce discussion in Sweden about the potential risks associated with the GM technology, and in particular the experiment with the winter flounder AFP protein was heavily criticized with the argument that such trees may escape and reproduce uncontrolled in the environment since they potentially could outcompete the native trees. The idea of moving "fish genes" to trees was not appreciated either. However, quite ironically it turned out that genotypes based on the T89 clone, including those containing the antifreeze gene, did not even survive the first winter in the field (in northern Sweden).

The second GM hybrid aspen field experiment, performed in 2004 and 2007 in Umeå, was a short-term experiment with trees in pots transferred to the field only for a short time in the summer, with a gene involved in the regulation of photosynthetic light harvesting (PsbS). The purpose was to see if the consequences of a lack of PsbS, engineered using an antisense transgene, were the same in hybrid aspen as in *Arabidopsis*.

The third experiment with GM hybrid aspen was established at two sites (in Halmstad and Laholm municipalities in southern Sweden) in 2010. In this experiment, 16 genes were selected from large gene finding projects performed by researchers at Umeå Plant Science Centre, in collaboration with the company SweTree Technologies. The basis of the selection was that each of these 16 genes had, upon overexpression or RNAi-mediated down-regulation, resulted in trees that under greenhouse conditions grew faster than the wild type. The aim of the field experiment was to see if the increased growth was transferrable to field conditions, and if this led to any trade-offs in terms of decreased hardiness or increased susceptibility to pests. Three independent lines transformed with each gene construct were included, in total 48 GM lines, and in total (including wild type) ca. 700 trees on each of the sites (Fig. 7). Many of the genes coded for transcription factors, while others had other more or less well defined functions. For all traits other than increased growth, the set could be regarded as a random set of 16 out of over 1000 genes whose expression levels had been modified in the large gene-finding projects. Therefore, this set of 48 clones could also be used to quantify unintentional changes to properties resulting from the transformation event, and for this—as well as demonstration purposes—a set of randomly selected natural genotypes of aspen was planted next to the GM trees on one site. This would allow for a comparison of all variation created unintentionally to the amount of natural variation of the same trait, as the same traits (e g growth, leaf shape, grazing by a wide variety of insects, pathogen infection, frost damage etc.) were measured on all trees. The establishment of the field experiment led to some media debate, NGOs were invited to participate in a discussion about the setup (but did not show up), background information was published (http://www.upsc.se/about-upsc/other-information/gmo-information.html), an information meeting was held and despite a few statements in local media, the experiments have been performed uninterrupted. The experiment has recently got a permit for a 5-year extension. The permit stipulates that a fence should surround the experiment (10 m from the trees), that native aspens 50 m from the fence should be removed as well as suckers between the trees and the fence.

Fig. 7 Swedish field trial. The GM hybrid aspen field trial established in Laholm, Sweden 2010. Ca. 800 trees (16 transgenes × 3 independent transformation events for each gene = 48 lines, plus wild type T89) with modified growth properties. *Photo* Stefan Jansson

The site should be regularly inspected for flowering and—if flowering is detected (which has not happened so far) all ramets of the flowering genotype should be removed.

Based on the experiences from this experiment, several others have been initiated at the same site (in Laholm). One rather similar experiment was initiated in 2011, led by researchers at the Swedish University of Agricultural Sciences, but with seven genes where modification of expression levels had resulted in altered wood properties under greenhouse conditions. Experiments led by SweTree Technologies, with 14 genes where altered expression levels had led to increased growth in the greenhouse, or which may influence drought tolerance, were initiated in 2011, and an experiment with 7 genes in 2012. Similarly, two experiments were started in 2014, one with trees with modified expression of 21 genes involved in the regulation of wood properties, and one with 27 genes potentially regulating phenology. At the Laholm site there are at present six experiments ongoing, in total covering almost 5 ha and with ca. 3000 trees with modified expression levels of over 90 different genes, the vast majority from *Populus* but a few originating from other species.

8 Field Tests in United Kingdom (2 Field Trials with GM Poplar)

Although there have been five separate notifications of GM tree trials in the United Kingdom (http://gmoinfo.jrc.ec.europa.eu/overview/GB.asp), only one has been reported on fully in the scientific literature (Pilate et al. 2002; Halpin et al. 2007). All of these trials were approved between 1993 and 1996 and involved eucalyptus, poplar or apple. Two approvals concerned the trialling of marker gene systems in trees, one involved the introduction of glyphosate tolerance, and two focused on the alteration of lignin biosynthesis. The lignin modification trials were planted one year apart at the same location at Syngenta's Jealott's Hill International Research Station in Berkshire (Fig. 8) in collaboration with similar trials by INRA in Orléans, France, within a research project with part-funding from the European Commission (FAIR CT95-0424). The trees grew for 3 and 4 years but both trials were ended abruptly in 1999 when the trees were vandalized and chain-sawed by activists. This likely explains why no further trials of trees have been performed in the UK, and research that the country was world-leading in has largely moved elsewhere.

Fig. 8 United Kingdom field trials. The lignin-modified GM hybrid aspen field trials were established at Jealott's Hill, Berkshire in 1995 and 1996 and grew for 4 or 3 years respectively until destroyed in 1999. Front trial block: 72 trees planted in 1995 in a randomized block design included two CAD antisense events and two COMT antisense events (12 ramets per event) plus 24 wild type 717-1B4 trees. Rear trial block: 80 trees planted in 1996 in a randomized block design included a CAD antisense event and a COMT antisense event (24 ramets per event) plus 40 wild type 717-1B4 trees. (*Photo* Claire Halpin)

Results of the longer four-year trial have been presented in several papers (Pilate et al. 2002; Halpin et al. 2007) and are discussed below.

Ironically, it was only evaluation of some of the ecological and environmental impacts of the trial, such as investigation of mycorrhizal associations that was prevented by the attack on the trees. The more applied objectives of the trial were unaffected and evaluation of wood from the trees for stability of gene suppression and lignin manipulation, and the utility in paper production, proceeded as planned (Pilate et al. 2002). Two types of transgenic plants suppressed in two different lignin biosynthesis genes and originally produced by INRA, were present in the trial. Some were suppressed in expression of caffeic acid/5-hydroxyferulic acid *O*-methyl transferase (COMT), while others were suppressed in expression of CAD (CAD). The trial was a sister trial to one performed at the same time by INRA near Orléans, France, on the same clonal trees. As has already been described earlier in this chapter for the trees in the French trial, the UK-grown trees grew normally but had modifications to lignin structure. The trees with reduced CAD expression were easier to delignify so that either less chemical could be used or more pulp could be recovered (Pilate et al. 2002). Gene suppression appeared to be stable over the 4 years of growth and at final harvest CAD enzyme activity was still reduced by 53–85 % of wild-type levels in the CAD-suppressed lines, while COMT activity was reduced by 56–68 % in the COMT-suppressed lines.

The UK trial was constantly monitored to determine whether the lignin-modified trees had altered biological or ecological impacts. Transgene escape was not possible since the trees were *P. tremula* × *P. alba* female clones (i.e. without pollen) and were harvested before any could flower. The incidence of visiting and resident insects was regularly monitored and visible damage and disease symptoms were recorded. All trees attracted a wide variety of insects but feeding damage to the trees was very modest and there was no significant incidence of disease other than occasional low frequency rust on individual leaves (Halpin et al. 2007). Neither insects nor disease affected the transgenic trees any more than the wild-type trees, and all trees remained very healthy throughout the four-year trial (Pilate et al. 2002).

The influence of modified lignin on soil microbiota and properties, and on CO_2 evolution and decomposition, were also considered using samples collected after the termination of the trial. When chopped roots were buried in a test soil over 5-months, slightly more CO_2 was emitted from the transgenic root samples compared to the wild type, indicative of a small increase in the rate of decomposition. The effect was greatest in the first month when labile polysaccharides, not lignin, would be decaying, suggesting that the modified lignin was less effective at protecting other cell wall components from microbial attack than wild type lignin is (Pilate et al. 2002). A longer study set up with trunk wood from the field trial harvest showed no statistically significant differences in CO_2 evolution or mass loss between the transgenic and wild-type samples (Tilston et al. 2004). Variation between samples of the same genotype was greater than variation between genotypes. There were no significant differences in C, N or microbial biomass between soil samples taken from beneath transgenic or wild type trees (Pilate et al. 2002).

Although microbial basal respiration was lower in soils from under two transgenic lines, this did not correlate consistently with transgene expression, and no significant differences in the substrates supporting microbial growth were evident between any of the soils. However, soil taken from a grassy area less than 4 m from the trees was significantly distinctive from the soil under the trees in all properties measured. The cumulative data from this trial suggests that plant residue decomposition and soil properties are more influenced by the type of plant grown and by field environmental variability than they are by genetic modifications to lignin (Hopkins et al. 2007).

9 Conclusion

Of nearly 800 GM tree field trials that have been approved to date worldwide, it is notable that only a very small number, perhaps fewer than 50, have been approved in Europe (Häggman et al. 2013). Indeed, most experiments involving GM trees—as for GM plants in general—do not go beyond the greenhouse stage as it is assumed that this is enough to assess the function of a gene in a biological process, at least for basic research purposes. Setting up any field trial requires time, money, space and technical skills, as well as the tenacity to navigate the considerable hurdle of assembling an application strong enough to gain regulatory approval. For GM trees, there are extra burdens as the evaluation of perennials needs to last for several years, and there are additional constraints related to the regulations on GM confinement to prevent dissemination of the transgene. Despite 25 years of safe field trials, these burdens only seem to be getting heavier (Walter et al. 2010). Taking all of these issues into account, it is easy to understand why most studies of transgenic trees are only performed in greenhouse conditions and European GM tree field trials remain the exception.

The GM tree field trials that have been set up in Europe have addressed numerous important scientific objectives, including evaluation of tree-specific traits (lignin and wood properties, flowering/sterility, transgene expression stability over several years), assessment of general agronomic traits (biomass, resistance to biotic and abiotic stress) and consideration of biosafety issues (horizontal gene transfer, possibility of vegetative dispersal, effect on biotic interactions) (see Table 1). Most experiments have been carried out on poplar, the model species for tree biology, and especially one clone that is easy to transform, INRA 717-1B4 (*P. tremula* × *P. alba*) and also to some extent on the clone T89 (*P. tremula* × *P. tremuloides*). These field trials can all be considered as confined trials as they were set up with strict controls on dissemination and usually terminated before sexual maturity. They were valuable tools for basic research, and for increasing our understanding of some tree-specific biological traits (dormancy, phase change, wood formation, etc.). Besides, they were also the first step towards more applied research and were, for example, essential to produce wood in sufficient amounts for industrial evaluation, under environmental conditions close to those of regular tree field trials. Field tests are an obligatory early step towards future commercial deployment, although, in Europe, no commercial deployment is foreseen in the short or mid-term. It is noteworthy that even though all experiments performed so far involved relatively small-scale field studies, a long way from any commercialization, none of the results detected any unforeseen risks or effects. Yet several trials were illegally destroyed by anti-GMO activists, and the burden to obtain authorization becomes heavier and heavier, despite the fact that virtually all of the concerns raised against GM trees are related to the commercial deployment stage, and contained field trials are the way to explore the evidence base for those concerns.

In this context, and to return to the question posed in the introduction, what did we really learn from these experiments?

A major outcome was gaining the knowledge that the expression of both silencing and gene-expression transgenes is generally stable over time even in long-lived perennial species such as trees, albeit sometimes variable in level. This has been most easily studied using marker genes such as bacterial *uid*A and *npt*II driven by a constitutive or tissue-specific promoter (Hawkins et al. 2003), or by using *rol*C, whose expression is associated with a very strong phenotype that is easy to monitor (Kumar and Fladung 2001). These studies led to the conclusions that transgene insertions were very stable as was transgene expression. Similarly, barnase expression from an inflorescence-specific promoter prevented flowering in silver birch over several years (Lännenpää et al. 2005), and poplars expressing glutamine synthetase grew taller over 3 years (Jing et al. 2004). Silencing transgenes appear to be equally stable, with the phenotypic effects of silencing a specific gene (e.g. modified lignin) persisting over several years of a trial and target enzyme activity remaining reduced (Pilate et al. 2002).

The trials also increased confidence that the phenotypic effects of transgenes initially evaluated in controlled greenhouse conditions, are frequently sufficiently robust to persist in fluctuating natural field environments, as evidenced by many of the examples described in this chapter. This validated several genes as real commercial targets for improving wood for industrial applications and enabled initial evaluation of the benefits likely to accrue. So, for example, down-regulation of CAD and consequent specific changes to lignin structure led to an increase in poplar pulp yield, or to an opportunity to make cost savings by using less pulping chemicals, with associated environmental benefits. Importantly, in this and many other field trials where growth modification was not the object of the modification, the GM trees showed no significant differences in growth compared to control trees. On the other hand, several deliberate strategies successfully led to valuable biomass increases in poplar. Although robustness of phenotype between greenhouse and field was generally the case, it was not always so. For example, transgenic poplars expressing a proteinase inhibitor were shown to tolerate insect attacks in greenhouse experiments (Leplé et al. 1995) but the effect was rather weak and was no longer significant in natural field conditions in France (unpublished results). In an example from the USA (see chapter by Strauss et al.), poplars down-regulated for 4-coumarate-CoA ligase (4CL, a key lignin biosynthesis gene) exhibited a sizeable growth increase in the greenhouse (Hu et al. 1999), but when similar plants were independently produced and grown in a field trial, they showed significant reduction in growth (Voelker et al. 2010). Crucially, the results obtained in the field can stimulate important new research directions. For example, poplars down-regulated for CCR produced wood that was more easily processed into ethanol, but the trees had reduced biomass yield, most likely because of xylem collapse and water conduction defects (Van Acker et al. 2014). Down-regulation of CCR would remain a valuable strategy if the yield penalty could be overcome, and this may be possible by targeting gene down-regulation to fibres only, using suitable promoters.

Although virtually all GM field trials enable evaluation of the robustness of genetic modifications over time, very few experiments have evaluated the robustness of the modifications and general field performance across different sites. Three duplicated field trials where the same lignin modified poplars were installed in France and in England (Pilate et al. 2002), in Belgium and in France (Van Acker et al. 2014) or in two sites in Sweden, allow such comparison. All revealed results to be predominantly and qualitatively similar at different sites, just varying occasionally in degree. Multisite field trials certainly deserve to be more frequently pursued, in just the same way that they are recommended for testing new tree varieties created through classical breeding.

In terms of new knowledge gained on the ecological, environmental and biosafety issues of growing GM trees, the European field trials have provided some of the best data to date from which conclusions can be drawn that are actually substantiated by evidence. Where microbial or fungal communities beneath GM trees were evaluated, they showed no significant differences to those under wild type trees, and the type of vegetation (Pilate et al. 2002) or the background genotype (Vauramo et al. 2006) was more important in determining the structure of soil communities than the transgenic status. In general, the degree of variation between GM trees and wild type trees is smaller than variations between different wild type genotypes when grown in the fluctuating conditions of the natural environment (Danielsen et al. 2012; Jansson et al. ms in preparation). Chitinase-expressing silver birch litter was not altered in decomposability (Vauramo et al. 2006) while even roots with modified lignin displayed only a small increase in the rate of decomposition (Pilate et al. 2002). No horizontal gene transfer could be detected between *rol*C expressing poplars and mycorrhiza (Kaldorf et al. 2004b; Nehls et al. 2006) and vegetative dispersal via root suckers either did not occur or was easily controlled by conventional means (Fladung et al. 2003). There was no evidence for significant unanticipated effects in any of the trials. We should emphasize that there has been no study in Europe on gene flow (vertical gene transfer), as in most cases, authorization for field trial was only given provided there was no dissemination of GM material into the environment from pollen, seeds, or suckers. Wind or insect pollinated species certainly have different requirements with respect to isolation distances. However, this will not be so great an issue for short rotation coppices where trees are harvested before being able to flower. The significance of these risks (of dissemination) remains an open question that deserves to be addressed in some cases when the genetic modification may confer an increased fitness to the GM tree (e.g. pest or pathogen increased tolerance). Paradoxically, conventionally bred commercial poplar clones are not assessed for risk in the same way, despite many being bred to have high resistance to rust, which is a clear fitness advantage. Given that many GM traits only confer an advantage under managed growth conditions or in subsequent industrial processing, only a few traits can be conceived of that might result in increased fitness compared to natural ecotypes. Given that (1) the anticipated variation generated in GM trees is often within the natural variation of the species, (2) any unanticipated effects are even smaller, and that (3) natural variation within a species is in no respect regarded as a risk, nor are traits introduced by conventional breeding irrespective of their influence on fitness, we may seriously wonder whether the huge effort and expense put into some risk evaluations of GM trees is fully rational or justified.

Looking to the future, new technologies are currently under development in plants, including trees (see chapter by Alburquerque et al.), involving nucleases associated with site-specific recognition systems: these technologies allow for specific targeted genes to be mutated, even knocked-out, avoiding the need for incorporation of silencing transgenes, and may enable targeted transgene insertion. These systems can avoid many of the issues that led to criticisms of conventional T-DNA transfer such as the randomness of insertion sites, and the need for integration of additional alien DNA such as selectable marker genes etc. Although, these new technologies (Meganucleases, ZFN, Talen, CrispCAS9) seem very promising, they also need to be carefully assessed, notably for the fidelity of their recognition mechanism. We believe that all new plant material modified for a given trait deserves to be evaluated in a similar way irrespective of whether it was produced by new plant breeding technologies, by classical gene transfer techniques or by traditional breeding (Strauss et al. 2009; Walter et al. 2010): indeed, what matters is the trait modification achieved and not the means of obtaining it.

References

Agerer R (1987–1993) Colour atlas of ectomycorrhizae. Einhorn-Verlag, Schwäbisch Gmünd D-73525, Germany

Aronen T, Nikkanen T, Häggman H (1998) Compatibility of different pollination techniques with microprojectile bombardment of Norway spruce and Scots pine pollen. Can J For Res 28:79–86

Aronen T, Nikkanen T, Häggman H (2003) The production of transgenic Scots pine (*Pinus sylvestris* L.) via the application of transformed pollen in controlled crossings. Transgenic Res 12:375–378

Barton KA, Binns AN, Matzke JM, Chilton M-D (1983) Regeneration of intact tobacco plants containing full length copies of genetically engineered T-DNA, and transmission of T-DNA to R1 progeny. Cell 32:1033–1043

Baucher M, Chabbert B, Pilate G, van Doorsselaere J, Tollier M-T, Petit-Conil M, Cornu D, Monties B, van Montagu M, Inzé D, Jouanin L, Boerjan W (1996) Red xylem and higher lignin extractability by downregulating cinnamyl alcohol dehydrogenase in poplar. Plant Physiol 112:1479–1490

Chen F, Dixon RA (2007) Lignin modification improves fermentable sugar yields for biofuel production. Nat Biotechnol 25:759–761

Clapham DH, Häggman H, Elfstrand M, Aronen T, von Arnold S (2003) Transformation of Norway spruce (*Picea abies*) by particle bombardment. In: Jackson JF, Linskens HF (eds) Molecular methods of plant analysis, vol 23. Genetic transformation of plants. Springer, Berlin. ISBN 3-540-00292-8, pp 127–146

Coleman HD, Cánovas FM, Man H, Kirby EG, Mansfield SD (2012) Enhanced expression of glutamine synthetase (GS1a) confers altered fibre and wood chemistry in field grown hybrid poplar (*Populus tremula X alba*) (717-1B4). Plant Biotech 10:883–889

Custers R (2009) First GM trial in Belgium since 2002. Nat Biotech 27:506

Danielsen L, Thürmer A, Meinicke P, Buee M, Morin E, Martin F, Pilate G, Daniel R, Polle A, Reich M (2012) Fungal soil communities in a young transgenic poplar plantation form a rich reservoir for fungal root communities. Ecol Evol 2:1935–1948

Danielsen L, Lohaus G, Sirrenberg A, Karlovsky P, Bastien C, Pilate G, Polle A (2013) Ectomycorrhizal colonization and diversity in relation to tree biomass and nutrition in a plantation of transgenic poplars with modified lignin biosynthesis. PLoS ONE 8:e59207

De Block M (1990) Factors influencing the tissue culture and the *Agrobacterium tumefaciens*-mediated transformation of hybrid aspen and poplar clones. Plant Physiol 93:1110–1116

El-Khatib R, Hamerlynck EP, Gallardo F, Kirby EG (2004) Transgenic poplar characterized by ectopic expression of a pine cytosolic glutamine synthetase gene exhibits enhanced tolerance to water stress. Tree Physiol 24:729–736

Ellis DD, McCabe DE, Mcinnis S, Ramachandran R, Russell DR, Wallace KM, Martinell BJ, Roberts DR, Raffa KF, McCown BH (1993) Stable transformation of *Picea glauca* by particle acceleration. Biotechnology 11:84–89

Filatti JJ, Sellmer J, McCown B, Haissig B, Comai L (1987) *Agrobacterium*-mediated transformation and regeneration of *Populus*. Mol Gen Genet 206:192–199

Fladung M (1999) Gene stability in transgenic aspen-Populus. I. Flanking DNA sequences and T-DNA structure. Mol Gen Genet 260:574–581

Fladung M, Hoenicka H (2012) Fifteen years of forest tree biosafety research in Germany. iForest 5:126–130

Fladung M, Kumar S (2002) Gene stability in transgenic aspen-*Populus*. III. T-DNA repeats influence transgene expression differentially among different transgenic lines. Plant Biol 4:329–338

Fladung M, Muhs HJ, Ahuja MR (1996) Morphological changes observed in transgenic *Populus* carrying the *rol*C gene from *Agrobacterium rhizogenes*. Silvae Genet 45:349–354

Fladung M, Nowitzki O, Ziegenhagen B, Kumar S (2003) Vegetative and generative dispersal capacity of field released transgenic aspen trees. Trees 17:412–416

Fladung M, Kaldorf M, Gieffers W, Ziegenhagen B, Kumar S (2004) Field analysis of transgenic aspen. In: Walter C, Carson M (eds) Plantation forestry of the 21st century, Research Signpost, pp 393–403

Fu J, Sampalo R, Gallardo F, Cánovas FM, Kirby EG (2003) Assembly of cytosolic pine glutamine synthetase holoenzyme in leaves of transgenic poplar leads to enhanced vegetative growth. Plant, Cell Environ 26:411–418

Gallardo F, Fu J, Cantón FR, García-Gutiérrez A, Cánovas FM, Kirby EG (1999) Expression of a conifer glutamine synthetase gene in transgenic poplar. Planta 210:19–26

GMO register European Commission http://gmoinfo.jrc.ec.europa.eu

Gallardo F, Fu J, Jing ZP, Kirby EG, Cánovas FM (2003) Genetic modification of amino acid metabolism in woody plants. Plant Physiol Biochem 41:587–594

Häggman H, Aronen T, Nikkanen T (1997) Gene transfer by particle bombardment to Norway spruce and Scots pine pollen. Can J For Res 27:928–935

Häggman H, Raybould A, Borem A, Fox T, Handley L, Hertzberg M, Lu MZ, Macdonald P, Oguchi T, Pasquali G, Pearson L, Peter G, Quemada H, Séguin A, Tattersall K, Ulian E, Walter C, McLean M (2013) Genetically engineered trees for plantation forests: key considerations for environmental risk assessment. Plant Biotechnol J 11:785–798

Halpin C, Thain SC, Tilston EL, Guiney E, Lapierre C, Hopkins DW (2007) Ecological impacts of trees with modified lignin. Tree Genet Genomes 3:101–110

Hawkins S, Leplé J-C, Cornu D, Jouanin L, Pilate G (2003) Stability of transgene expression in poplar: a model forest tree species. Ann For Sci 60:427–438
Herrera-Estrella L, Depicker A, van Montagu M, Schell J (1983) Expression of chimaeric genes transferred into plant cells using a Ti plasmid-derived vector. Nature 303:209–213
Hopkins DW, Webster EA, Boerjan W, Pilate G, Halpin C (2007) Genetically modified lignin below ground. Nat Biotech 25:168–169
Hu WJ, Harding SA, Lung J, Popko JL, Ralph J, Stokke DD, Tsai CJ, Chiang VL (1999) Repression of lignin biosynthesis promotes cellulose accumulation and growth in transgenic trees. Nat Biotechnol 17:808–812
Jing ZP, Gallardo F, Pascual MB, Sampalo R, Romero J, Torres de Navarra A, Cánovas FM (2004) Improved growth in a field trial of transgenic hybrid poplar overexpressing glutamine synthetase. New Phytol 164:137–145
Kaldorf M, Fladung M, Muhs HJ, Buscot F (2002) Mycorrhizal colonization of transgenic aspen in a field trial. Planta 214:653–660
Kaldorf M, Renker C, Fladung M, Buscot F (2004a) Characterization and spatial distribution of ectomycorrhizas colonizing aspen clones released in an experimental field. Mycorrhiza 14:295–306
Kaldorf M, Zhang C, Nehls U, Hampp R, Buscot F (2004b) Interactions of microbes with genetically modified plants. In: Varma A, Abbott L, Werner D, Hampp R (eds) Plant surface microbiology. Springer, Heidelberg, pp 179–196
Kang SH, Singh S, Kim JY, Lee W, Mulchandani A, Chen W (2007) Metabolically engineered for enhanced phytochelatin production and cadmium accumulation. Appl Environ Microbiol 73:6317–6320
Kontunen-Soppela S, Sillanpää M, Luomala E-M, Sutinen S, Kangasjärvi J, Vapaavuori E, Häggman H (2010) Photosynthetic characteristics in genetically modified sense-*Rbc*S silver birch lines. J Plant Physiol 167:820–828
Kumar S, Fladung M (2000a) Transgene repeats in aspen: molecular characterisation suggests simultaneous integration of independent T-DNAs into receptive hotspots in the host genome. Mol Gen Genet 264:20–28
Kumar S, Fladung M (2000b) Determination of T-DNA repeat formation and promoter methylation in transgenic plants. Biotechniques 28:1128–1137
Kumar S, Fladung M (2001) Gene stability in transgenic aspen (*Populus*). II. Molecular characterization of variable expression of transgene in wild and hybrid aspen. Planta 213:731–740
Lännenpää M, Hassinen M, Ranki A, Hölttä-Vuori M, Lemmetyinen J, Keinonen K, Sopanen T (2005) Prevention of flower development in birch and other plants using a BpFULL1: BARNASE construct. Plant Cell Rep 24:69–78
Lapierre C, Pollet B, Petit-Conil M, Toval G, Romero J, Pilate G, Leplé J-C, Boerjan W, Ferret V, de Nadaï V, Jouanin L (1999) Structural alterations of lignin in transgenic poplars with depressed cinnamyl alcohol dehydrogenase or caffeic acid O-methyltransferase activity have an opposite impact on the efficiency of industrial Kraft pulping. Plant Physiol 119:153–163
Lemmetyinen J, Järvinen P, Pasonen H-L, Keinonen K, Lännenpää M, Keinänen M (2008) Birches. In: Kole C, Hall TC (eds) A compendium of transgenic plants, vol 9: Forest tree species. Blackwell Publishing, Oxford, UK
Leplé J-C, Bonadé-Bottino M, Augustin S, Pilate G, Dumanois Lê Tân V, Delplanque A, Cornu D, Jouanin L (1995) Toxicity to *Chrysomela tremulae* (*Coleoptera*: *Chrysomelidae*) of transgenic poplars expressing a cysteine proteinase inhibitor. Mol Breed 1:319–328
Leplé J-C, Dauwe R, Morreel K, Storme V, Lapierre C, Pollet B, Naumann A, Kang KY, Kim H, Ruel K, Lefèbvre A, Joseleau J-P, Grima-Pettenati J, de Rycke R, Andersson-Gunnerås S, Erban A, Fehrle I, Petit-Conil M, Kopka J, Polle A, Messens E, Sundberg B, Mansfield SD, Ralph J, Pilate G, Boerjan W (2007) Downregulation of cinnamoyl-coenzyme a reductase in poplar: multiple-level phenotyping reveals effects on cell wall polymer metabolism and structure. Plant Cell 19:3669–3691

Levée V, Lelu M-A, Jouanin L, Cornu D, Pilate G (1997) *Agrobacterium tumefaciens*-mediated transformation of hybrid larch (*Larix kaempferi* × *L. decidua*) and transgenic plant regeneration. Plant Cell Rep 16:680–685

Lohtander K, Pasonen H-L, Aalto MK, Palva T, Pappinen A, Rikkinen J (2008) Phylogeny of chitinases and implications for estimating horizontal gene transfer from chitinase transgenic silver birch (*Betula pendula*). Environ Biosafety Res 7:227–239. doi:10.1051/ebr:2008019

Moreno-Cortés A, Hernández-Verdeja T, Sánchez Jiménez P, González-Melendi P, Aragoncillo C, Allona I (2012) CsRAV1 induces sylleptic branching in hybrid poplar. New Phytol 194:83–90

Nehls U, Zhang C, Tarkka M, Hampp R, Fladung M (2006) Investigation of horizontal gene transfer from transgenic aspen to ectomyccorhizal fungi. In: Fladung M, Ewald D (eds) Tree transgenesis—recent developments. Springer, New York, pp 323–333

Pascual MB, Jing ZP, Kirby EG, Cánovas FM, Gallardo F (2008) Response of transgenic poplar overexpressing cytosolic glutamine synthetase to phosphinothricin. Phytochemestry 69:382–389

Pasonen H-L, Seppänen S-K, Degefu Y, Rytkönen A, von Weissenberg K, Pappinen A (2004) Field performance of chitinase transgenic silver birches (Betula pendula): resistance to fungal diseases. Theor Appl Genet 109:562–570

Pasonen H-L, Degefu Y, Brumos J, Pappinen A, Timonen S, Seppänen S-K (2005) Transgenic silver birch (*Betula pendula*) expressing an antifungal sugar beet chitinase IV gene forms normal ectomycorrhizae with *Paxillus involutus in vitro*. Scand J For Res 20:385–392

Pasonen H-L, Vihervuori L, Seppänen S-K, Lyytikäinen-Saarenmaa P, Ylioja T, von Weissenberg K, Pappinen A (2008) Field performance of chitinase transgenic silver birch (*Betula pendula* Roth): growth and adaptive traits. Trees-Struct Funct 22:413–421. doi:10.1007/s00468-007-0202-7

Pasonen H-L, Lu J, Niskanen AM, Seppänen S-K, Rytkönen A, Raunio J, Pappinen A, Kasanen R, Timonen S (2009) Effects of sugar beet chitinase IV on root-associated fungal community of transgenic silver birch in a field trial. Planta 230:973–983

Peuke AD, Rennenberg H (2005) Phytoremediation: molecular biology, requirements for application, environmental protection, public attention, and feasibility. EMBO Rep 6:497–501

Peuke AD, Rennenberg H (2006) Heavy metal resistance and phytoremediation with transgenic trees. In: Fladung M, Ewald D (eds) Tree transgenesis—recent developments. Springer, New York, pp 137–155

Pilate G, Ellis D, Hawkins S (1997) Transgene expression in field-grown poplar. In: Klopfenstein N, Chung YW, Kim M-S, Ahuja MR (eds) Micropropagation, genetic engineering and molecular biology of *Populus*, USDA Forest Service, pp 84–89

Pilate G, Guiney E, Holt K, Petit-Conil M, Lapierre C, Leplé J-C, Pollet B, Mila I, Webster EA, Marstorp HG, Hopkins DW, Jouanin L, Boerjan W, Schuch W, Cornu D, Halpin C (2002) Field and pulping performances of transgenic trees with altered lignification. Nat Biotechnol 20:607–612

Report WIV-ISP (2010) The Scientific Institute of Public Health, Belgian focal point for Biosafety "1990–2010: 20 years of risk assessment of GMOs and pathogens". ISBN 9789074968287 (NUR-code: 884). http://www.biosafety.be/Book/PDF/SBB_20yearsBiosafety_EN_LR.pdf

Ryynänen L, Sillanpää M, Kontunen-Soppela S, Tiimonen H, Kangasjärvi J, Vapaavuori E, Häggman H (2002) Preservation of transgenic silver birch (*Betula pendula* Roth) lines by means of cryopreservation. Mol Breed 10:143–152

Seppänen S-K, Pasonen H-L, Vauramo S, Vahala J, Toikka M, Kilpeläinen I, Setälä H, Teeri T, Timonen S, Pappinen A (2007) Decomposition of the leaf litter and mycorrhiza forming ability of silver birch with a genetically modified lignin biosynthesis pathway. Appl Soil Ecol 36:100–106

Strauss SH, Tan H, Boerjan W, Sedjo R (2009) Strangled at birth? Forest biotech and the convention on biological diversity. Nat Biotechnol 27:519–527

The Poplar Files http://www.vib.be/en/news/Pages/The-Poplar-Files.aspx

Tilston EL, Halpin C, Hopkins DW (2004) Genetic modifications to lignin biosynthesis in field-grown poplar trees have inconsistent effects on the rate of woody trunk decomposition. Soil Biol Biochem 36:1903–1906

Valve H, McNally R, Pappinen A (2010) Doing research, creating impact: Using "PROTEE" to learn from a GM tree field trial. Sci Public Pol 37:369–379. doi:10.3152/030234210X501216

Van Acker R, Vanholme R, Storme V, Mortimer JC, Dupree P, Boerjan W (2013) Lignin biosynthesis perturbations affect secondary cell wall composition and saccharification yield in *Arabidopsis thaliana*. Biotechnol Biofuels 6:46–63

Van Acker R, Leplé JC, Aerts D, Storme V, Goeminne G, Ivens B, Piens K, Van Montagu M, Santoro N, Foster C, Ralph J, Soetaert W, Pilate G, Boerjan W (2014) Improved saccharification and ethanol yield from field-grown transgenic poplar deficient in cinnamoyl-CoA reductase. Proc Natl Acad Sci USA 111:845–850

Vauramo S, Pasonen H-L, Pappinen A, Setälä H (2006) Decomposition of leaf litter from chitinase transgenic silver birch (*Betula pendula*) and effects on decomposer populations in a field trial. Appl Soil Ecol 32:338–349

Vihervuori L, Lyytikäinen-Saarenmaa P, Tuomikoski E, Luoma M, Niemelä P, Pappinen A, Pasonen H-L (2012) Palatability of transgenic birch and aspen to roe deer and mountain hare. Biocontrol Sci Technol 22:1167–1180. doi:10.1080/09583157.2012.716393

Voelker SL, Lachenbruch B, Meinzer FC, Jourdes M, Ki CY, Patten AM, Davin LB, Lewis NG, Tuskan GA, Gunter L, Decker SR, Selig MJ, Sykes R, Himmel ME, Kitin P, Shevchenko O, Strauss SH (2010) Antisense down-regulation of 4CL expression alters lignification, tree growth, and saccharification potential of field-grown poplar. Plant Physiol 154:874–886

Walter C, Fladung M, Boerjan W (2010) The 20-year environmental safety record of GM trees. Nat Biotechnol 28:656–658

Lessons from Two Decades of Field Trials with Genetically Modified Trees in the USA: Biology and Regulatory Compliance

Steven H. Strauss, Cathleen Ma, Kori Ault and Amy L. Klocko

Abstract We summarize the many field trials that we have conducted in the USA beginning in 1995 and continuing to this day. Under USDA APHIS federal regulatory notifications and permits, we have planted nearly 20,000 trees derived from approximately 100 different constructs in more than two dozen field experiments. The large majority of the trials were in *Populus* and included hybrid white poplars (*P. tremula* × *alba* INRA 717-1B4 and *P. tremula* × *tremuloides* INRA *353-53*), but also included diverse hybrid cottonwoods such as *P. trichocarpa* × *deltoides* and *P. deltoides* × *nigra*. One field trial used transgenic sweetgum (*Liquidambar*). Most trials were conducted on Oregon State University (OSU) land, but several were also conducted on the land of industry collaborators in Oregon, Washington, and other states. The main traits we have studied are floral sterility and flowering time modification; size and growth rate modification by gibberellin perturbation; activation-based gene tagging; stability of reporter gene expression and RNAi suppression; herbicide and pest resistance gene impacts on plantation productivity; lignin modification and its impacts on physiological processes; and effects of isoprene reduction on growth and stress tolerance. The most significant lessons from these years of trials are: (1) Visual abnormalities in form or growth rate due to the transformation and in vitro regeneration (somaclonal variants) have been observed in several experiments, but are extremely rare (below 1 % of events produced). (2) Gene expression and RNAi-induced gene suppression have been highly stable —with a virtual absence of transgene silencing—over many years for virtually all transgenic trees whether assayed by a visual phenotype (reporter gene, flowering time, sexual sterility, herbicide or pest tolerance), or by molecular measures of transgene expression (e.g., quantitative RT-PCR). (3) The regulatory process has largely been efficient and workable, though it imposes significant biological constraints, costs, and risks that are very difficult for an academic laboratory to bear when trials span several years. It is most difficult where flowering is

---

S.H. Strauss (✉) · C. Ma · K. Ault · A.L. Klocko
Department of Forest Ecosystems and Society, Oregon State University, Corvallis, OR 97331, USA
e-mail: steve.strauss@oregonstate.edu

C. Vettori et al. (eds.), *Biosafety of Forest Transgenic Trees*,
Forestry Sciences 82, DOI 10.1007/978-94-017-7531-1_5

needed. (4) Field environments invoke complex and largely unpredictable changes to expression and associated phenotypes when studying physiology-modifying transgenes, including those affecting wood properties, suggesting the need to study several field sites, genetic backgrounds, and gene insertion events over many years, similar to common practices of conventional breeding. However, regulatory requirements make this very difficult to do for transgenic trees. (5) Collaborative field trials with industry have shown that common transgenic traits, such as herbicide and insect resistance, can have large productivity benefits in near-operational plantation conditions (e.g., two-year volume growth improvements of ~20 %)—suggesting that it could be highly beneficial to incorporate transgenic traits into production programs. Regulatory reforms to focus on product benefits as well as risks, and that do not assume harm from the use of recombinant DNA methods, are needed if transgenic technology is to provide significant benefits in forestry.

1 Introduction

During the span of our research program, society has gone from a position of great enthusiasm for use of transgenic plants in agriculture and forestry, to one where regulatory, market, and social barriers have grown to the point that transgenic studies are increasingly difficult to fund and carry out (e.g., Viswanath et al. 2012). Very few academic research programs conduct field studies with transgenic trees anymore as a result of these barriers. We have planted more hectares of transgenic trees than any public sector research program in the USA (Biotechnology 2014). However, our program has nearly been shut down several times due to a lack of adequate funding to support the substantial costs of regulatory compliance. Whether the pendulum will swing back or not is unclear, especially as regulations and international market barriers are very slow to change. But, our experience and insights from field studies with transgenic trees, including problems and opportunities missed, may be of value to inform society about whether and how to ease restrictions in the future. Providing these lessons in a single, easily accessible place is the main reason that we have written this chapter.

Our previous summaries of field experience, and the reasons that we believe extensive field research is essential for progress in tree molecular biology and biotechnology, can be found in several review and analysis papers published earlier (Bradshaw and Strauss 2000; Brunner et al. 2004; Strauss et al. 2004; Busov et al. 2005a, b; Valenzuela and Strauss 2005; Wei et al. 2006; Strauss et al. 2009a, b; Strauss et al. 2010; Voelker et al. 2010; Elorriaga et al. 2014). The different types of studies we have carried out over the last two decades are summarized in Table 1, and illustrated in Figs. 1, 2, 3 and 4.

Table 1 List of field trials

Trial name and clones used	Promoter/transgene/terminator	Years	No. trees	Publications
Flowering modification				
First generation sterility				
717, 353	TA29::Barnase::NOS	1995–2009	228	Elorriaga et al. (2014)
	35S::LEAFY::NOS			Skinner et al. (2000), Rottmann et al. (2000)
	APETALA1::GUS::NOS			
	APETALA1::DTA::G7			
	TTS-1::Barnase::35S			
	TTS-1::GUS::35S			
	DTA::NOS::35S			
	SLG::DTA::NOS			
	TTS-1::DTA::35S			
Second Generation sterility—PTD				Elorriaga et al. (2014)
717, 353	PTD::GUS::NOS	2000–2009	229	
	AP3::DTA::NOS			
	35S::GUS::ST-LS1::RUBP			
	ACTIN2::GUS-ST-LS1::RUBP			
	ACTIN11::GUS-ST-LS1::RUBP			
	35S::PTLF::NOS			
Second generation sterility—DNM				
353	ACT11::PTD-1::E9	2001–2009	280	–
	ACT11::PTD-2::E9			
	ACT11::PTAG2-1::E9			
	ACT11::PTAP1.1b-1::E9			
	ACT11::PTAP1.1b-2::E9			
Sterility trial: overexpression/suppression				
717	35S::PMFT-IR::OCS (delay)	2003–2009	202	Mohamed et al. (2010)
	35S::PMFT::35S			
	35S::PCENL1::35S			
	35S::PCENL1-IR::OCS (promote)			
Attenuation sterility trial				
717	PTLF::GUS::G7, MARs	2003–2009	588	Wei et al. (2007)
	35SBP::BARSTAR::E9, MARs			
	35SBPW::BARSTAR::E9, MARs			
	NOS::BARSTAR::E9, MARs			

(continued)

Table 1 (continued)

Trial name and clones used	Promoter/transgene/terminator	Years	No. trees	Publications
	PTLF::BARNASE::G7 35SBP::BARSTAR::E9, MARs			
	PTLF::BARNASE::G7 35SBPW::BARSTAR::E9, MARs			
	PTLF::BARNASE::G7 NOS::BARSTAR::E9, MARs			
	35S::GUS::E9, MARs			
	35SOmega::GUS::E9, MARs			
	NOS::GUS::E9, MARs			
Third generation sterility				
717, 353, 6K10	35S::AG::E9, MARs	2011—current	3,539	Klocko et al. (2014b)
	35S::AP1-M2::E9, MARs			
	35S::AP1-M3::E9, MARs			
	35S::PAGL24-IR::OCS			
	35S::PTFT1-IR::OCS			
	35S::PFT1/PAGL20-IR::OCS			
	35S::PtFT-PAGL20-IR::OCS/35S::PfPFL1-IR::OCS			
	35S::PTAG-IR::OCS			
	35S::PTAG-IR::OCS, MARs			
	35S::PTAP1-IR::OCS			
	35S::PTAP1-PTAG-IR::OCS			
	35S::PTAP1-PTLF-IR::OCS			
	35S::PTD-IR::OCS			
	35S::PTLF-PTAP1-PTAG-IR::OCS			
	35S::PTLF-IR::OCS			
	35S::PTLF-PTAG-IR::OCS			
	35S::PAGL20-IR::OCS			
	35S::PAGL24-IR::OCS			
	35S::PFPFL1-IR::OCS			
	35S::PFPFL2-IR::OCS			
	35S::PSVP::OCS			
	35S::PTLF-IR::OCS/35S::PTAG-IR::OCS			
Sweetgum sterility trial				
Liquidambar styraciflua CV *Worplesdon*	35S::LAG/LSAG-IR::NOS	2007—current	328	–
	En35S::AG-M3::E9, MARs			
	PTD::BARNASE/35S::BARSTAR, MARs			

(continued)

Table 1 (continued)

Trial name and clones used	Promoter/transgene/terminator	Years	No. trees	Publications
Management				
Herbicide resistance stability trial				
717, 353	pTA29::BARNASE::NOS, pSSUARA-TP::BAR::G7	1997–2006	384	Li et al. (2008)
Glyphosate-resistance screening trail				
50-197, 189-434, 195-529, 311-93	FMV::CP4::T9, FMV::GOX::NOS	1996–1999	1,176	Ault et al. (2016) in press
Glyphosate-resistance management trail				
95-529, 311-94	FMV::CP4::T9, FMV::GOX::NOS	2000–2003	1944	Ault et al. (2016) in press
BT screening trial				
24-305, 50-197, OP-367, 189-434	35S::cry3Aa::orf25	1998–2002	502	Klocko et al. (2014a), Meilan et al. (2000a, b)
BT large-scale trial				
OP-367	35S::cry3Aa::orf25	1999–2002	402	Klocko et al. (2014a), Meilan et al. (2000b)
Form and growth rate				
Semi-dwarfism trial				
717	35S::ptaGA2ox::OCS 35S::2-Oxidase::NOS	2003–2008	882	Etherington et al. (2007), Zawaski et al. (2011)
	GAI::GAI (from Arab cDNA, wt-gene)::GAI GAI::gai (from Arab cDNA,51-bp in-frame deletion)::GAI			
	35S::GAI (from Arab genomic, wt-gene)::35S			
	35S::gai (from Arab genomic, 51-bp deletion)::35S			
	35S::rgl1 (51-bp, in-frame deletion)::NOS			
GA competition and yield				
717	35S::GA2-oxidase::NOS	2006–2008	2400	Elias et al. (2012)
	35S::Atrgl-1::NOS			
	35S::AtGAI::35S			
	35S::PtaGA2-ox::OCS			
	NativeGAI::AtGAI::nativeGAI			
	NativeGAI::Atgai::nativeGAI			
GA growth enhancement				
717	GA20ox7::GA20ox7::GA20ox7	2008–2010	429	Lu et al. (2015)
GA signaling modification				
	Empty vector			
	35S::AtSPY::OCS			

(continued)

Table 1 (continued)

Trial name and clones used	Promoter/transgene/terminator	Years	No. trees	Publications
	35S::HvSPY::OCS			
GA modification growth trial				
717	35S::GA2oxidase2,7::OCS	2009–2011	502	Lu et al. (2015)
	35S::GA2oxidase1,6::OCS			
	35S::GA2oxidase3::OCS			
	SHI1::SHI1::SHI1			
	PHOR1::PHOR1::PHOR1			
	GA2ox1::GA2oxidase2::NOS			
	RGL1-1::GA20oxidase2::NOS			
	Empty vector			
Phytochrome trial				
717	35S::PHYB1::OCS	2004–2013	220	–
Activation tagging				
Activation tagging trial				
717	35S::none::none	2003–2009	2564	Busov et al. (2003)
Activation tagging trial				
717	35S::none::none	2007–2009	1783	Busov et al. (2010)
Tools and stability				
Alcohol inducible				
717	alcA::GUS	2005–2009	40	–
	35S::PHYB2::OCS			
Transgene stability trials				
717, 353	35S::BAR-IR::OC	2004–2007	340	Li et al. (2008)
	35S::BAR-IR::OC, MARs			
	35S::rbcSp-IR::OC			
	35S::rbcSp-IR::OC, MARs 35S::GFP::35S rbcs::TP::BAR::G7 35S::GFP::35S rbcs::TP::BAR::G7, MARs	2003–2006	3416	Li et al. (2008, 2009)
Other physiology modifications				
Lignin modification				
717	Pt4CL1P::4CL1::NOS	2005–2009	97	Voelker et al. (2010, 2011a, b)
Isoprene reduction trial				
717	35S::PcISPS-RNAi::OCS	2012—current	504	–
All trials	**Total trees**	22,979		

Fig. 1 Sexual sterility field trials. **a** 9-acre sterility trial (photo taken in spring 2013), **b** Dr. Steve Strauss with the same sterility trial showing a block with transgenic clone INRA 353-53 (*P. tremula* × *P. alba*; photo taken spring 2014), **c** Program Manager Liz Etherington inside a coppiced clone bank used to produce cuttings for the same sterility trial, **d** collecting catkins in an earlier sterility trial in ~1995, and **e** seven-year-old sterility trial of trangenic sweetgum trees showing their fall color

Fig. 2 Gene tagging. **a** Program Manager Liz Etherington with field trial of activation-tagged trees in 2009. A variety of phenotypic alterations were observed in these field trials, including, **b** early bud flush, **c** early leaf senescence, **d–f** altered wood color and bark texture (**d** is wild type), and **g–h** changes in tree form (more upright branches in **h** compared to wild type in **g**)

(a)
(b)
(c)
(d)
(e)
(f)
(g)
(h)

◀ **Fig. 3** Management trials. **a–e** Herbicide resistance field trails of poplars showing **a** rows which have been sprayed with glyphosate. A variety of insertion events with high resistance are shown on the *left*, and the row of dead non-transgenic control trees are on the *right* indicated by the *red arrow*. **b** Glyphosate application directly onto trees. **c** Glyphosate-resistant trees growing well after multiple direct sprays and nearly 2 years in the field. **d** Conventionally grown poplar showing weed proliferation. **e** Glyphosate-resistant poplar with much less weed competition from same experiment as in **d**. **f–g** Insect-resistant poplars showing **f** Bt poplar with no insect damage from cottonwood leaf beetle. **g** Non-transgenic control poplar tree from the same experiment as (**f**) showing extensive insect damage, and **h** comparable growth and morphology between Bt trees (*right*) and unmodified trees (*left*) after two growing seasons

Fig. 4 Form and architecture modification. Dramatic alterations in **a** tree shape (bushy vs. narrow, *back row*) and size (dwarf, *front row*) and **b** stature of GA-modified poplars

Our laboratory has focused on poplars, mostly using model genotypes obtained from collaborators at INRA in France (G. Pilate and L. Jouanin provided hybrids 717-1B4 and 353-53) that are easy to transform and perform well in the field in the Pacific Northwest of the USA. We have also been successful in transforming many other poplar genotypes, but with greater difficulty. We have been able to use transformation as a routine tool, enabling us to produce and test tens of thousands of transgenic gene insertions, called "events." However, for many other tree species and most genotypes, including the closely related willows (*Salix*), transformation remains a costly or extremely difficult tool to use, requiring optimization and trial-and-error protocol development in each laboratory and for each genotype. Unfortunately, as a consequence of the stigma over transgenic methods there has been very little public investment in development of science-based, generalized transformation methods. In contrast, great and often proprietary advances have been made in the private sector. With the extraordinary advances in developmental biology, it is likely that transformation barriers can be much reduced to support a next generation of transgenic tree biology and field testing, if society chooses to reinvest in the area.

In this review we will first summarize the mechanics of high throughput transformation and field trials, including the management challenges and surprises we have run into. Many of these are common to any field trials, but take on additional importance given the high cost and regulatory oversight of transgenic materials.

2 The Lab to Field Pipeline

Because of the ease and reliability in response of the in vitro propagated poplar genotypes that we have used, it has been easy to standardize media and steps in transformation, subculture, and propagation. As a result, most of the steps can be performed by high school or undergraduate students after a modest period of training. The methods we employ are variations of the well known "leaf disc" type of organogenic transformation (Horsch et al. 1985) where sterile in vitro leaf disks, internode sections, and sometimes petioles are cocultivated with *Agrobacterium*, then sequentially placed in callus induction medium, shoot induction medium, and rooting medium in the presence of a selective antibiotic or herbicide. Further propagation to produce the number of clonal replicate trees (ramets) needed for field trials are developed by further shoot development and rooting of nodal segments. Once roots are produced, plants are transplanted to soil and exposed to ambient conditions in a greenhouse slowly over a period of one to two months. After a further one to two months of growth and acclimation in the greenhouse, they are either planted directly in the field, grown out of doors for some weeks to enable further hardening, or induced to go dormant in a greenhouse or out of doors then planted as dormant materials (whole plants or stem sections). While poplar trees can be established using dormant plants or cuttings, most of our trials have been planted with continuously growing materials; the degree of hardening prior to planting depends on the time of year and harshness of the planting site. The summer drought can make planting in Oregon very challenging for planting after early June. Often trees are pruned to a standard size (e.g., ~30 cm) or transpiration inhibitors applied to the leaves to help them survive in the field. High quality weed control and irrigation are essential to obtain a high rate of survival and growth when using poplar transplants that are actively growing, especially when planting occurs after spring.

3 Management and Its Challenges

At the start of our field research, most management activities were carried out by staff of the Colleges of Forestry or Agriculture as part of the research infrastructure. This setup was commonplace throughout the USA. However, because of rapidly declining support for public agricultural and forest research at OSU and most other academic research universities in the USA, we have had to obtain external research grants to fund, direct, and often personally undertake most of the management activities needed to conduct field research (Fig. 5). This includes obtaining regulatory permits from USDA; basic site preparation and weed control; fencing to exclude deer (>3 m high); planting and fertilization; irrigation management and associated water rights permits; post-planting mowing and weed control; tree harvest and herbicide-assisted devitalization (killing); and site monitoring to kill root sprouts

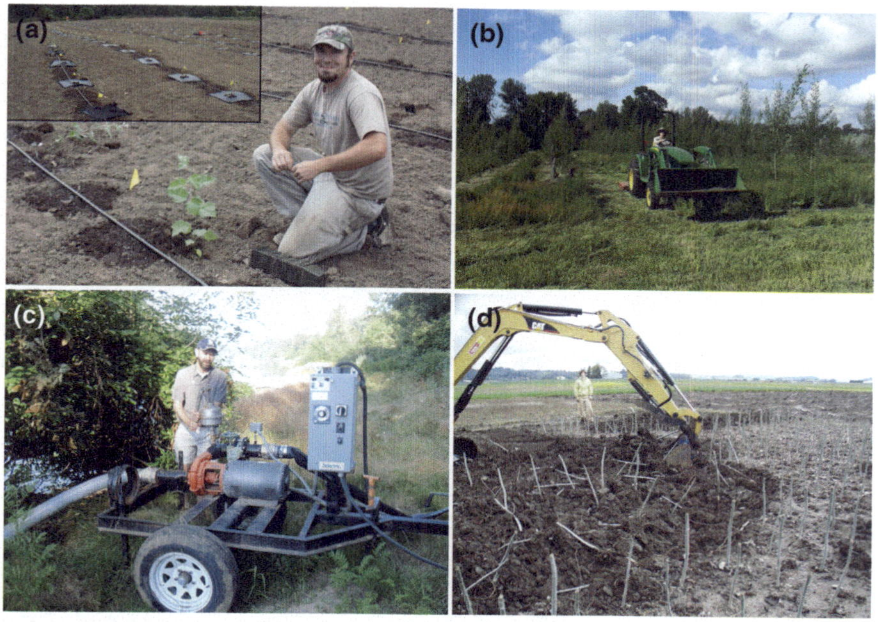

Fig. 5 Management of field sites. **a** Student Chad Washington planting the ~3 ha third generation sterility trial. **b** Program Manager Kori Ault mowing weeds. **c** Dr. Steve Strauss priming the irrigation pump. **d** Removal of tree roots at the end of a trial for biomass assessment and devitalization

which can appear for several years after harvest and initial attempts to kill trees. Complete killing of all trees, and an absence of root sprouts, is essential for regulatory compliance. Large trees often have extensive root systems that are particularly difficult to completely kill, especially with our model poplar genotypes (which are vigorous natural sprouters). In fact, poplar trees re-sprout so well that they can be coppiced (cut off at ground level) and regrown on a regular basis. While this trait is advantageous for maintaining trees at a smaller size, it is a great nuisance for tree removal. Until there is a complete absence of living sprouts for at least two full years, the field sites remain a regulated piece of land that must be monitored and reported on to the USDA. Individual citizens, not institutions, are the responsible organizations under USDA regulations, making full compliance especially important to the responsible individuals, usually the science director (Strauss).

Any significant deviations from expected conditions in USDA permits must be promptly reported to the USDA. For us, these have included multiple instances where our field sites were partially inundated by the nearby Willamette River after severe winter rain storms, often destroying parts of fences and depositing soil, logs, and other debris onto field sites (Fig. 6).

Weed control is a continual problem and nuisance, and methods for weed suppression such as the use of "shade clothes" around trees (Fig. 4a) have often

Fig. 6 Field site disturbances. **a** Flooding of a field site by a nearby river, **b** large debris brought in by flood waters. All trees were monitored, accounted for, and the event promptly reported to the USDA

been helpful, though sometimes these provide protection from predators for rodents who can girdle trees. In one field planting during a year of high vole populations, nearly all trees that were protected by shade clothes were girdled, requiring that we abandon and replant the entire experiment (with many hundred trees) the following year when populations had begun to collapse in the area. Weeds can rapidly grow to overtop young trees, especially during our mild and wet winters in Oregon. They also impede access to the trees, making it more difficult to collect samples and data. In addition, weeds provide cover and food for rodents that can damage trees. Mowing provides only a short respite from competition, thus herbicides are often used. However, most herbicides, even when applied with sheltered sprayers, risk damage to poplars even when the trees are dormant. Our experience with weed management has shown us firsthand why herbicide resistant crops are considered of such high value to farmers (who have employed them on a massive scale in agriculture). As discussed below, the superior weed control they afford can significantly increase tree growth.

4 Regulatory Experiences and Shocks

In contrast to many other countries, it is relatively straightforward to get a permit for a field trial of transgenic trees in the USA. One does not need to do extensive research to characterize each insertion and its physiology as is required in some countries (Viswanath et al. 2012); the USDA accepts your claim that it is transgenic then regulates it at a stringency that is proportional to the risk of broad classes of phenotypes, such as if it produces a biopharma product or not, and if it is perennial or not. For multiple year trials of transgenic trees, in recent years the USDA has required full permits rather than the faster and easier "notifications" that were once allowed, thus increasing the burden of permit applications, field trial establishment, and reporting. The USDA is also requiring more stringency in monitoring and reports. Because all transgenic materials must be contained and then removed from the site (and a feasible plan for this must be presented), beyond the broad categories cited above the specific nature of the transgenic trait is not very important when research trials are considered.

For most situations, trees have wild or planted relatives nearby within pollination distance, or have the ability to disperse viable seeds a long distance (e.g., the cottony seeds of poplars, bird dispersed fruits). Thus for most trials trees are not allowed to flower. A benefit of our transgenic poplar models is that they are sexually incompatible with the wild cottonwoods near to our research sites (they are from different sections of the genus *Populus*); this has enabled us to allow flowering in some trials where that trait is important to the research (e.g., to study sterility transgene effects). For many forest trees flowering occurs after the trees are too large to practically remove or bag all flowers (Fig. 1d), which is a major impediment to the conduct of the ecologically and economically most desirable full-rotation trials (Strauss et al. 2010). The USA biotechnology regulatory system is a complex mixture of a trait-risk and method-trigger system that was adapted from prior laws governing plant pests, pesticides, and food safety. In the USDA system, if a plant pest such as *Agrobacterium* is used for gene transfer, or there is any DNA from a plant pest, such as a promoter or terminator, the transgenic plant is putatively a plant pest and assumed potentially harmful until deregulated. Thus, even cisgenic, intragenic, and domesticating traits (Strauss 2003; Bradford et al. 2005) must be fully contained—severely restricting the length and thus the relevance of field trials of transgenic trees. This has made the development of containment mechanisms an extremely high research priority in the USA (Strauss et al. 2009a).

The regulations have resulted in very costly and scary incidences for us. As discussed above, all transgenes produced with a plant pest or DNA from them are presumed to be potentially hazardous, regardless of their real risk or benefit, until "proven" otherwise by a full regulatory petition and review. In today's world where there are groups who will challenge any and all GMO trees in court, this means that USDA must obtain extensive data on all submissions so that they can produce a defensible Environmental Impact Statement. We did not realize how important the distinction between biological hazard and legal burden of proof is until we found that some poplars in a field trial that flowered abnormally—producing upright catkin-shoot hybrid structures (versus normal drooping catkins), and doing so in the middle of summer (poplars only flower in spring in nature, prior to leaf-out, months earlier; Fig. 7). As required by our permit, we reported this to the USDA, but also indicated that such female catkins in summer—when there is no compatible pollen anywhere—are not a biological hazard. Moreover, all of the trees had genes for semi-dwarfism in them; they would not be competitive with wild trees were any released. We wished to allow them to continue flowering to observe their behavior, and to avoid the significant cost of removal of hundreds of catkins from many dozens of trees. In addition, some of the flowers had unusual transitional phenotypes,

Fig. 7 Collection of unexpected flowers. Trees in one trial flowered unexpectedly in the summer, and our permit requirements (which did not allow for flowering) necessitated removal of every single flower. **a** Trees with fully expanded leaves began to develop flowers. **b** Unusually, flowers grew in an upright conformation instead of the typical hanging orientation. **c** Some flowers had abnormal phenotypes that appeared to be a catkin-vegetative shoot hybrid, with leaf-like organs rather than flowers. **d** Trees flowered again the following spring, again requiring hand-removal of all flowers to comply with permit conditions. The dwarf size of the trees made it feasible to complete the collection without the use of bucket trucks or other machinery. **e** However, trees had numerous catkins, which made the task extremely time-consuming

appearing to be part leaf and part flower. This phenotype was both scientifically interesting and likely to be sterile, further decreasing any possible gene flow. Although the USDA scientists we conferred with agreed with us about the lack of significant risk, and although we had many other trees of the same background genotype for which a permit for flowering on that site had been obtained from USDA, these trees were not intended for flowering thus no such permit had been obtained for them. Finally, after much discussion the USDA scientists felt that they had no choice but to report this as a possible permit violation to their compliance branch (essentially, federal police). Fortunately for us, they informed us of this, and gently but persuasively recommended that we might wish to avoid a legal confrontation over this, with whatever those outcomes might be (e.g., fines, permit revocation, jail time, public embarrassment). Thus we immediately put every student we had available to work in the field removing the catkins before they matured, and reported this action to USDA. We had to undertake a second round of catkin removal the following spring, when these trees flowered at the typical time of year. A possibly cataclysmic legal violation was avoided, one that was prompted by a regulatory system focused on method versus trait, and on legal technicalities vs. science. This is a major reason that we believe that reform of the regulatory system governing field trials is essential if we are to move forward in developing transgenic solutions to speed tree breeding (as we have argued elsewhere; e.g. Strauss et al. 2009b, 2010).

5 Vandalism and Its Impacts

There have been several successful acts of vandalism against transgenic tree research in the USA. The most significant were in 2001 when sites at OSU in Corvallis, the University of Washington in Seattle (UW), and GreenWood Resources (GWR) in Portland, Oregon were attacked. In the latter two cases this caused millions of dollars in damage. The vandalism at our research sites near to Corvallis had very little consequence, as the trees damaged were either non-transgenic, in experiments that were completed, or were recently planted young "seedlings" that easily sprouted after their tops were cut. However, the arson conducted at UW and GWR caused very serious damage, and prompted investments in alarms and magnetic card entry systems at OSU to reduce the risk of similar events. It also caused the movement of our plots to a place where vandalism is likely to have caused lesser damage. Perhaps due to the "9–11" attack on the World Trade Center in New York in fall 2001, the classification of "eco" terrorism as a form of terrorism in the USA (within similar legal consequences), and the FBI's successful pursuit and jailing of many "eco" terrorists, there has been very little further vandalism against biotechnology in the USA, trees or otherwise. Although we have not seen any signs of vandalism since 2001, it remains an ongoing concern, especially as anti-GMO activism seems to be on the rise in the USA.

6 Transformation, Mutation, and Stability

A striking result from our studies of thousands of transgenic poplars is how rarely we observe unintended changes in tree morphology or growth due to the gene transfer and regeneration process, or due to mutagenic effects from where the gene of interest is inserted (Busov et al. 2005a, b). In contrast, there is a great deal of variation in phenotype due to the extensive variation in transgene expression that occurs as a result of the "random" insertion locus and perhaps the unique epigenetic state that is imparted to transgenes with each insertion. We found that dwarf or visible mutants occur at a frequency of around 0.1–1 %. Two rare mutations of note (Fig. 8) were events that did not express their mutations until after dormancy and resumption of growth in the field (they had been planted after continuous in vitro propagation following transformation)—suggesting that dormancy might have triggered epigenetic changes in gene expression due to a somaclonal mutation. One mutant showed much reduced field growth and another both reduced growth and mottling of leaves (Ault et al. 2016). However, for most transgenic poplar the process of producing transgenic poplars does not seem to impact their growth. In a recently published study, we compared the growth of transgenic poplars containing various GUS reporter constructs to wild type controls after 3 years in the field; there were no significant or even modest differences in growth rate (Elorriaga et al. 2014). Although it is difficult to compare species that are grown and transformed differently, and have distinctive morphology, it is our impression that poplars suffer an unusually low rate of somaclonal and transformation-related variation compared to many other species.

Fig. 8 Rare tree abnormalities observed in transformation events with hybrid cottonwoods. **a** Leaves with variable shapes and mottling (Ault et al. 2016: control in *center*), and **b** a dwarfed mutant (*right*) compared to majority of transgenic trees (*left* and rest of plantation)

We have tested hundreds of events in field trials that showed a visibly detectable phenotype, mostly herbicide resistance or insect resistance under insect pressure. We have not observed any cases of gene silencing over years that resulted in a sudden loss of phenotype. Li et al. (2009) showed that herbicide resistance was stable over 8 years and multiple coppicing (cutting and regrowth), and Klocko et al. (2014a) showed that *Bt* gene expression was stable over 14 years despite multiple coppice cycles. Likewise, the early flowering observed by Mohamed et al. (2010) due to RNAi suppression was stable over multiple years, as was RNAi against an herbicide resistance gene (based on qRT-PCR; Li et al. 2009) Reporter gene expression has also been remarkably stable over time in our multi-year field studies (Meilan et al. 2001; Li et al. 2009). Male-sterility due to tapetal ablation was very high and essentially complete over 4 years in the field (Elorriaga et al. 2014). It appears that poplar, either because of its low rate of somaclonal mutation or because (in contrast to annual species) it is not subjected to rounds of sexual propagation after transformation (meiosis and related processes might trigger gene silencing at a higher rate than vegetative development), has a very low rate of instability in gene expression.

7 Biological Lessons

We have conducted a number of field studies with the goal of testing biological hypotheses under conditions of physiological relevance to trees, or ascertaining if transgenic modifications could provide value in a plantation forestry context. The results of these studies are mostly published; we highlight a few below.

7.1 Acceleration of Flowering

We have found that reducing the expression of a poplar homolog of the *Terminal Flower 1* gene using RNAi gave rise to trees of normal form and growth rate, but which flowered 2–4 years earlier than normal (Mohamed et al. 2010). As a dominant gene it could therefore be used to accelerate flowering in poplar breeding programs, then segregated away in progeny during further selection and propagation. In contrast, overexpression of poplar Flowering Locus T generally does not lead to viable pollen and seed (but see Hoenicka et al. 2014) and causes large pleiotropic effects that would impair growth and survival in the field (Zhang et al. 2010). Many other early flowering transgenes that are effective in *Arabidopsis* do not function at all in transgenic poplar (Rottmann et al. 2000; Strauss et al. 2004).

7.2 Sexual Sterility

We have found that a male-sterility transgene functions well in poplar over four growing seasons, preventing virtually all pollen production (Elorriaga et al. 2014). In addition, by field testing and comparison we also found that the barnase gene we employed impaired the growth rate of nearly all transgenic lines, stimulating the development of attenuated forms of barnase in subsequent commercial constructs used in pine and eucalypts (Zhang et al. 2014). Recently, Klocko et al (2014b), reported that RNAi suppression of the poplar *LEAFY* gene prevented catkin maturation while allowing normal vegetative growth. It was shown in a female genotype, but because of the function of *LEAFY* is expected to also work in male genotypes, giving complete sexual sterility in a field-grown tree for the first time. In a greenhouse experiment, a barnase construct controlled by the poplar *LEAFY* promoter and attenuated to various degrees by its specific inhibitor barstar gave normal growth and development of poplar in the greenhouse. However, in the field many of the same trees expressed diverse malformations and much slowed growth, providing a vivid demonstration of the importance of field testing to observe pleiotropic effects (Wei et al. 2006).

7.3 Gene Tagging

Although great strides have been made in recent years in high precision genomic mapping, it remains very difficult to definitively link allelic variants within tree populations to specific quantitative traits. Such linkages were essentially impossible to make at the time we began work on gene tagging in about 1995. The goal of our work in gene tagging was to create allelic variants from natural genes that were large and strong enough to be definitively associated with the affected trait, and to be expressed when hemizygous as primary transgenic plants (i.e., genetically dominant, as poplars cannot be selfed to produce loss of function homozygotes, and inbreeding depression in trees creates additional confounding phenotypic variation). Activation tagging, a recent innovation, seems to be an answer (reviewed in Busov et al. 2005a, b). In this method, a strong enhancer is ~randomly inserted into the genome such that, when it lands near enough to a gene to affect its promoter and cause abnormal upregulation of the gene, it is easily identified by methods such as inverse- or TAIL-PCR (reviewed in Busov et al. 2010a, b). This approach has been used to identify several genes whose functions in tree biology were previously unknown or poorly understood (e.g., Busov et al. 2003; Yordanov et al. 2014). One problem is that the large majority of activation-tagged transgenic trees does not show visible trait modifications in the laboratory or greenhouse, and many of those seen, such as modified leaf morphology, are not of particular interest to tree biologists. We therefore created two hectare-scale field trials of our activation-tagged populations (Fig. 2), which we showed led to the identification of many more, and more biologically interesting, trait modifications than had been observed in the greenhouse alone (Busov et al. 2010a, b). The modified traits included timing of bud flush and bud set, timing of leaf senescence, bark morphology, wood color, crown form, wood chemistry, phototropism, and leaf pubescence (examples in Fig. 2). After the affected genes are identified, RNAi is typically used to understand the gene's role in the absence of overexpression (e.g., Yordanov et al. 2014).

7.4 Form and Architecture Modification

Genetic modification of gibberellin (GA) synthesis and signaling have been used extensively in agriculture to produce semi-dwarf plants and are the basis of the green revolution that revolutionized cereal yields in many parts of the world. With the identification of the underlying genes, we used transgenic methods to study whether similar modifications would be effective in trees, and if they are useful for trees grown as ornamentals (where dwarfism is often highly desirable) or for bioenergy or short rotation forestry (where semi-dwarfism could improve wind-firmness, wood quality, and stress resistance: (Klocko et al. 2013). The transgenes produced a burst of morphological variability in leaf size and color, stature, and crown structure that would be impossible to fully appreciate without growth in the field (Zawaski et al. 2011). This method could clearly be of great value for ornamental horticulture if GMO trees were not so stigmatized and costly to market (Etherington et al. 2007). We also conducted high density field trials at different spacings to test if, as with agricultural crops, the relative performance of semi-dwarfs was most expressed at high planting density, and if semi-dwarfs would have a significant disadvantage in competition with wild type (dwarfism genes have been suggested as a means to mitigate ecological impacts from transgene release). Both hypotheses were supported by the field results, and the semi-dwarf poplars shown to have much higher allocation of biomass to roots compared to shoots, suggesting that they could promote drought tolerance and wind-firmness (Elias et al. 2012). Finally, we have studied the potential for overexpression of genes that induce synthesis of active GAs to improve growth rate in the greenhouse and the field. The transgenes were very successful in many experiments, including in modifying the allocation of biomass among plant organs, but growth rate improvement was highly variable and the correlation between field and greenhouse growth extremely poor (Viswanath et al. 2011). Were the trees not transgenic, the next step would have been much expanded field plantings, similar to that of a conventional breeding program. However, we have been unable to obtain funds for such work, mainly because of the national disinvestment in transgenic technology by grant agencies.

7.5 Ecophysiology and Lignin Perturbation

In the 1990s and early 2000s, the benefits of transgenic lignin modification for improved growth rate and improved pulp or biofuel yields had been widely touted. Other than some high quality field trials in Europe, many of these claims were based on poorly designed greenhouse or field evaluations, or lacked field data entirely. One widely acclaimed transgenic modification was *4CL* downregulation (Hu et al. 1999). To see if these benefits were real, we produced transgenic poplars with an antisense *4CL* gene provided by the Chiang laboratory at North Carolina State University and grew them in the field in Oregon for two years. In contrast to expectations based on casual observation or greenhouse experiments, the growth and drought tolerance of those trees in the field were disappointing; the traits of interest were either unaffected or impaired, and there was no benefit for bioethanol production. Instead, the most strongly downregulated trees had badly malfunctioning and collapsed xylem cells which made the wood stiffer and less able to transport water (Voelker et al. 2010, 2011a, b). This case is perhaps the most striking demonstrations of the need, early in research, for field trials over several years, with different genotypes, and at multiple sites when transgenes are used to significantly modify fundamental aspects of plant structure and physiology.

Economic value. After promising results in obtaining highly pest or herbicide tolerant trees in field screens (Meilan et al. 2000a, b) we worked with industry partners to establish "management trials" on their land. In these trials, large numbers of trees were planted and managed under near-operational conditions. In the case of insect resistance, we used a modified gene from *Bacillus thuringiensis* (a variant of Bt cry3a) provided by Mycogen that we had shown earlier should produce a toxin that is lethal to a major pest of poplar, the cottonwood leaf beetle (*Chrysomela*) (James et al. 1998). The results of the screening and growth trial revealed a very strong and stable benefit from the expression of the gene that—even under conditions of low insect pressure (damage that was not visually obvious in most trees)—led to a 10–20 % improvement of volume growth over 2 years in the field (Klocko et al. 2014a). In a management trial of highly glyphosate tolerant trees, (Ault et al. 2016) found that in a weed control regime designed to take advantage of herbicide tolerance, weed populations were greatly reduced in density and tree volume growth was increased by approximately 20 % over two growing seasons. There are likely to be underestimates of the benefits that would accrue when pest damage is high, and when weed management has been more fully adapted to take advantage of herbicide resistance. These results suggest that, if combined with a genetic containment option to reduce management trade-offs and facilitate regulatory and social approval, these traits could be of considerable value in the management of short rotation poplar plantations.

8 Regulatory Lessons and Conclusions

As biotechnologists, there is nothing more gratifying than seeing a transgenic concept based on biological research in a model organism, or when following upon a lab or greenhouse experiment, express itself in the field. It is also instructive when a trait does not show what you expected, as it teaches you about how traits depend intimately on the ecophysiology of the organism—which is influenced by many factors. These include species and genotype; transgene expression intensity and pattern; stage of tree development; and environment. The unpredicted outcomes become probes of specific developmental and physiological processes. To produce a useful modified organism with complex traits requires plant breeding-scale studies in many genotypes and environments, over many years of growth. However, these are rarely done in public sector research due to high costs and regulatory risks or preclusions. As a consequence, there is a dearth of information on long-term performance of transgenic trees in the scientific record (private sector studies are rarely and selectively published). Old myths, including of the unworkable instability, dramatic danger, or magical performance of transgenic trees persist much longer than necessary as a consequence.

Our experience has shown that the transgenic trees we have worked with–mainly poplars, eucalypts, and sweetgum–perform reliably and stably. Unfortunately, the most valuable traits for production and tolerance of environmental stresses, especially for trees under climate associated stresses, are likely to be those that modify physiology, including for pest resistance. Thus, as discussed above, field studies in many genotypes and environments are needed. Unfortunately, under current process-based regulatory regimes (which implicitly assume all transgenic trees are hazardous), it is nearly impossible to conduct the kind of wide-ranging trials needed to test these genes and incorporate them into breeding programs (Strauss et al. 2010, 2015). Such regulations would seem to be a violation of the Precautionary Principle, if you believe, as we do, that it is foolish not to develop transgenic options for coping with growing stresses on forest and agricultural systems. Our experience has suggested that transgenic trees are a valued and reliable tool that could, if unlocked from its strong regulatory and market restraints in most countries, make large and extensive contributions. But fundamental regulatory reforms will be needed if the required biology and breeding studies, in the field, are to occur beyond that in a very few companies, academic laboratories, and countries. Given the extensive political resistance to all kinds of GMOs, this will take leadership, communication, and investment at the highest levels of government, business, and civil society.

References

Ault K, Viswanath V, Jayawickrama J, Ma C, Eaton JA, Meilan R, Hohenschuh W, Murthy G, Strauss S (2016) Improved growth and weed control of glyphosate-tolerant poplars. New Forests (in press)

Biotechnology, Information Systems for (2014) USDA field trials of GM crops
Bradford KJ, Van Deynze A, Gutterson N, Parrott W, Strauss SH (2005) Regulating transgenic crops sensibly: lessons from plant breeding, biotechnology and genomics. Nat Biotechnol 23 (4):439–444
Bradshaw, HD Jr, Strauss SH (2000) Breeding strategies for the 21st century: Domestication of poplar. Poplar Culture in North America, Part 2, Chapter 14. Dickmann DI, Isebrands JG, Eckenwalder JE, Richardson J, National Research Council of Canada, Ottawa, ON K1A 0R6, NRC Research Press, Canada, pp 383–394
Brunner AM, Busov VB, Strauss SH (2004) Poplar genome sequence: functional genomics in an ecologically dominant plant species. Trends Plant Sci 9(1):49–56
Busov VB, Meilan R, Pearce DW, Ma C, Rood SB, Strauss SH (2003) Activation tagging of a dominant gibberellin catabolism gene (GA 2-oxidase) from poplar that regulates tree stature. Plant Physiol 132(3):1283–1291
Busov VB, Brunner AM, Meilan R, Filichkin S, Ganio L, Gandhi S, Strauss SH (2005a) Genetic transformation: a powerful tool for dissection of adaptive traits in trees. New Phytol 167(1):9–18
Busov V, Fladung M, Groover A, Strauss S (2005b) Insertional mutagenesis in populus: relevance and feasibility. Tree Genet Genomes 1(4):135–142
Busov V, Strauss SH, Pilate G (2010a) Transformation as a tool for genetic analysis in populus. In: Genetics and genomics of populus, pp 113–133
Busov V, Yordanov Y, Gou J, Meilan R, Ma C, Regan S, Strauss SH (2010b) Activation tagging is an effective gene tagging system in populus. Tree Genet Genomes 7:91–101
Elias AA, Busov VB, Kosola KR, Ma C, Etherington E, Shevchenko O, Gandhi H, Pearce DW, Rood SB, Strauss SH (2012) Green revolution trees: semidwarfism transgenes modify gibberellins, promote root growth, enhance morphological diversity, and reduce competitiveness in hybrid poplar. Plant Physiol 160(2):1130–1144
Elorriaga E, Ma C, Etherington E, Meilan R, Skinner JS, Brunner A, Strauss SH (2014) A tapetal ablation transgene induces stable male sterility and slows field growth in poplar. Tree Genet Genomes 10:1583–1593
Etherington E, Gandhi H, Busov V, Meilan R, Ma C, Kosola KR, Strauss SH (2007) Dwarfism genes for modifying the stature of woody plants: a case study in poplar. Landscape Plant News 18:3–6
Hoenicka H, Lehnhardt D, Nilsson O, Hanelt D, Fladung M (2014) Successful crossings with early flowering transgenic poplar: interspecific crossings, but not transgenesis, promoted aberrant phenotypes in offspring. Plant Biotechnol J 12:1066–1074
Horsch RB, Fry JE, Hoffmann NL, Eichholtz D, Rogers SG, Fraley RT (1985) A simple and general-method for transferring genes into plants. Science 227(4691):1229–1231
Hu WJ, Harding SA, Lung J, Popko JL, Ralph J, Stokke DD, Tsai CJ, Chiang VL (1999) Repression of lignin biosynthesis promotes cellulose accumulation and growth in transgenic trees. Nat Biotechnol 17:808–812
James RR, Croft BA, Strauss SH (1998) Susceptibility of the cottonwood leaf beetle (Coleoptera: Chrysomelidae) to different strains and transgenic toxins of *Bacillus thuringiensis*. Environ Entomol 28:108–115
Klocko A, Stanton BJ, van Oosten C, Strauss SH (2013) Green revolution plantations: could short trees be a big thing? Information systems for biotechnology news report. 7–9 May
Klocko AL, Meilan R, James RR, Viswanath V, Ma C, Payne P, Miller L, Skinner JS, Oppert B, Cardineau GA, Strauss SH (2014a) Bt-Cry3Aa transgene expression reduces insect damage and improves growth in field-grown hybrid poplar. Can J Forest Res 44(1):28–35
Klocko AL, Ault K, Ma C, Dow M, Robertson S, Strauss SH (2014b) Targeting floral development genes for genetic containment of forest trees: PtLFY-RNAi causes sterility while allowing vigorous tree growth in field-grown poplar. American Society of Plant Biologists Annual Meeting, Portland, Oregon: Poster P06034-B
Li J, Brunner AM, Meilan R, Strauss SH (2009) Stability of transgenes in trees: expression of two reporter genes in poplar over three field seasons. Tree Physiol 29(2):299–312

Li J, Brunner AM, Shevchenko O, Meilan R, Ma C, Skinner JS, Strauss SH (2008) Efficient and stable transgene suppression via RNAi in field-grown poplars. Transgenic Res 17(4):679–694
Lu H, Viswanath V, Ma C, Etherington E, Dharmawardhana P, Shevchenko O, Pearce DW, Rood SB, Busov V, Strauss SH (2015). Transgenic modification of gibberellic acid modifies growth rate and biomass allocation in *Populus*. Tree Genet Genomes 11:127
Meilan R, Han KH, Ma CP, James RR, Eaton JA, Stanton BJ, Hoien E, Crockett RP, Strauss SH (2000a) Development of glyphosate-tolerant hybrid cottonwoods. Tappi J 83(1):164–166
Meilan R, Ma C, Cheng S, Eaton JA, Miller LK, Crockett RP, DiFazio SP, Strauss SH (2000b) High levels of Roundup® and leaf-beetle resistance in genetically engineered hybrid cottonwoods. In: Blatner KA, Johnson JD, Baumgartner DM (eds) Hybrid poplars in the pacific northwest: culture, commerce and capability. Washington State University Cooperative Extension Bulletin MISC0272, Pullman, WA, pp 29–38
Meilan R, Auerbach DJ, Ma C, DiFazio SP, Strauss SH (2001) Stability of herbicide resistance and GUS expression in transgenic hybrid poplars (Populus sp.) during several years of field trials and vegetative propagation. HortScience 37:1–4
Mohamed R, Wang CT, Ma C, Shevchenko O, Dye SJ, Puzey JR, Etherington E, Sheng XY, Meilan R, Strauss SH, Brunner AM (2010) Populus CEN/TFL1 regulates first onset of flowering, axillary meristem identity and dormancy release in populus. Plant J 62(4):674–688
Rottmann WH, Meilan R, Sheppard LA, Brunner AM, Skinner JS, Ma C, Cheng S, Jouanin L, Pilate G, Strauss SH (2000) Diverse effects of overexpression of LEAFY and PTLF, a poplar (populus) homolog of LEAFY/FLORICAULA, in transgenic poplar and Arabidopsis. Plant J Cell Mol Biol 22(3):235–245
Skinner JS, Meilan R, Brunner A, Strauss SH (2000) Options for genetic engineering of floral sterility in forest trees. In: Jain SM, Minocha SC (eds) Molecular biology of woody plants, pp 135–153
Strauss SH (2003) Regulating biotechnology as though gene function mattered. Bioscience 53 (5):453–454
Strauss SH, Brunner AM, Busov VB, Ma CP, Meilan R (2004) Ten lessons from 15 years of transgenic Populus research. Forestry 77(5):455–465
Strauss SH, Schmitt M, Sedio R (2009a) Forest Scientist views of regulatory obstacles to research and development of transgenic forest biotechnology. J Forest 107(7):350–357
Strauss SH, Tan H, Boerjan W, Sedjo R (2009b) Strangled at birth? forest biotech and the convention on biological diversity. Nat Biotechnol 27(6):519–527
Strauss SH, Kershen DL, Bouton JH, Redick TP, Tan HM, Sedjo RA (2010) Far-reaching deleterious impacts of regulations on research and environmental studies of recombinant DNA-modified perennial biofuel crops in the united states. Bioscience 60(9):729–741
Strauss SH, Costanza A, Seguin A (2015) Genetically engineered trees: paralysis from good intentions. Science 349:794–795
Valenzuela S, Strauss SH (2005) Lost in the woods. Nat Biotechnol 23(5):532–533
Viswanath V, Albrectsen BR, Strauss SH (2012) Global regulatory burden for field testing of genetically modified trees. Tree Genet Genomes 8(2):221–226
Viswanath V, Ma C, Etherington E, Dharmawardhana P, Pearce DW, Rood SB, Busov VB, Strauss SH (2011) Greenhouse and field evaluation of transgenic poplars with modified gibberellin metabolism and signaling genes. BMC Proceedings 5:O22
Voelker SL, Lachenbruch B, Meinzer FC, Jourdes M, Ki CY, Patten AM, Davin LB, Lewis NG, Tuskan GA, Gunter L, Decker SR, Selig MJ, Sykes R, Himmel ME, Kitin P, Shevchenko O, Strauss SH (2010) Antisense down-regulation of 4CL expression alters lignification, tree growth, and saccharification potential of field-grown poplar. Plant Physiol 154(2):874–886
Voelker SL, Lachenbruch B, Meinzer FC, Kitin P, Strauss SH (2011a) Transgenic poplars with reduced lignin show impaired xylem conductivity, growth efficiency and survival. Plant, Cell Environ 34(4):655–668
Voelker SL, Lachenbruch B, Meinzer FC, Strauss SH (2011b) Reduced wood stiffness and strength, and altered stem form, in young antisense 4CL transgenic poplars with reduced lignin contents. New Phytol 189(4):1096–1109

Wei H, Meilan R, Brunner AM, Skinner JS, Ma CP, Gandhi HT, Strauss SH (2007) Field trial detects incomplete barstar attenuation of vegetative cytotoxicity in Populus trees containing a poplar LEAFY promoter: barnase sterility transgene. Mol Breeding 19(1):69–85

Wei H, Meilan R, Brunner AM, Skinner JS, Ma C, Gandhi HT, Strauss SH (2006) Field trial detects incomplete barstar attenuation of vegetative cytotoxicity in populus trees containing a poplar LEAFY promoter:barnase sterility transgene. Mol Breeding 19:69–85

Yordanov YS, Ma C, Strauss SH, Busov VB (2014) EARLY BUD-BREAK 1 (EBB1) is a regulator of release from seasonal dormancy in poplar trees. Proc Natl Acad Sci USA 111(27): 10001–10006

Zawaski C, Kadmiel M, Pickens J, Ma C, Strauss S, Busov V (2011) Repression of gibberellin biosynthesis or signaling produces striking alterations in poplar growth, morphology, and flowering. Planta 234:1285–1298

Zhang C, Norris-Caneda KH, Rottmann WH, Gulledge JE, Chang S, Kwan BY-H, Thomas AM, Mandel LC, Kothera RT, Victor AD, Pearson L, Hinchee MA (2014) Control of pollen mediated gene flow in transgenic trees. Plant Physiol 159:1319–1334

Zhang H, Harry DE, Ma C, Yuceer C, Hsu CY, Vikram V, Shevchenko O, Etherington E, Strauss SH (2010) Precocious flowering in trees: the FLOWERING LOCUS T gene as a research and breeding tool in Populus. J Exp Bot 61(10):2549–2560

Part II
Environmental Impact Assessment and Monitoring of GMTs

Specific Environmental Considerations for GM Trees and Guidance on Their Risk Assessment and Monitoring

Marja K. Ruohonen-Lehto, Katileena M.M. Lohtander and Jeremy B. Sweet

Abstract This chapter introduces the specific biological and ecological elements of trees that need to be considered in the risk assessment of GM (genetically modified) trees, and the basics of environmental risk assessment. Trees have perennial growth habits with long lifespan, and many pathways for dispersal of propagules. Moreover, trees are often key species in ecosystems, they are present from temperate to tropical climates and have complex and multi-level interactions with other organisms. The possible environmental impacts of GM trees are discussed, and examples of scenarios where the GM trees may disperse and have environmental impacts are given. The chapter describes different environmental risk assessment frameworks. The basics of any risk assessment are hazard identification, estimation of potential consequences, estimation of the likelihood of these consequences to occur, estimation of the overall risk and deciding about risk management measures. Carrying out risk assessment usually includes a problem formulation step, definition of assessment endpoints from the broad protection goals and identification of those attributes that will be measured (measurement endpoints). As examples the EFSA guidance on risk assessment of GM plants and the Cartagena Biosafety Protocol guidance of GM trees are described. Moreover, both the comparative and per se risk assessment approaches are discussed.

M.K. Ruohonen-Lehto (✉) · K.M.M. Lohtander
Finnish Environment Institute, PO Box 140, 00251 Helsinki, Finland
e-mail: marja.ruohonen-lehto@ymparisto.fi

K.M.M. Lohtander
e-mail: katileena.lohtander-buckbee@ymparisto.fi

J.B. Sweet
Sweet Environmental Consultants, 6 Green Street, Willingham, CB CB24 5JA, UK
e-mail: jeremysweet303@aol.com

C. Vettori et al. (eds.), *Biosafety of Forest Transgenic Trees*,
Forestry Sciences 82, DOI 10.1007/978-94-017-7531-1_6

1 Introduction

GM technology is expected to be part of the toolbox for the future breeding of trees for agriculture and forestry use. The value to breeders is that novel traits can be introduced without having to go through the long processes associated with hybridisation and regeneration involved in conventional breeding methods (see Chap. 3). Traits introduced to GM Trees (GMTs) include modification of lignin and cellulose composition in order to improve efficiency of processing, improved biomass production for timber and biofuel, resistance to pests and diseases and resistance to abiotic stresses such as temperature, drought and mineral salts (Harfouche et al. 2011, 2012; Häggman et al. 2013). The benefits of these traits for production are described in Chaps. 2, 4 and 5, and also include social, economic and environmental benefits in relation to substitution or offsetting deforestation and improving the economics of forestry and hence encouraging more tree cultivation. Two GMTs have been commercialised in Asia and the application of genetic modification techniques to trees is currently at an advanced stage of research in many countries, with experimental field trials in several countries including European Union (EU) member states (Walter et al. 2010; Fladung 2011; Häggman et al. 2013).

All GM plants, including GMTs, are regulated at experimental and development stages and in relation to commercial cultivation. GMTs are thus required to be risk assessed prior to both experimental and commercial releases (Hoenicka and Fladung 2006; Fladung et al. 2012). The cultivation of GMTs may have environmental impacts where the GMT and its products are released. These impacts can be caused by the new trait conferred to the GMT and/or by the genetic modification. For instance, if the transgenes are not part of the gene pool of the tree species and they transfer to wild counterparts, relatives or to other species they could impact other biota and ecosystems (see below and Chaps. 8 and 9). In addition, genetic modification might lead to unintended changes in the characteristics of modified trees that could also affect the environment. If the intended use of a GMT is agricultural (e.g. fruit trees) there is also a potential for changes to the food safety characteristics of the products derived from it.

Different national and international authorities have established regulations to ensure the safe use of GMOs (GM Organisms), including GMTs (Kikuchi et al. 2008). In addition, there are internationally agreed frameworks (OECD1986, 1992; Codex Alimentarius 2009) on how to assess the potential health and environmental risks caused by such organisms. In the EU, different legal instruments regulate the use of GMOs including their risk assessment, labelling, marketing, transport, sampling and detection and risk management as well as post-market environmental monitoring (EC 2001, 2002, 2003). In accordance with relevant legislation, the European Food Safety Authority (EFSA2009, 2010a, 2011a) assesses all GMO applications for commercial release and cultivation in the EU. To date, no applications for commercial release of trees have been made in the EU and Aquilera et al. (2013) considered that the current risk assessment frameworks of EFSA would be applicable to GMTs, though additional studies may be required for some applications. In this chapter, we discuss aspects of the existing international regulatory frameworks and describe studies and activities related to the development and risk assessment of GMTs in the EU. We highlight the characteristics of GMTs that may require particular consideration in their environmental risk assessment (ERA) and discuss future research needs.

2 Basics of Environmental Risk Assessment

The basics of any ERA are to identify the possible hazards/adverse effects, estimate their potential consequences and estimate the likelihood/probability of these consequences to occur. The stressor can be a chemical, a GMO or an alien species. During the assessment process, it is vital to have good information (baseline data) of the surrounding environment: the characteristics of the stressor, the level and route of exposure, including the areas that will be affected and the duration of the exposure (i.e. spatial and temporal exposure parameters) and what are the most valued entities that we want to protect in the given environments (e.g. species, habitats and natural resources).

The basics of different environmental/ecological risk assessment frameworks have been well described by Suter II (2007). To structure the assessment process, the following concepts have been introduced.

- Hazard identification is done in a so-called "problem formulation" step (see also Wolt et al. 2009; EFSA 2009). In simple terms, problem formulation means defining the goals of the assessment and specifying the methods for achieving these goals. Problem formulation thus includes, in addition to the above-mentioned hazard identification, integrating available information, defining assessment endpoints and developing a description of the hypothesised relationships between the stressor and the assessment endpoints. The purpose of problem formulation is to focus on those hazards that are most relevant for those entities that we want to protect.
- Assessment endpoints are defined as environmental values that we want to protect. They must include an entity and specific attribute(s) of that entity. An entity can be an organism, population, species, community or ecosystem. The attributes include survival, behaviour or abundance.
- Measurement endpoint (measure of effect) refers to a measurable or estimable ecological characteristic that is related to the valued characteristic chosen as the assessment endpoint. Measurement endpoints provide tools to measure the possible hazards/harm caused by the stressor. Measurement endpoints are related to the above-mentioned attributes (e.g. numbers of individuals as a measure of abundance).
- The next steps are characterization of exposure and characterization of ecological effects (hazards) and risk characterization.
- Finally, risk management and estimation of overall risk is carried out.

Many ERA frameworks have been developed by different stakeholders. In the EU, Directive 2001/18/EC (EC 2001) and an EFSA Guidance (EFSA 2010a) describe a framework that is based on the same principles described by Suter II (2007). Another framework, the so-called Roadmap, has been developed under the Cartagena Protocol on Biosafety (UNEP/CBD/BS 2012). This Roadmap consists of five steps and each step includes points to consider-sections, a type of check list that assists in taking all relevant issues and information into account when performing the ERA.

3 Comparative Risk Assessment

The basic risk assessment framework by Suter II (2007) introduces what can be called a per se ERA. This means that the environmental impacts/risks are evaluated as such and not compared to effects caused by other stressors or other man-driven activities. This has been the case in, e.g. traditional chemicals risk assessment. In a comparative ERA, the assessment always includes a comparator, i.e. the possible adverse effects of one stressor are compared to another stressor.

All so-far developed ERA frameworks for GMOs are based on a comparative approach. Possible adverse effects associated with a GMO should be assessed in the context of the adverse effects posed by the non-modified counterparts in the likely receiving environment. In ideal cases the comparator is the conventional, isogenic counterpart of the modified organism (EFSA 2011c).

In the comparative ERA of GM plants conventional agriculture or forestry is often regarded as the baseline or "the terms of reference", i.e. the possible adverse effects of GM cultivation are compared to conventional practices. The comparative approach is practical in many cases, as it can be considered to prioritise the ERA while focusing only on those adverse effects that are different from those of the comparator, but it has certain limitations. It is not always clear what is to be considered as "the terms of reference" while, e.g. agricultural or forestry practices differ in different countries and regions. Moreover, a comparator may not always be available, for instance, in case of a GM plant with an abiotic stress tolerance trait (e.g. cold tolerance, salinity tolerance).

Assessing consequences of invasiveness and gene flow in a comparative approach can be challenging. Invasiveness as such can be a serious problem and considered a hazard to receiving environments where the plant species is not established. The addition of a transgene to an invasive species could change the invasiveness of the species and cause different or additional adverse impacts. When invasiveness is already an identified problem in an introduced conventional cultivar, the GM cultivar should be assessed per se (i.e. as a novel invasive species) and not in a comparative approach.

Gene flow is a route of dissemination and exposure and, as such, is not or has not been considered a hazard. Gene flow should always be assessed in connection with the genes or characteristics that are transferred to other areas and to other species through gene flow/hybridization. Genes move to different cultivars, wild relatives and even to other species (horizontal gene flow). Wild relatives can form "transgene reservoirs" through which the transgenes move to different cultivars. Also in this case, the exposure assessment should be a per se assessment but the consequences of gene flow for recipient plants can be assessed in a comparative assessment.

In principle, problem formulation and defining assessment endpoints do not include a comparison step. Comparing environmental impacts of GM plants to conventional agriculture forms part of political decision making as it places impacts of GM plants into context with other impacts of agriculture.

EFSA Guidance (2011c) recognises the technical limitations of comparative ERA vs per se ERA in the case of applications where a comparator is not available. This can occur where the GM construct only occurs in vitro and is then combined with other GM traits, or when a GM plant is adapted to environments in which the comparator cannot grow. In the former case, the GM plant containing the stacked events is risk assessed and compared with appropriate comparators. An example of the latter case would be the GM cold tolerant *Eucalyptus* and, as indicated previously, the ERA would require examination of the impact of the whole novel plant per se on cooler environments not previously exposed to *Eucalyptus* introductions.

4 Specific Issues to Consider in Relation to the Environmental Risk Assessment and Risk Management of GM Trees

In order to assess the environmental impacts of the introduction of any new tree type (GM or non-GM) into new regions, it is important to understand the ecosystem functions of trees in each geographic area and their interrelation with other biota, microclimate, soil structure, water courses, etc. Trees form a succession of roles in environments from early pioneers to developing different stages of the evolution of a woodland or forest up to the climax community. Within woods and forests, different species form different layers of the understorey and main canopy. In temperate climates, rapid establishing and growing trees such as *Salix* and *Corylus* will form the first levels of tree vegetation but will later tend to be displaced by the trees that form the climactic upper canopy such as *Fagus*, *Fraxinus*, *Quercus* and *Tilia* species. Each tree species directly supports a diversity of different biota, including soil organisms, which in turn support a matrix of other biota. In addition, mixed and solid stands of trees will support different populations and diversities of species. For example, the seed and fruits of trees support a diversity of birds and small mammals with different tree species supplementing each other in different seasons. These birds and small mammals also feed on a range of invertebrates and are themselves food for a range of predators.

Native and introduced trees are the key components of established semi-natural and plantation woodlands and so replacing or displacing them with newly introduced or GMTs is likely to have impacts on their ecological equilibrium and their associated biota. However, many of the plantation trees currently grown in Europe are either alien introduced species or are non-native to the areas in which they are grown. This is particularly the case with many of the pine trees grown in temperate regions. Replacing indigenous woodlands or planting landscapes with these exotic plantation species already has considerable impact on these environments and can change the biodiversity that is present. It is against this background that the environmental impacts and risks associated with the introduction of GMTs need to be considered.

Aquilera et al. (2013) and Häggman et al. (2013) indicate that the ERA of trees should follow the normal steps of the ERAs used to assess agricultural and horticultural crops (e.g. EFSA 2009, 2010a). However, the problem formulation step will rapidly indicate that many cultivated tree species either also occur as wild types or have compatible closely related wild relatives in the areas where they will be cultivated. In addition, if the GMT is also an alien species to a region (e.g. *Eucalyptus* outside Australia) then it also has a potential for ferality and invasion. Thus consideration needs to be given as to whether a GMT will become established outside its cultivated areas or whether the transgene(s) will introgress into wild types and related species. If the problem formulation step indicates that this is likely to be the case, then the consequences of the invasion and/or introgression need to be characterised. The consequences will depend on the characteristics of the tree and the specific traits conferred by the transgene(s), the levels of exposure and on the nature of the different receiving environments.

If the GM traits are likely to confer a fitness advantage in terms of reproduction, dissemination and survival, then it is likely that the genes will introgress into wild populations and/or that the GMT will have a higher potential for ferality and therefore could spread outside its cultivation areas. The consequences of such introgression and invasion will need to be considered. Different scenarios can be envisaged:

- The GMTs replace existing plantation trees of same species: the environmental consequences will depend on the GM trait and the tree species. If the GM trait confers pest or disease resistance then there could be effects on non-target organisms, such as predators or symbionts, and these should be assessed. For example, Burgess et al. (2014) demonstrated the adverse effects of *Bacillus thuringiensis* (Bt) toxins on some native Lepidoptera and other fauna in a study simulating the release of a Bt *Pinus radiata* in New Zealand. If the GM trait is considered environmentally neutral, e.g. herbicide tolerance in the absence of the use of the herbicide, then replacing native trees with GMTs is likely to have the same effects as replacement with other non-GM cultivated types.
- The GMTs (or relatives hybridised with GMTs) displace other trees (related or unrelated) in semi-natural or managed environments: this could lead to much greater environmental consequences since the GMTs could ultimately dominate

certain receiving environments. The impacts would depend on the GMT species, the introduced trait(s) and the tree species that are displaced and removed from their niches. Take two hypothetical examples: (i) A GM elm (*Ulmus*) tree is introduced with resistance to elm disease (*Ophiostoma ulmi*) which replaces current elm trees and the trees occupying niches previously occupied by native elm. It may be considered that the environment has reverted back to that which existed before the disease eradicated most of the native elm trees. (ii) A GM *Eucalyptus* with cold tolerance is introduced into Europe. We know that *Eucalyptus* is already an alien invasive species in some areas and that it supports a very limited biodiversity. Therefore, we would have to consider whether this GM trait would extend the invasive areas of *Eucalyptus* and what the consequences of this might be for native flora and fauna.

The environmental impacts of GMTs will also be influenced by their management and how they are exploited. The impacts of short rotation coppice or plantations of juvenile GMTs will be mainly confined to biota associated with these cultivation systems. The trees will produce few or no flowers and set very little seed and so any dispersal will be confined mostly to vegetative spread. Traits which affect the fitness of the tree such as growth rate and wood quality will have little or no influence outside the plantation areas. Traits which affect biotic interactions such as pest and disease control will mostly affect target and non-target populations within the cultivation area but may also have some broader consequences on species with wide dissemination characteristics.

By contrast GMTs, which are allowed to reach sexual maturity, will be able to disperse pollen and seed over large distances, depending on the tree species. Therefore, the potential receiving environment for these propagules will extend far beyond the cultivation areas and will need to be considered in the ERA. For further information on gene flow in trees and its consequences, see papers by DiFazio et al. (2004), Brunner et al. (2007), Firbank (2008), Ahuja (2009) and Bialozyt (2012).

Herbicide tolerance traits will allow applications of broad spectrum herbicides on young tree plantations to control weeds. This will facilitate rapid establishment of the trees in the absence of competitors, especially perennial weeds. The herbicides will only be applied in the early years of establishment until the trees are large enough to successfully compete with the weeds. Thus botanical diversity will be reduced in early years, but there will be some re-establishment later from the seed bank. Similarly, effects on biodiversity will occur while weed vegetation is destroyed but will rapidly re-establish when herbicide treatment is stopped.

5 Focusing Risk Assessment/Recent Developments

Assessment endpoints commonly derive from protection goals which are based on policy goals and societal values that are often described in the legislation of a particular country or region. Protection goals are broad and include such environmental values as protection of biodiversity, protection of water resources or protection of large nature areas.

Choosing appropriate and comprehensive assessment endpoints makes it possible to assess impacts on sets of protection goals. It is also important to choose the correct measurement endpoints, i.e. how to best determine (measure) the possible adverse effects to a certain species or population?

Recently, a new approach has been introduced to ERA. Protection goals are still seen as very general and broad, and difficult to implement in practice. In order to move forward, and divide/translate these general protection goals into more specific protection goals (assessment endpoints), the ecosystem services concept has been introduced to the ERA of plant protection products (PPPs), alien species and GMO risk assessment. These discussions are especially advanced in the PPP risk assessment (see, e.g. EFSA 2010b). During the development of ecosystem services-based framework for PPP risk assessment, those ecosystem services were identified that could potentially be harmed by the use of PPPs in agriculture. Further, key drivers (main indicators) for each ecosystem service were identified and specific protection goals (assessment endpoints) defined. For further reading on the ecosystem services concept see EFSA Scientific Colloquium Summary Report 19 (2014) and Nienstedt et al. (2012).

6 Guidance for Risk Assessment of GM Trees

6.1 The EU

The EFSA guidance (2010a) provides general risk assessment guidance to assess all GM plants, including tree species. In a recent review paper on risk assessment and regulation of GMTs in the EU Aguilera et al. (2013) describe the legal background and discuss the usefulness of the EFSA guidance in this context. In their conclusions, they state that EFSA guidance provides useful information to the ERA of GMTs, but due to the biology of GM fruit and forest trees and their management practices some changes in data types, collection and analysis may be needed. The specific considerations for GMTs have been discussed earlier in this paper.

6.2 The OECD

The OECD working group on harmonisation of regulatory oversight on biotechnology has produced several biology documents on different tree species. These documents have proven to be very useful in the ERA of GMTs as they provide baseline information on the biology and phenology of tree species. They support both the comparative, and per se approach to the ERA. So far, there are documents published of the conifers *Picea abies* (OECD 1999a), *P.glauca* (OECD 1999b), *P.mariana* (OECD 2010), *Pinus banksiana* (OECD 2006), *P.contorta* (OECD 2008a), *P.monticola* (OECD 2008b), *P.sitchensis* (OECD 2002a), *P.strobus* (OECD 2002b), North American larches (*Larixlyallii, L. occidentalis* and *L. laricina*; OECD 2007) and *Pseudotsuga Menziesii* (OECD 2008c), as well as the broadleaved trees *Populus* (OECD 2000), *Prunus* spp. (OECD 2002c) and *Betula pendula* (OECD 2003).

7 The Cartagena Protocol on Biosafety

The only internationally available specific ERA guidance on GMTs has been developed under the Cartagena Protocol on Biosafety (UNEP/CBD/BS 2012). The specific section on trees complements the Roadmap (generic risk assessment framework) of the same guidance document.

The Roadmap and the specific guidance are under revision at the moment but are and have been actively used by different stakeholders. The specific guidance starts with *Background*, *Scope* and *Introduction*. These chapters introduce many specific characteristics of trees that have to be taken into consideration when assessing GMTs. Trees are big in size, have often a perennial growth habit with a long lifespan, and a delayed onset of reproductive maturity. Many characteristics differentiate trees from annual crop plants. This aspect has been discussed earlier in this paper, but the list below gives an idea of the specificity of the Cartagena guidance.

According to the guidance, the special features compared to the crop plants include high fecundity, seed dormancy and high seed viability as well as many pathways for dispersal of propagules. Their root system is associated with different microorganisms and fungi. Furthermore, tree species and genotypes are highly diverse, and they have wide range of distribution and complex associations. Trees are present from temperate to tropical climates and have complex and multi-level ecological interactions with other organisms, e.g. from decomposers to birds and from insect pollinators to large wild animals. Trees also have significant ecological, economic, environmental, climatic and socio-economic values. The non-managed forests, such as tropical rain forests or boreal forests have high conservation value.

The next step is the *Planning phase of the ERA*. The guidance follows the comparative approach, and lists a number of points to be considered in the planning step. In the *Choice of comparators*, the following issues should be taken into account: long lifespan, high potential for dispersal, outcrossing and establishment beyond the intended receiving environment (into natural or less managed ecosystems). Moreover, well-adapted provenances are often used in forestry. They may show better adaptive capabilities and better performance than unselected germplasm. These regional provenances may provide appropriate comparators. However, when little information is available on ecological functions and interactions, the comparative approach is sometimes challenging. Points to consider when choosing comparators are the following: information and knowledge of the biology and ecological interactions of the suitable comparators and whether one or more comparators are available. The planning phase can also include field trials design where important information can be obtained from experience with non-modified trees.

Conducting the ERA follows the steps of the Roadmap (Step 1: Identification of adverse effects; Step 2: Evaluation of the likelihood; Step 3: Evaluation of the consequences; Step 4: Estimation of overall risk and Step 5: Recommendation as to whether or not the risks are acceptable or manageable and Identification of management strategies). The GM trees ERA guidance lists specific points to be taken into account under each step of the Roadmap.

For Step 1 presence of genetic elements and propagation methods and long lifespan, genetic and phenotypic characterization and stability of modified genetic elements need special attention. For both Step 1 and Step 2, dispersal mechanisms should be carefully considered.

For Steps 1, 2 and 3, the likely potential receiving environments must be carefully assessed, taking into consideration different environments and degree of their management, presence and proximity of compatible species, proximity of protected areas and centres of origin, genetic diversity, ecologically sensitive regions, ecosystem functions and services and change in landscape patterns and sensitivity to human activities.

Exposure of the ecosystem to living modified trees and potential consequences must be considered in both Step 2 and Step 3.

For Steps 4 and 5, risk management strategies include certain specific issues when dealing with GM trees. When there is a need to limit or prevent dispersal of forest or plantation GM trees, strategies may include delaying or preventing flowering and biological confinement (e.g. male sterility or flower ablation).

8 Specific Data Requirements for GM Trees: Research Needs

In all of the above ERA systems, data on impacts of GMTs on biota and the whole ecosystem in potential receiving environments are required, as for other GM plants. However, as trees are perennial long lived woody species that can form dominant and climactic populations, environmental impacts of GMTs are very different from those of the mostly annual GM crop species that have been assessed by regulators and risk assessors. Research is therefore needed to determine what types of data are required in order to make meaningful comparisons between GM and non-GM trees which inform risk assessors not only of any differences in phenology and impact but also the consequences of the differences for receiving environments and their biota. In addition, due to the longevity of trees and the need to make risk assessments proportionate and realistic, modelling, scenario analysis and other predictive methods may need to be researched and developed in order to quantify impacts on protection goals and assessment endpoints.

9 Environmental Monitoring

In *the EU* there is a requirement for post-market environmental monitoring of all commercialised GM plants (EC 2001; EFSA 2011b). Applications for release of GMTs will require post-market environmental monitoring (PMEM) plans, which take account of any identified potential risks or uncertainties as well as allow for unanticipated effects. All commercialised GM trees will therefore be monitored and the results of the monitoring reported to the relevant authorities and placed in the public domain. The monitoring should be part of the general stewardship and conditions of release and management of the GM tree. If results of the monitoring indicate that adverse effects are being caused as a consequence of the cultivation of the GM tree, then action can be taken to prevent or mitigate these effects.

The Cartagena Protocol guidance (UNEP/CBD/BS 2012) includes also guidance on environmental monitoring. The guidance is divided into sections *Introduction*, *Objective and scope*, Monitoring and its purposes and *Development of a Monitoring Plan.*

A monitoring plan may include the following steps. The first step is choice of indicators and parameters for monitoring (What to monitor?). Monitoring for potential effects of a GMO involves the observation of changes to *indicators* (e.g. species, populations, soil, environmental processes, etc.) and/or *parameters* (e.g. species abundance and soil organic matter).

The second step is monitoring methods, baselines including reference points, and duration and frequency of monitoring (How to monitor?). The selection of monitoring methods should take into account the level of sensitivity and specificity needed to detect changes in the indicators and parameters. Monitoring methodology includes the means for sampling and observing indicators and parameters, and for the analysis of the resulting data. Appropriate methods for collecting data may include, e.g. observations, descriptive studies and questionnaires.

The third step is choice of monitoring sites (Where to monitor?). The sites are selected on a case-by-case basis depending on the geographical location of the release, the used indicators and parameters, the intended use of the GMO, and taking into account the associated management practices.

In the fourth step guidance is given on how to report of monitoring results (How to communicate?).

The Cartagena Protocol guidance on monitoring is a general guidance but includes many relevant issues to be taken into account when planning for monitoring of GM trees. The planning should be done hand-in-hand with the specific ERA guidance.

References

Aguilera J, Nielsen KM, Sweet JB (2013) Risk assessment of GM trees in the EU: current regulatory framework and guidance. iForest: http://www.sisef.it/iforest/contents/?id=ifor0101-006

Ahuja RM (2009) Transgene stability and dispersal in forest trees. Trees 6:1125–1135

Bialozyt R (2012) Gene flow in poplar—experiments, analysis and modeling to prevent transgene outcrossing. iForest 5:147–152

Brunner A, Li J, DiFazio SP, Shevchenko O, Montgomery BE, Mohamed R, Wei H, Ma C, Elias AA, Van Wormer K, Strauss S (2007) Genetic containment of forest plantations. Tree Genet Genomes 3:75–100

Burgess EPJ, Kean AM, Barraclough EI, Poulton J, Wadasinghe G, Markwick NP, Malone LA (2014) Evaluation of the field impacts of simulated *Bacillus thuringiensis*-transgenic *Pinus radiata* on nontarget native Lepidoptera and their natural enemies in a New Zealand plantation forest. Agric Entomol 16:63–74

Codex Alimentarius (2009) Foods derived from modern biotechnology. Codex Alimentarius Commission, Joint FAO/WHO Food Standards Programme, Rome. http://www.fao.org/docrep/011/a1554e/a1554e00.htm

DiFazio SP, Slavov GT, Burzyk J, Leonardi S, Strauss SH (2004) Gene flow from tree plantations and implications for transgenic risk assessment. In: Carson M (ed) Plantation Forest Biotechnology for the 21st Century, (Walter C. Research Signpost, India, pp 405–422

EC (2001) Directive 2001/18/EC of the European Parliament and of the Council of 12 March 2001 on the deliberate release into the environment of genetically modified organisms and repealing Council Directive 90/220/EEC. Official J Eur Union L 106:1–39

EC (2002) Commission Decision (2002/623/EC) of 24 July 2002 establishing guidance notes supplementing Annex II to Directive 2001/18/EC of the European Parliament and of the Council on the deliberate release into the environment of genetically modified organisms and repealing Council Directive 90/220/EEC. Official J Eur Union L 200:22–33

EC (2003) Regulation (EC) No 1829/2003 of the European Parliament and of the Council of 22 September 2003 on genetically modified food and feed. Official J Eur Union L 268:1–23

EFSA (2009) Scientific opinion on Guidance for the risk assessment of genetically modified plants used for non-food or non-feed purposes. EFSA J 1164:1–42

EFSA (2010a) Guidance on the environmental risk assessment of genetically modified plants. EFSA J 8:1879

EFSA (2010b) Scientific Opinion on the development of specific protection goal options for environmental risk assessment of pesticides, in particular in relation to the revision of the Guidance Documents on Aquatic and Terrestrial Ecotoxicology (SANCO/3268/2001 and SANCO/10329/2002). EFSA J 8:1821. doi:10.2903/j.efsa.2010.1821

EFSA (2011a) Scientific Opinion on Guidance for risk assessment of food and feed from genetically modified plants. EFSA J 9:2150

EFSA (2011b) Guidance on the Post-Market Environmental Monitoring (PMEM) of genetically modified plants. EFSA J 9:2316

EFSA (2011c) Guidance on selection of comparators for the risk assessment of genetically modified plants and derived food and feed. EFSA J 9:2149

EFSA Scientific Colloquium Summary Report 19 (2014) Biodiversity as protection goal in environmental risk assessment for EU agro-ecosystems. EFSA publication office. 72 pp. ISBN: 978-92-9199-588-2. doi:10.2805/57358

Firbank LG (2008) Assessing the ecological impacts of bioenergy projects. Bioenergy Res 1:12–19

Fladung M (2011) Genetically modified trees for a sustainable, environmentally responsible and resource-saving production of wood for energy production. Gesunde Planzen 63:101–110

Fladung M, Altosaar I, Bartsch D, Baucher M, Boscaleri F, Gallardo F, Häggman H, Hoenicka H, Nielsen KM, Paffetti D, Séguin A, Stotzky G, Vettori C (2012) European discussion forum on transgenic tree biosafety. Nat Biotechnol 30:37–38

Häggman H, Raybould A, Borem A, Fox T, Handley L, Hertzberg M, Lu M-Z, Macdonald P, Oguchi T, Pasquali G, Pearson L, Peter G, Quemada H, Seguin A, Tattersall K, Ulian E, Walter C, McLean M (2013) Genetically engineered trees for plantation forests: key considerations for environmental risk assessment. Plant Biotechnol J 11:785–798

Harfouche A, Meilan R, Altman A (2011) Tree genetic engineering and applications to sustainable forestry and biomass production. Trends Biotechnol 29:9–17

Harfouche A, Meilan R, Kirst M, Morgante M, Boerjan W, Sabatti M, Mugnozza GS (2012) Accelerating the domestication of forest trees in a changing world. Trends Plant Sci 17:64–72

Hoenicka H, Fladung M (2006) Biosafety in *Populus* spp. and other forest trees: from non-native species to taxa derived from traditional breeding and genetic engineering. Trees 20:131–144

Kikuchi A, Watanabe K, Tanaka Y, Kamada H (2008) Recent progress on environmental biosafety assessment of genetically modified trees and floricultural plants in Japan. Plant Biotechnol 25:9–15

Nienstedt KM, Brock TCM, van Wensem J, Montforts M, Hart A, Aagaard A, Alix A, Boesten J, Bopp SK, Brown C, Capri E, Forbes V, Koepp H, Liess M, Luttik R, Maltby L, Sousa JP, Streissl F and Hardy AR (2012) Development of a framework based on an ecosystem services approach for deriving specific protection goals for environmental risk assessment of pesticides. Sci Total Environ 415:31–38. http://www.sciencedirect.com/science/article/pii/S0048969711005821

OECD (1986) Recombinant DNA safety considerations. Organisation for Economic Cooperation and Development (OECD), Paris, France

OECD (1992) Safety considerations for biotechnology. Organisation for Economic Cooperation and Development (OECD), Paris, France

OECD (1999a) Consensus Document on the Biology of *Picea abies* (L.) Karst (*Norway Spruce*). Organisation for Economic Cooperation and Development (OECD), Paris, France

OECD (1999b) Consensus Document on the Biology of *Picea glauca* (Moench) Voss (White Spruce). Organisation for Economic Cooperation and Development (OECD), Paris, France

OECD (2000) Consensus Document on the Biology of *Populus* L. (Poplars). Organisation for Economic Cooperation and Development (OECD), Paris, France

OECD (2002a) Consensus Document on the Biology of *Picea sitchensis* (Bong.) Carr. (Sitka Spruce). Organisation for Economic Cooperation and Development (OECD), Paris, France

OECD (2002b) Consensus Document on the Biology of *Pinus strobus* L. (Eastern White Pine). Organisation for Economic Cooperation and Development (OECD), Paris, France

OECD (2002c) Consensus Document on the Biology of *Prunus* spp. (Stone Fruits). Organisation for Economic Cooperation and Development (OECD), Paris, France

OECD (2003) Consensus Document on the Biology of European White Birch (*Betula pendula* Roth). Organisation for Economic Cooperation and Development (OECD), Paris, France

OECD (2006) Consensus Document on the Biology of *Pinus banksiana* (Jack Pine). Organisation for Economic Cooperation and Development (OECD), Paris, France

OECD (2007) Consensus Document on the Biology of the Native North American Larches: Subalpine Larch (*Larix lyallii*), Western Larch (*Larix occidentalis*), and Tamarack (*Larix laricina*). Organisation for Economic Cooperation and Development (OECD), Paris, France

OECD (2008a) Consensus Document on the Biology of Lodgepole Pine (*Pinus contorta* Dougl. *ex*. Loud.). Organisation for Economic Cooperation and Development (OECD), Paris, France

OECD (2008b) Consensus Document on the Biology of Western White Pine (*Pinus monticola* Dougl. ex D. Don). Organisation for Economic Cooperation and Development (OECD), Paris, France

OECD (2008c) Consensus Document on the Biology of Douglas-Fir (*Pseudotsuga Menziesii* (Mirb.) Franco (2008)

OECD (2010) Consensus Document on the Biology of *Picea mariana* [Mill.] B.S.P. (Black spruce). Organisation for Economic Cooperation and Development (OECD), Paris, France

Suter II GW (2007) Ecological risk assessment. CRC Press, Taylor and Francis Group. 2nd edn. 643 pp. ISBN 1-56670-634-3

UNEP/CBD/BS (2012) Guidance on risk assessment of living modified organisms. UNEP/CBD/BS/COP-MOP/6/13/Add. 1

Walter C, Fladung M, Boerjan W (2010) The 20-year environmental safety record of GM trees. Nat Biotechnol 28:656–658

Wolt JD, Keese P, Raybould A, Fitzpatrick JW, Burachik M, Gray A, Olin SS, Schiemann J, Sears M, Wu F (2009) Problem formulation in the environmental risk assessment for genetically modified plants. Transgenic Res 19:425–436

Field Trials with GM Trees: A Step-by-Step Approach

Debora C.M. Glandorf and Didier Breyer

Abstract Field trials are an important step in the experimental research with and commercial development of genetically modified (GM) plants, including GM trees. Field trials with GM plants in the European Union (EU) are subject to authorisation granted on the basis of an environmental risk assessment (ERA). Data requirement for the ERA varies depending on the purpose of the field trial and the level of knowledge on the GM plant and its environmental impact. In the Netherlands a step-by-step approach has been developed for the categorisation of field trials. Under this approach the confinement of GM plants in a field trial can be gradually decreased and the scale of the trial increased in a step-wise manner at the same time that knowledge on the GM plant and its environmental interactions increases. Very few other countries seem to apply a similar classification of field trials. We argue that a formal step-by-step approach may be a helpful tool to facilitate the approval process for field trials of GM plants and the collection of relevant data/material for the ERA without compromising the environmental safety, and that this approach is also applicable to field trials with GM trees.

1 Introduction

Genetically modified (GM) plants, including trees, intended for commercial cultivation in the European Union (EU) shall be authorised according to Directive 2001/18/EC (EU 2001) or Regulation (EC) 1829/2003 (EU 2003). The latter Regulation is only applicable in case of plants also intended for food and/or feed

D.C.M. Glandorf (✉)
GMO Office, National Institute of Public Health and the Environment,
P.O. Box 1, 3720 Bilthoven, BA, The Netherlands
e-mail: boet.glandorf@rivm.nl

D. Breyer
Scientific Institute of Public Health, Biosafety and Biotechnology Unit,
Rue J. Wytsmanstraat 14, 1050 Brussels, Belgium
e-mail: didier.breyer@wiv-isp.be

C. Vettori et al. (eds.), *Biosafety of Forest Transgenic Trees*,
Forestry Sciences 82, DOI 10.1007/978-94-017-7531-1_7

use. Before an application for commercial growth can be filed, sufficient data should be obtained on the safety of the GM plant itself (e.g. molecular, phenotypic and agronomical characterisation) as well as on its interactions with the receiving environment, both intended and unintended. Data requirements are described in Directive 2001/18/EC and have been complemented by additional Guidance from the European Food Safety Authority (EFSA).

Such data can be obtained from various sources. However, to obtain reliable data on phenotypic and agronomic characteristics and on environmental interactions, field trials are a necessity. Indeed, plant behaviour and the plant's interaction with the environment can be only realistically studied in an open environment including the full set of biotic and abiotic interactions.

Field trials with GM plants can be performed with different objectives in mind: (1) scientific research or demonstration purposes, (2) regulatory purposes, (3) selection of the best lines for future commercial release (product development) or (4) variety registration of selected lines (in the EU).

(1) Field trials for scientific research are often small scale. They are generally meant to study the stability and expression of the trait and the phenotypic characteristics of the plant under field conditions or to demonstrate "proof of principle" of the introduced trait. Such field trials can also be done as part of biosafety research, to generate independent data on the biosafety of GM plants. GM plants cultivated in these field trials will not end up as a commercial product. The same accounts for trials for demonstration purposes. These trials are especially set up for the public, mainly for educational purposes. They are also often small scale.

(2) Trials for regulatory purposes are performed to generate data that are required to complete the environmental risk assessment of GM plants (potential effects on target and non-target organisms, the soil ecosystem, the abiotic environment or human and animal health, see Sect. 3) or the food and feed safety evaluations (compositional analysis, potential toxicity or allergenicity).

(3) Trials for product development are performed by breeders who need to evaluate new germplasm or new crosses for their agronomic performance (efficacy of the modification, yield, etc.). These trials can take many years, during which promising lines are selected and tested on a larger scale to be able to collect enough data on performance and for analyses.

(4) Field trials can also be performed for variety registration on a national or European catalogue and for breeder rights. Variety registration is a precondition for the certification and commercialisation in the European Union of seed of agricultural plant species, vegetable species and fruit trees (for more information, see http://ec.europa.eu/food/plant/plant_propagation_material/plant_variety_catalogues_databases/index_en.htm).

Except for field trials for scientific research or demonstration purposes, data and material to obtain these data are usually collected from trials in multiple locations, representing a range of growing conditions, and over multiple years.

In many cases GM plants that are tested in the field are not fully characterised from a molecular, phenotypic and agronomical viewpoint and/or there is incomplete knowledge on their interactions with the receiving environment. In this chapter we discuss how these uncertainties could be taken into account during the risk assessment and how field trials with GM plants can be performed in an efficient way to obtain all relevant data/material without compromising the environmental safety. The approach towards field trials with GM plants of several EU member states, Canada and the USA is described and discussed with a special focus on field trials with GM trees.

2 Methodology and Data Requirement for the Environmental Risk Assessment of GM Plants

The purpose of the environmental risk assessment (ERA) is to assess if the introduction of a GM plant into the environment would have adverse effects (direct or indirect, immediate or delayed) on human and animal health and the environment. The ERA of GM plants is a comparative exercise, i.e. it involves generating, collecting and assessing information on a GM plant in order to determine its potential adverse impact relative to its non-GM comparator. The ERA should be carried out in a scientifically sound and transparent manner. It should also be carried out on a case-by-case basis, meaning that the required information may vary depending on the species of GM plants concerned, the introduced genes, their intended use(s) and the potential receiving environment(s), taking into account specific cultivation requirements and the presence of other genetically modified organisms (GMOs) in the environment.

Different national and international legal instruments have been established to regulate the environmental release of GM plants. Most of them follow the same general principles and methodology for the ERA. In the European Union (EU), these principles and methodology are laid down in Directive 2001/18/EC, in particular its Annex II. These principles and methodology are supplemented by guiding instruments, in particular Commission Decision 2002/623/EC (EC 2002) and the Guidance on the ERA of GM plants developed by the European Food Safety Authority (EFSA 2010). Supporting tools to assess the potential environmental risks associated with GMOs have also been developed at international level, in particular by the (OECD 2014) and in the frame of the Cartagena Protocol on Biosafety (CPB 2012).

In accordance with the above principles and methodology, an ERA should be conducted in six steps, in an integrated process and in an iterative manner, as follows (see also Fig. 1):

1. Problem formulation in which all important questions for the risk characterisation are identified. This step includes the identification of any characteristics of the GM plant which may cause adverse effects (hazards), of the nature of these effects, and of pathways of exposure through which the GM plant may adversely affect the environment;

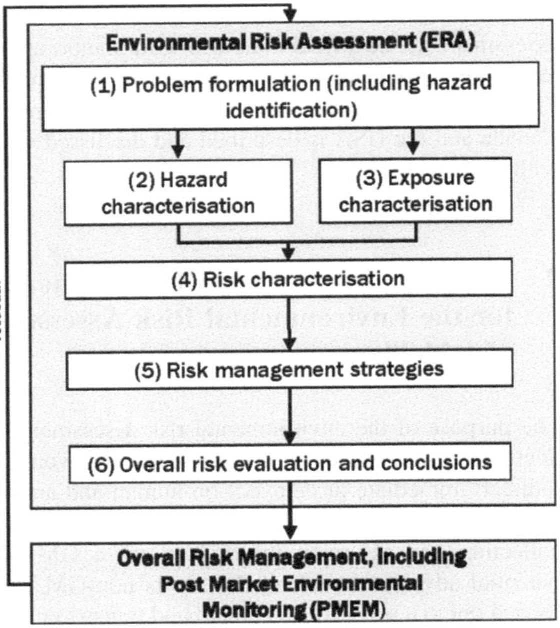

Fig. 1 Six steps in the environmental risk assessment (ERA) of GM plants (*Source* EFSA 2010)

2. Hazard characterisation, i.e. the evaluation of the potential consequences of each adverse effect;
3. Exposure characterisation, i.e. the evaluation of the likelihood of the occurrence of each identified potential adverse effect;
4. Risk characterisation, which is an estimation of the risk posed by each identified characteristic of the GM plant which has the potential to cause adverse effects;
5. Identification of management strategies to reduce potential identified risks associated with the GM plant to a level of no concern, and to address the uncertainties;
6. Evaluation of the overall risk of the GM plant, taking into account the results of the ERA and associated levels of uncertainty and the risk management strategies proposed.

Although quite some experience and knowledge has already been gained with (GM) crops worldwide, in many ERAs, both hazard and exposure are measured with a certain level of uncertainty. Uncertainty is an inherent and integral element of scientific analysis and risk assessment of any organism, whether it is GM or not. It can arise from: (i) lack of information, (ii) incomplete knowledge, and (iii) biological or experimental variability, for example due to inherent heterogeneity in the population being studied or to variations in the analytical assays (CPB 2012). Uncertainty resulting from lack of information includes, for example, information that is missing and data that are imprecise or inaccurate (e.g. due to study designs, model systems and analytical methods used to generate, evaluate and analyse the information).

Although it may be impossible to identify all the uncertainties, the ERA shall include a description of the types of uncertainties encountered and considered during the different risk assessment steps. Their relative importance and their influence on the assessment outcome shall be described (EFSA 2009). Uncertainties originating from lack of information may be addressed by requesting further information on the specific issues of concern.

Addressing uncertainties originating from lack of information is especially relevant in the context of field trials and of the step-by-step approach foreseen in Directive 2001/18/EC. According to this Directive the containment of GMOs is reduced and the scale of field trials increased gradually, step by step, but only if evaluation of the earlier steps in terms of protection of human health and the environment indicates that the next step can be taken. In this approach, it is anticipated that by progressively increasing knowledge on the characterisation and potential adverse effects of the GMO, uncertainties can be reduced.

It is recognised that an ERA can only be carried out based on scientific and technical data available at the time it is conducted. The ERA may not always result in definitive answers to all the questions considered because of lack of such data. Therefore appropriate risk management, including monitoring, has to be considered in accordance with the precautionary principle in order to prevent adverse effects on human health and the environment.

3 Information Requirements for the ERA

For each element of an ERA, information must be compiled. In the EU, the information requirements are specified in Annex III of Directive 2001/18/EC. It includes information relating to the recipient or parental organism(s), to the genetic modification, to the resulting GM plant (for instance factors that may affect its release, presence and persistence in the environment), to the characteristics of the release and of the receiving environment and to the interactions between the GM plant and the receiving environment. Relevant information can be obtained from a variety of sources such as scientific literature, expert opinions, monitoring reports, or experimental data obtained during or prior to the risk assessment process.

As stated in Directive 2001/18/EC, not all the information elements set out in the Annex III have to be delivered in every case. Only the particular subset of considerations which is appropriate to individual situations should be addressed (case-by-case). The level of detail required in response to each subset of considerations is likely to vary according, among other things, to the biology/ecology of the recipient organism, the intended use of the GM plant and its likely potential receiving environment, the (biological) containment of the trial, and the scale and duration of the environmental exposure (e.g. whether it is for field testing or for commercial use). Obviously, the level of detail required is lower for field trials than for an unconfined commercial release.

Defining the appropriate set of information to be provided for a specific ERA is not straightforward and can give rise to different interpretations amongst risk assessors and evaluators.

For small-scale field trials, especially at early experimental stages or in the early steps of environmental releases of GM plants that are conducted in a step-wise manner, limited information may be available to identify or characterise some of the potential hazards at the time the risk assessment is conducted. Whereas sufficient information will be generally available on the biology on the parent species and the characteristics and consequences of the genetic modification (from published data and laboratory experiments or from other sources), less information may be available with regards to the phenotypic and agronomic properties of the GM plant and its (un)anticipated interactions with the receiving environment. As mentioned above, in such cases the resulting uncertainties may be addressed through the implementation of risk management measures, aiming in particular at mitigating impacts on the receiving environment (e.g. spatial isolation, planting of non-GM border rows, prevention of pollen and seed dissemination). When additional information is requested in order to decrease uncertainty, one should keep in mind that this information should always be considered in terms of its relevance and contribution to the identification and evaluation of potential adverse effects of the GM plant in the context of its intended use(s), or how far it can affect the outcome of the risk assessment. Providing more information will not always contribute to reduce uncertainty. Data requirements should always consider the issue of "need to know" versus "nice to know". Only data useful for coming to a conclusion on the risk assessment should be generated, otherwise the study moves into more basic research (EFSA 2008) or may actually give rise to new uncertainties (CPB 2012).

For large scale or commercial environmental releases where the intent is widespread introduction of the GM plant in the environment, usually with few or no restrictions, the potential hazards should be characterised more completely, meaning that more detailed and comprehensive information should be considered for the risk assessment. Such information may be obtained from different sources including the scientific literature, laboratory studies, past risk assessments where relevant and also results from experimental field trials for the same or similar GMOs introduced in similar receiving environments.

Determining whether, when and how field studies should be set up to gather data to further inform the risk assessment of a (pre-)commercial release and contribute to reduce uncertainty is a complex matter.

First, environmental risk assessment for regulatory purposes is commonly organised through a tiered (step-wise) approach, whereby hazards are evaluated within different tiers that progress from worst-case scenario conditions framed in highly controlled laboratory environments to more realistic conditions in semi-field and subsequently field experiments (EFSA 2010). In this context the question whether field trials are always necessary or could be avoided when worst-case studies in the lab show the absence of a hazard with high enough certainty is subject to debate (Romeis et al. 2006; Lang et al. 2007).

Second, as mentioned in the previous section, adequate problem formulation is a critical step in the ERA process. If field trials must generate data useful for the risk assessment, their design should comply with a science-based hypothesis-driven approach focusing on appropriate assessment endpoints and relevant exposure pathways that can measure what is really at harm.

Third, field trial designs should be adequate to provide relevant and statistically valid data. This is not always evident to achieve in small-scale trials which rarely have the statistical power to detect low effect sizes due to their insufficient sample size. When sample size is low, meta-analysis may improve statistical power by combining several trials and assuming a common measure of effect size. Extrapolating the results of such trials is also an issue, since effects assessed at one scale, period or climate may not pertain at other scales, periods or climates, and since it is hardly conceivable (from time and resources reasons) to perform multiple testing tailored to all potential receiving environments relevant for the GM plant in question. Some models are available for making predictions for larger scales based on results obtained from smaller scale experiments ("scaling up of results") but their routine application to GMO risk assessment is still a matter of discussion (EFSA 2008).

Fourth, another important aspect is how to gather reliable data for the ERA from more or less confined field trials. Due to applied confinement measures, the interactions between the GM plant and its receiving environment are often limited. Results (including monitoring data) of these field trials may be therefore less informative with regard to the full potential of interactions that may occur between the GM plant and its receiving environment in large scale and unconfined trials and may not generate scientifically robust data.

To help applicants and risk evaluators determining when and how field trials and data gathering for an ERA should be performed, some countries have established specific procedures and requirements mainly based on a tiered (step-wise) approach. As an example, the approach as taken in the Netherlands is detailed in the next section.

4 A Step-by-Step Approach for the Design and Evaluation of Field Trials with GM Plants: The Dutch Model

In the Netherlands testing of GM crops in field trials is performed in a step-wise manner, based upon the step-by-step principle described in Directive 2001/18/EC which states that confinement of GMOs in field trials can be gradually decreased and the scale of the introduction can subsequently be increased in a step-wise manner, under the condition that the conclusion of the ERA of the former steps allows the next step.

In line with this principle, the Dutch Commission on Genetic Modification (COGEM) proposed a system of field trials in which GM plants can be tested in the

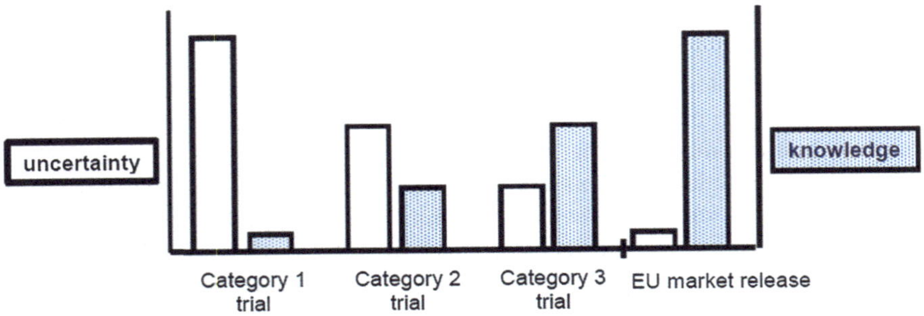

Fig. 2 Step-by-step approach for field trials with GM plants in the Netherlands. An increase in data results in a decrease of uncertainty. Due to this decreased uncertainty, the scale of the trials can increase and containment measures are no longer deemed necessary in category 2 and 3 trials (from COGEM 2008)

field on an increasing scale and with decreasing confinement measures on the condition that increasing knowledge on the GM plant and its environmental interactions is available (COGEM 2005, 2008). This system means to ensure that breeders can obtain sufficient data on their GM plants during the selection process in the field, without compromising the environmental safety of those trials. In general, the selection process of new plant varieties starts with a number of plants exhibiting the desired phenotype that are not fully characterised (phase 1). Thereafter a subset will be further tested and characterised (phase 2), resulting in one or two fully characterised events that will be commercialised (phase 3).

Following these 3 phases in the selection process of plant breeding, COGEM proposed a system in which three categories of field trials can be distinguished (Fig. 2) (Glandorf 2014).

The first category is meant to test the many uncharacterised plant lines in the start of the selection process. Due to the uncertainty related to the low level of characterisation of the plants, confinement measures can be applied such as isolation distances or removal of flower buds. The scale of the introduction is in this stage limited to maximum five locations of not more than one hectare per year. Thereafter, one can apply for the second category of field trials. This is only possible when there are sufficient data available to conclude, based on risk assessment, that potential adverse effects on human health and the environment are unlikely. This conclusion can be based on data obtained from the first category field trial, or based on data from other field trials or literature. This second category of field trials is meant—if applicable—to further characterise a subset of earlier tested plants which are of interest for the breeder. Based on the outcome of the risk assessment, the use of confinement measures to prevent outcrossing is considered no longer necessary. However, there still remains uncertainty since the GM plants are not fully characterised. Therefore the scale of the field trials is limited to a maximum of 10 ha per year.

The third and last category is meant for pre-commercial field trials with a single fully characterised event, which has proven in earlier trials in the Netherlands (or in other EU countries) not to have adverse effects on human health and the environment. There is no limitation to the scale of the field trials and no confinement measures to prevent outcrossing are deemed necessary.

For each trial measures have to be taken to prevent spreading from GM material from the trial site. In addition, applicants have to comply with the Dutch coexistence distances between GM and non-GM plants.

GM plants should not necessarily be subject to all categories of field trials in a step-wise manner. For example, a trial with GM plants could be initiated directly in the second or third category based on information already available on the GM plant from other (EU) trials or from the literature. If the information is not considered sufficient, the trial will be classified as a lower category of field trial that fits with the level of GM plant characterisation. In that case also the maximum size and number of locations will be adjusted accordingly and confinement measures may apply if necessary. Alternatively, a trial with GM plants could be performed under category 1 only without any need to scale up, for example in case of scientific research without any intention to follow this up for commercial purposes. This applies also in cases where there is sufficient data available on the specific GM plant for category 2 but there is no need for a large size trial. Applicants are free to choose for a lower category, despite the fact that mandatory confinement measures may apply (category 1) or the size or number of location will be limited as a consequence of this choice (Glandorf 2014).

Data requirements and other requirements for each category of field trials are laid down in COGEM advices (COGEM 2005, 2008). Ideally, the scientific advice issued by COGEM for a field trial in a certain category indicates which specific data requirements are considered necessary for the next category of field trials so that applicants can gather the relevant data before applying for a next category of field trials.

Monitoring data are an important pillar in data gathering with regard to expected and potential unexpected effects of GM plants on the environment. A monitoring plan is therefore mandatory for all categories of field trials. Monitoring consists of general monitoring (phenotype, general agronomic features, any unexpected effect) and, depending on the crop/trait combination, specific monitoring. Examples are specific monitoring to study persistence in the soil of an antimicrobial protein expressed by GM apple trees, or to study a potential increase in frost tolerance in low amylose starch potatoes.

5 Approaches in Other Countries for the Evaluation of Field Trials with GM Plants

Several EU and non-EU countries having experience with field trials involving GM plants, including GM trees, were contacted to provide information on the way such field trials were assessed and managed depending on their purpose and the level of knowledge on the GM plant and its environmental impact. Belgium, Finland, France, Romania, Spain, Sweden, the United Kingdom, as well as Canada and the USA provided some feedback based on the following questions:

1. Do you have in your country a similar system as the "Dutch system", in which field trials are upscaled in a step-wise manner? If so, do you have examples or documentation on this?
2. If such an approach is applied, do you indicate to the applicants which information is necessary to perform an upscaled field trial? If so, do you have examples or documentation on this?
3. If such an approach is not applied, do you use any other specific approach for dealing with experimental versus pre-commercial field trials, and for the gathering of biosafety data informing the environmental risk assessment? If so, do you have examples or documentation on this?

Belgium, Finland, Romania, Spain, Sweden, the United Kingdom and the USA indicated that they do not have a formal notification system similar as the Dutch system, in which field trials with GM crops or trees are—or can be—upscaled in a step-wise way. In these countries field trials are assessed on a case-by-case basis without making distinction between experimental and pre-commercial field trials, and without any pre-defined requirements on confinement measures, size or number of locations of the field trials. Belgium, Spain and Sweden reported cases where an authorisation for a small-scale research field trial was granted on the condition that specific additional studies are performed during the trial to increase knowledge on the potential environmental impacts of the GM plant. It was recognised however that small-scale trials may not always be adequate to provide scientifically robust data. Most countries indicated that management measures accompanying field trials were important to address uncertainties resulting from the risk assessment. It was noted that because such measures focus on minimising environmental exposure, they are rarely adequate to generate new data on the environmental impact of the GM plants, which could potentially be used if subsequent field trials with the same GM plants are considered.

France indicated that the High Council for Biotechnologies (HCB) has established in a recommendation an indicative classification of field trials in three categories corresponding to three stages of development of a GM plant. Indications are given on the level of detail of information expected by the HCB depending on the type of experimentation. This classification has no regulatory value. It is only indicative and does not prejudge the additional information that may be requested by the Ministry of Agriculture or the HCB during the examination of the application on a case-by-case basis.

In Canada the system is binary. An applicant can apply for a confined research field trial of plants with novel traits (PNTs) or an unconfined commercial release but there is no prescriptive "scale-up" procedure (CFIA 2014; Finstad et al. 2007). Confined field trials are limited in size as a risk management measure and terms and conditions are imposed. These terms and conditions provide for reproductive isolation of the plants within the trial from plants outside it, provide for physical separation of plant material from the trial from food and feed supply chains, and mitigate persistence of the PNTs in the environment post-harvest. Developers can make a justification for larger size field trials based on their research, experimental design, the biology of the species in the trial or other relevant criteria. Applications for unconfined release usually follow several years of confined trials. Approvals are given following a thorough environmental safety assessment based on extensive high-quality, statistically sound data and/or valid scientific rationale provided by the applicant to demonstrate the environmental safety of the PNT.

6 Discussion: How a Step-by-Step Approach May Be Applied to Field Trials with GM Trees

The ERA of GM plants should be carried out based on the most complete set of available data relevant to the GM plant to be assessed. In that respect, a challenging aspect is whether and how experience gained from a field trial with a specific GM plant can be taken into account in the ERA of subsequent trials with the same or similar GMO plants. In the Netherlands, this aspect has been addressed through the application of a formal and binding step-by-step approach. Under this approach the confinement of GM plants in a field trial can be gradually decreased and the scale of the trial increased in a step-wise manner at the same time that knowledge on the GM plant and its environmental interactions increases.

Our survey, even if it provides only partial and indicative information, shows that very few other countries have applied specific procedures or approaches to assess and manage fields trials with GM plants depending on their purpose (experimental vs. pre-commercial) and the level of knowledge on the GM plant and its environmental impact. France reported the use of an approach similar to the Dutch one, although not being prescriptive. In Canada, environmental releases are categorised in "confined" or "unconfined" levels but without prescriptive "scale-up" procedure.

A formal step-by-step approach may provide several interesting features.

First, it provides clarification with regards to the necessary data requirements for the ERA. For research field trials involving uncharacterised GM plants, the risk assessment will be based on information relevant to the implementation and effectiveness of conditions of use (scale, duration, types of activities) and management measures to ensure risk mitigation, while allowing knowledge gaps with regards to the GM plant and its potential environmental impact (category 1 in the Dutch approach, or "confined" level in Canada). For pre-commercial environmental releases, the risk assessment will require exhaustive and scientifically sound information demonstrating the environmental safety of the GM plant (category 3 in the Dutch approach, or "unconfined" level in Canada).

Second, it can facilitate the collection of useful data for the ERA. Indeed considerably more information is necessary for the risk assessment of unconfined field trials. Laboratory and greenhouse studies may be helpful in generating specific case-specific data for the risk assessment but, as mentioned earlier, such studies do not allow testing the GM plant against the full set of biotic and abiotic conditions. The information needed to feed the risk assessment may be difficult to collect from confined field trials (category 1) due to the application of strict confinement measures avoiding or limiting interactions between the GM plants and the environment. The necessary replication required to generate ecological effects data from confined field studies would also make such studies extremely difficult (Häggman et al. 2013). In that respect, the step-by-step approach allows for an intermediate phase (category 2 in the Dutch approach) between complete confinement and unconfinement in which relevant data can be collected for the ERA in open-field conditions. Examples are field trials without confinement measures, but with size limitations or spatial separation from sexual compatible species, in which interactions between the GM plant and the environment can be studied. Monitoring is an essential mean of generating data on those interactions.

Third, a step-by-step approach is applicable to all plants, including GM trees. It has been helpful as a tool to aid the approval process for GM trees, as illustrated by the following examples in the Netherlands.

The first example is a small-scale category 1 trial with GM poplar (*Populus* × *canescens*) for bioethanol production, grown in a short rotation coppice (COGEM 2010a). The tree was modified with the *ccr* gene, coding for cinnamoyl coenzymeA reductase, resulting in low lignin content, and a marker gene *hpt*, coding for hygromycin resistance. No flowering was expected. The application complied with the criteria for a category 1 trial. To prevent spreading of the poplar outside the field location, removal of inflorescence, root suckers and falling branches was required. Monitoring on root suckers should take place for at least two years.

An example of a category 2 trial is a small-scale field trial with flowering, scab-resistant apple trees (*Malus pumila*) (COGEM 2010b). The trees were modified with a resistance gene *HcrVf2* obtained from apple (*M. floribunda*), which is already present in commercial apple varieties and natural apple populations. The absence of vector backbone sequences was confirmed. The field plot was located 150 m from any apple tree and 500 m from commercial apple orchards. This distance was not considered enough to prevent outcrossing. However, among others, an important consideration in the ERA was that the *HcrV* gene is already present in commercial apple varieties and in natural apple populations. Therefore no environmental risks were foreseen as a result of outcrossing and no (additional) confinement measures were necessary.

The risk assessment of GM trees can already rely on existing knowledge on the biology of the corresponding non-GM tree species and their interaction with the receiving environment (see e.g. OECD 2014). This is especially true for intensively managed systems such as plantation forest trees. Additionally, commercial application of fruit trees or other woody perennial species may also provide useful information relevant to the ERA (Häggman et al. 2013). Many genes used to genetically modify trees and their resulting traits are not new and experience exists with regard to the assessment of those traits. In that respect the compilation of existing information performed in the frame of the EU COST Action FP0905 on the biosafety of forest transgenic trees (Fladung et al. 2012) will be of great value. As mentioned before, laboratory or greenhouse studies may also be a valuable tool in generating data on ecological effects.

With the application of a step-by-step approach, the ERA of GM trees might benefit from additional data generated from unconfined field trials that would complement already existing information and studies about the host, the genetic modification and the receiving environment. When designing field trials with GM trees one should however cope with some safety and methodological considerations specific to GM trees. There are some differences between trees and crops, for example their longevity and ability to disperse, that are no new aspects in the ERA of GM plants but may need more emphasis in the ERA of field trials with GM trees, in particular trees for plantation forests (Aguilera et al. 2013). Trees are generally perennial, woody, long lived species with long life cycles taking several years to reach sexual maturity and commence reproduction. When mature they can produce large amounts of seed and pollen that can disperse over long distances (Hoenicka and Fladung 2006). The choice of appropriate non-GM comparators may be also more limited. All these considerations may require some changes in data types, collection and field design. For example, extremely long field trials with these long lived species may be challenging. Although it seems difficult to avoid field trials with GM trees to generate data on possible unexpected/unintended effects resulting from the genetic modification, useful data may be obtained in field trials with non-GM trees of the same species exhibiting natural variation in the trait of interest.

As indicated by Aguilera et al. (2013) and Fladung et al. (2012) further discussions are still needed on the types of studies required for providing safety data to be used in the risk assessment of GM trees. Further clarifications may also be necessary with regards to the current EFSA guidance documents (that might not be readily applicable to GM trees) and the guidance developed under the Cartagena Protocol (that does not differentiate between confined and unconfined field trials). However, it is broadly recognised that field trials are important to collect data that are relevant to the specific characteristics of GM trees. A step-by-step approach involving a categorisation of field trials, such as the one developed in the Netherlands, provides a useful mean to frame and facilitate the design, evaluation and regulation of these field trials.

References

Aguilera J, Nielsen KM, Sweet J (2013) Risk assessment of GM trees in the EU: current regulatory framework and guidance. iForest 6:127–131

CFIA (2014) Directive Dir2000-07: Conducting confined research field trials of plant with novel traits in Canada. Available online: www.inspection.gc.ca/plants/plants-with-novel-traits/applicants/directive-dir2000-07/eng/1304474667559/1304474738697

COGEM (2005) Advies indeling veldwerkzaamheden genetisch gemodificeerde planten CGM/050929-03. Available online: www.cogem.net

COGEM (2008) Herziening advies indeling veldwerkzaamheden met genetisch gemodificeerde planten. CGM/081125-02. Available online: www.cogem.net

COGEM (2010a) Kleinschalige veldproef met genetisch gemodificeerde populieren met een verminderd ligninegehalte. CGM/081205-01. Available online: www.cogem.net

COGEM (2010b)Veldproef met genetisch gemodificeerde schurftresistente appelbomen. CGM/101214-01. Available online: www.cogem.net

CPB (2012) Guidance on risk assessment of Living Modified Organisms. Available online: www.bch.cbd.int

EC (2002) Commission Decision 2002/623/EC of 24 July 2002 establishing guidance notes supplementing Annex II to Directive 2001/18/EC of the European Parliament and of the Council on the deliberate release into the environment of genetically modified organisms and repealing Council Directive 90/220/EEC. Official J Eur Commu L200:22–33

EFSA (2008) Environmental risk assessment of genetically modified plants—challenges and approaches. IN colloquium Es (Ed.) EFSA scientific colloquium—summary report

EFSA (2009) Guidance of the scientific committee on transparency in the scientific aspects of risk assesments carried out by EFSA. Part 2: general principles. EFSA J 1051:1–22

EFSA (2010) Guidance on the environmental risk assessment of genetically modified plants. EFSA Panel on Genetically Modified Organisms (GMO). EFSA J 8(11):1879–1990

EU (2001) Directive 2001/18/EC of the European Parliament and of the Council of 12 March 2001 on the deliberate release into the environment of genetically modified organisms and repealing Council Directive 90/220/EEC. Official J Eur Commun L106:1–38

EU (2003) Regulation (EC) No 1829/2003 of the European Parliament and of the Council of 22 September 2003 on genetically modified food and feed. Official J Eur Commun L 268:1–23

Finstad K, Bonfils A-C, Shearer W, Macdonald P (2007) Trees with novel traits in Canada: regulations and related scientific issues. Tree Genet Genomes 3:135–139

Fladung M, Altosaar I, Bartsch D, Baucher M, Boscaleri F, Gallardo F, Häggman H, Hoenicka H, Nielsen K, Paffetti D, Séguin A, Stotzky G, Vettori C (2012) European discussion forum on transgenic tree biosafety. Nat Biotechnol 30:37–38

Glandorf DCM (2014) Categorisation of field trials with GM plants in the Netherlands: applicable to field trials with GM forest trees? iForest 8:222–225 doi:10.3832/ifor1311-008. Available online: www.sisef.it/iforest/contents/?id=ifor1311-008

Häggman H, Raybould A. Borem A, Fox T, Handley L, Hertzberg M, Lu M, MacDonald P, Oguchi T, Pasquali G, Pearson L, Peter G, Quemada H, Séquin A, Tattersall K, Ulian E, Walter C, McLean M (2013) Genetically engineered trees for plantation forests: key considerations for environmental risk assessment. Plant Biotechnol J:1–14

Hoenicka H, Fladung M (2006) Biosafety in *Populus* spp. and other forest trees: from non-native species to taxa derived from traditional breeding and genetic engineering. Trees 20:131–144

Lang A, Lauber E, Darvas B (2007) Early-tier tests insufficient for GMO risk assessment. Nat Biotechnol 25(1):35–36; Author reply 36–37

OECD (2014) Safety assessment of transgenic organisms. OECD Consensus Documents. Available online: www.oecd.org/env/ehs/biotrack

Romeis J, Meissle M, Bigler F (2006) Transgenic crops expressing *Bacillus thuringiensis* toxins and biological control. Nat Biotechnol 24(1):63–71

Soil Effects of Genetically Modified Trees (GMTs)

Fernando Gallardo, Conchi Sánchez, Marcin Grabowski, Juan Jesús Molina-Rueda, Nieves Vidal and Matthias Fladung

Abstract The activity of the root and the dead material from genetically modified trees (GMTs) might potentially alter soil features and turn into an impact on soil ecosystem. Several greenhouse and field studies of forest transgenic trees including poplars, silver birch, white spruce, American chestnut and *Eucalyptus* engineered for lignin biosynthesis and other relevant traits have addressed a potential impact on the receiving environment. Most of the available studies have considered effects on mycorrhizal fungi because of their intimate relationship with trees, and their support for the plants' acquisition of water and nutrients. Futhermore, changes in fungal community may also affect other fungal or bacterial communities and be thus indicative of more complex changes to soil ecosystem. To the knowledge of the authors, significant changes in bacterial, fungal community or mycorrhizal plant colonization have not been reported in peer-review of GMTs to date. However, some studies reported effects on indicators species. Similar observations have been

F. Gallardo (✉) · J.J. Molina-Rueda
Departamento de Biología Molecular y Bioquímica, Facultad de Ciencias, Universidad de Málaga, 29071 Málaga, Spain
e-mail: fgallardo@uma.es

J.J. Molina-Rueda
e-mail: juanjesusmolina@gmail.com

C. Sánchez · N. Vidal
Instituto de Investigaciones Agrobiológicas de Galicia, Consejo Superior de Investigaciones Científicas, 15780 Santiago de Compostela, Spain
e-mail: conchi@iiag.csic.es

N. Vidal
e-mail: nieves@iiag.csic.es

M. Grabowski
Department of Applied Entomology, Faculty of Horticulture, Biotechnology and Landscape Architecture, Warsaw University of Life Sciences-SGGW, 02-776 Warsaw, Poland
e-mail: grabo55@yahoo.pl

M. Fladung
Thünen-Institute of Forest Genetics, 22927 Grosshansdorf, Germany
e-mail: matthias.fladung@ti.bund.de

C. Vettori et al. (eds.), *Biosafety of Forest Transgenic Trees*, Forestry Sciences 82, DOI 10.1007/978-94-017-7531-1_8

reported in bioremediation studies with GMTs. The lack of baseline data on the diversity and natural variability of the soil microbiota, including fungi, in silvicultural practices limits the evaluation of the ecological relevance of the observed changes. Some studies suggest that plant stage, type of soil and other environmental factors may have a greater influence on the soil microbiota, as seen with indicator species, than the effect of GMTs.

1 Introduction

In order to optimize growth and productivity of tree plantations, active management of the crop soil and other vegetation is required. Knowledge about the genetics and phenotype of the tree and its interaction with the environment is important, as is managing resistance to pests and diseases affecting both yield and value of the final product. Environmental impact assessment of new genetically modified tree (GMT) clones is a prerequisite to their possible and deliberate release into the environment. The European directive 2001/18/EC describes some potential adverse effects of GMOs that require assessment. Due to the diversity of potential modifications and unique properties of GM plants such as assessment of adverse effects must be case-specific. For GMTs, due to their longevity, it is important to determine effects occurring throughout their long life cycle and to consider effects on populations of species in their receiving environment (see Häggman et al. 2013 for a recent review). Among the possible long-term effects of GMTs to be considered are the impacts on soil functions.

Adverse effects of GMTs on the soil biota and a modified composition of transgenic material could potentially alter the decomposition rates of their organic material, and thereby alter particularly the biogeochemical cycling of carbon and nitrogen at the receiving ecosystems. Effects on soil water relations and on its capacity for filtering and buffering, or on its physical and anchoring capacity to support the root system, may profoundly affect soil ecosystems in the long term. Therefore, the use of GMTs for economic and societal needs raise questions about the use of indicators to determine already at early state effects on the soil environment. A number of field and greenhouse works have studied the effects of GMTs on soil ecosystem on which we report in this chapter.

2 Potential Soil Effects and Indicators

The impacts on soil ecosystem of trees are associated with effect of decaying plant material that reach soil surface, or caused by the root growth and root exudates. Together, these materials contribute to the organic compounds in the soil and may affect existing soil microorganisms and fauna, including decomposers, plant pathogens and symbionts. Cultivation practices also affect the organisms and their activities. The tilling of the soil, the use of plant protection products and fertilizers, and rotational coppicing, all may potentially have an impact on the soil environment. A hazard identification of GMT plantations must therefore keep in mind that generally cultivation practices have strong effects on soil ecosystems, and those should serve as a baseline in a comparative approach for identifying possible ecological impacts of a GMT plantation.

The study of plants on soil functions has received special attention because of increasing number of hectares dedicated to transgenic crops in recent years. General indicators for a potential impact on soil ecosystem include analysis of abiotic and biotic factors that later encompass microbial communities, including mycorrhizal fungi colonization of roots, the study of degradation of organic matter and changes in invertebrate community (Smith et al. 2010).

The effects of transgenic crops such as corn, potato, sugar beet and oil seed rape, on soil microbial communities have been considered in relation to GM plants expressing Bt toxins and glufosinate-tolerant herbaceous crops (Smith et al. 2010). In some cases, an effect on microbial community structure was revealed. Most field studies, however, could not demonstrate significant differences between GM and non-GM, or effects produced by transgenic products released into soil. This conclusion is also shared in other works (Griffiths et al. 2007; Dunfield and Germida 2004) that showed that factors like tillage, the type of crop, the state of plant development and other environmental parameters, as well as the location of the plantations, have more impact on soil microbial communities than the introduction of a transgenic crop. In addition, another main problem in the assessment of potential effect on microbial communities is that different methods are used to evaluate the microbial structure in the different studies. These different methods include those based on degradation in vitro of different substrates, the analysis of phospholipid fatty acid composition, or the analysis of molecular data based on differences in ribosomal gene sequences. The use of different methods creates difficulties for comparison of the studies (Icoz and Stotzky 2008) and their interpretation in terms describing effects on microbial ecology. It has been suggested that other indicators may be more indicative of potential impacts than general analysis of soil microbial communities. Analysis of changes in arbuscular mycorrhizal fungi (AM fungi) appears to be a more preferred method for assessing potential impacts of a GM crop (Smith et al. 2010). AM fungi symbiosis increases the uptake of nutrients from the root system and it substantially varies across many environmental gradients. In addition, AM fungi affects the diversity of other organisms in soil through the exudates produced, altering other microbial communities. Although it is not easy to study AM fungi, it seems to be a very reliable indicator to determine a possible environmental impact on soil ecosystem (Smith et al. 2010).

Other informative indicators of potential impacts caused by transgenic crops are the degradation rate of organic matter and changes in the invertebrate community. The analysis of some Bt transgenic crops has revealed that degradation rates of plant material are different from non-transgenic crops. These results are probably due to changes in plant constituents as lignin, glucosamine and in ergosterol content associated to the plant material (revised in Smith et al. 2010). The interpretation of the data obtained from decomposition of organic matter is problematic as changes in the decomposition rate can be considered as a positive or negative effect depending on the organic content of the soils being studied and the crops being cultivated. The ecological significance of potential changes observed on the decomposition rate of organic material is very context dependant. However, experts consider that relevant changes in microbial communities should be detectable at metabolic level and have an effect on microbial respiration in soil. Changes in organic compounds and microbial respiration are two indicators that could be used to define baseline for different soils. This conclusion is emphasized by Smith et al. (2010) and they recommend, along with other indicators, the above-mentioned parameters could be considered for general surveillance of transgenic plants.

Soil invertebrates have also been considered as potential indicators of environmental impacts. Several studies include analysis of collembola, earthworm and nematode populations in transgenic crops. Nematode populations are considered a useful indicator for an impact since it is described as the unique group of invertebrate that may be affected on a small effect of different crop lines in different types of soil (revised in Smith et al. 2010).

3 Studies on the Effect of GMTs in Soil Ecosystem

A number of publications about GMTs describe the potential effects of transgenic lines on general indicators to analyse their possible impact on the receiving environment. These studies include several field and greenhouse experiments using transgenic poplar, silver birch, white spruce, *Eucalyptus* and American chestnut plants.

3.1 Decomposition of DNA from GMTs

A concern about GMTs is the persistence of the transgenic material in the soil, and the possibility of a horizontal gene transfer to other organisms which could contribute to the spread of the transgene and eventually alter the receiving ecosystem. Hay et al. (2002) addressed the persistence of transgenic DNA from decaying material. The work was focused on the stability of a neomycin phosphotransferase II gene, a common resistance marker use in plant transgenesis. Transgenic poplar leaves were placed in permeable bags on weeds, and on soil or below the soil under natural conditions for up to 12 months. The presence of the transgenic gene was then analysed by PCR in the samples. Under the natural conditions, Hay et al. (2002) observed that the modified DNA was not detectable in dead plant material after four months. Although many factors can contribute to the degradation of DNA in different soils, this study highlights the fact the rapid degradation of DNA from transgenic material. Therefore, the authors concluded that a possible transfer of the transgenic material to microorganisms is unlikely to occur under natural field conditions (Hay et al. 2002).

3.2 Trees with Altered Lignification

Alteration of lignin composition is a main area of research in plant biology. The biosynthesis of lignin has received a relevant attention to better understand the biosynthesis of this complex polymer. The composition of the polymer is associated to wood properties and quality, and it is also a relevant factor to consider in the production of pulp. Lignin content can be altered by changing the composition of monolignols participating in its synthesis. This can result in the production of new or altered wood, and also to modify the pulp yield for cellulose or biocombustible production. Lignin has also a significant ecological role in ecosystems. It participates in sequestering atmospheric carbon, and its degradation from dead plant material is relatively slow and contributes to the formation of stable organic matter, including humic material, and to moisture retention in soil. Therefore, GMTs with altered lignin content could potentially alter the characteristics of the receiving soil considering the full long life cycle of trees.

Relevant target genes for modification and alteration of lignin content are those encoding cinnamyl alcohol dehydrogenase (CAD) or caffeic acid/5-hydroxyferulic acid *O*-methyltransferase (COMT). These enzymes participate in the production of monolignols, the brick material for the production of the lignin polymer. Several studies of transgenic poplar (*Populus tremula* × *P. alba*) expressing an antisense gene for CAD or COMT. Pilate et al. (2002) describe field yield studies over 4 years in different locations in France and England. This work described the stable modification in lignin produced by alteration in CAD or COMT expression. Associated to the gene alteration, they observed easier delignification of the GMT material, producing a more high-quality pulp. However, the plantation of GMTs at the two different locations had no effect on microbial communities in soil. Some changes in microbial respiration were observed but these did not correlated with transgene expression. These results were related to spatial variability or fossil properties in the field. The study of decomposition of material from the same GMTs was described by Tilston et al. (2004). After studying the decomposition of truck sections from GMTs, including total CO_2 production and effect on microbial biomass, they concluded that environmental variability during the field assay had a greater effect on wood decomposition than the alteration of lignin biosynthesis of GMTs (Tilston et al. 2004). A similar conclusion is described in the review of Halpin et al. (2007) after comparison of results from GMTs and transgenic tobacco also altered in the expression of genes involved in lignin biosynthesis and composition. The review of Halpin et al. (2007) also emphasizes that no apparent effect on leaf-feeding insects, microbial pathogens and soil organisms of lignin-altered transgenic plants was observed, although short-term decomposition of transgenic roots was slightly enhanced but this was considered to be without apparent ecological significance.

Altered lignin biosynthesis and effect on soil microbial community was also described for transgenic *Populus tremuloides* lines (Bradley et al. 2007). This work was carried out in greenhouse using three soils from different locations with significant differences in chemical composition and physical properties. Reduced lignin content in stem, up to 50 %, and increased syringlyl-type lignin was observed in certain lines. Different effects on root biomass in comparison to control lines were observed in the different soil types. The response of the soil microbial communities was investigated using phospholipid fatty acids, neutral lipid and extracellular enzyme assays. Besides the changes observed, the responses of lines across the three soils were not consistent. The authors concluded that soil type had a large impact on the microbial community composition, and that it was difficult to assess the potential ecological impacts of the transgenic lines studied on soil microbial communities and their functions.

Alteration in the expression of 4-coumarate:coenzyme A ligase (4CL), involved in the lignin biosynthetic pathway, was also described by using an antisense construct in silver birch (*Betula pendula*; Seppänen et al. 2007). In this case, the genetic transformation did not affect wood chemistry but it had a relevant influence on root biomass and morphology. The number of root tips was lower than in control plants, and also the GMT exhibited less root biomass. Besides the significant effect on the root system, the genetic modification had no relevant effect on the development of ectomycorrhizal fungi, and no differences in the mycorrhizal colonization were observed *in vitro* when compared with control plants. Similarly, no relevant differences were observed in total microbial biomass and basal respiration. Moreover, similar degradation rates of leaf litter were observed (Seppänen et al. 2007). Together these studies on GMTs with different species and lines suggest that no relevant short-term environmental impact is detectable in lines with altered expression of genes involved in lignin biosynthesis.

3.3 Aspen Overexpressing the rolC Gene from Agrobacterium Rhizogenes

Extensive research and analysis of the mycorrhizal status of GMT lines was carried out in the frame of the first release experiment in Germany by the Thünen-Institute of Forest Genetics (Fladung and Muhs 2000). The study was initiated in 1996 and lasting for 5 years. Several aspen transgenic lines including *Populus tremula* and *P. tremula* × *P. tremuloides* were prepared for the expression of the *rolC* gene from *Agrobacterium rhizogenes* using the 35S (35S::*rolC*) and a rubisco small subunit promoter (rbcS::*rolC*). The aim of the field experiment was related to the mode of integration of the foreign gene and the stability of the transgene expression under natural environmental conditions and during the long life span of trees. The *rolC* gene from *A. rhizogenes* was selected in this study because it produces a remarkably dwarf-like phenotype characterized by stunted growth and modified leaf size and colour (Schmülling et al. 1988; Fladung 1990; Fladung and Ballvora 1992), allowing the phenotypic selection without tissue destruction (Fladung 1990, 1996; Fladung and Ballvora 1992). The *rolC* expression in rbcS::*rolC* lines was restricted to the leaf and resulted in a slightly reduced size of the plant with no remarkable reduction in apical dominance, but with light green leaves compared with darker green one of control plants. The expression was also detected in the root in the lines plants expressing the 35S::*rolC* transgene (Schmülling et al. 1993 Fladung et al. 1997).

In a first study, the mycorrhizal status of transgenic aspen grown in the field was checked regularly on two transgenic lines and the respective untransformed control clone. The status of ectomycorrhizal fungi (EM) was determined by anatomical and morphological characters (Agerer 1991) as well as by PCR-RFLP (Pritsch et al. 2000). The majority of fine roots (67%) were colonized by EM. No significant differences in the quantity of mycorrhizal colonization of the different aspen lines were detected. Five morphotypes represented 94% of all EM and up to eleven further EM types were found occasionally (Kaldorf et al. 2002); but differences in the morphotypes were not statistically significant with the exception of one line which showed a specific pattern of EM not linked with the expression of the transgene. Thus, in general, the mycorrhizal diversity did not differ between transgenic and non-transgenic trees, and the EM population was relatively uniform in the whole experimental field site (Kaldorf et al. 2002). On the other hand, arbuscular mycorrhiza (AM) was rare but could be identified in root samples from all aspen lines, with an average of less than 10% of the root length colonized (Kaldorf et al. 2002). Differences between the transgenic and non-transgenic aspen lines were small and not significant.

In a further study, the community structure and spatial distribution of EM from field grown transgenic and control aspen clones were investigated by morphotyping, restriction analysis and internal transcribed spacer (ITS) sequencing (Kaldorf et al. 2004). In this second study, a total of 23 morphotypes were observed. Among these, six EM types dominated, roughly forming 90% of all ectomycorrhizas, namely *Phialocephala fortinii*, *Laccaria sp.*, *Cenococcum geophilum*, two different *Thelephoraceae*, and one member of the *Pezizales* (Kaldorf et al. 2004). The three most common EM had an even spatial distribution, confirming the high degree of homogeneity of the experimental field. The distribution of three other predominant EM was correlated with the distances to the spruce forest and deciduous trees bordering the experimental field. The succession of the EM community colonizing the genetically modified aspen grown in the field was monitored over four years (Kaldorf et al., unpublished). The early stage of succession was characterized by the dominance of only three EM fungi, forming EM with short or medium distance exploration strategy, as did the other twelve EM morphotypes found at this stage. During the third and fourth year of investigation, EM diversity of the normally growing wild type and rbcS::*rolC* transgenic trees increased significantly. In contrast, EM diversity of the slow growing, dwarf 35S::*rolC* transgenic aspen did not change (Kaldorf et al., unpublished). Most striking, the 35S::*rolC* transgenic aspen were only poorly colonized by complex EM morphotypes with rhizomorphs, compared to the wild type (Kaldorf et al., unpublished). This demonstrates that the severe phenotypic changes induced by 35S::*rolC* expression have a significant impact on EM community development under field conditions.

3.4 Other Studies of GMTs Expressing Genes with a Potential Effect on Soil Microbiota

In addition to the alteration of lignin content, the tolerance to pests is also a trait being introduced into GMTs. Several trees have been modified for the expression of chitinase. Chitinases are pathogenesis-related enzymes that are activated upon pathogen infection or wounding of leaves. The enzyme hydrolyzes chitin, an important cell wall component of many fungi and exoskeleton of arthropods. Therefore, GMTs with altered expression of chitinase may potentially have an effect on soil microbiota, existing soil fungal communities and invertebrates.

Decomposition of leaf litter has been studied in a field experiment of transgenic silver birch expressing a chitinase gene (Vauramo et al. 2006). The study included the effect on soil fungal and microbial biomass, basal respiration and effects on soil nematodes abundance. After 8 and 10 months from the establishment of the study, no effects were observed in fungal or total microbial biomass between the lines used in the study. Only one out of fifteen transgenic lines differed in the number of nematodes compared to control plants, indicating a potential indirect effect on microbial populations in the litter. Therefore, the main conclusion raised was that changes in litter quality due to gene transformation were absent or too small to affect the decomposition of the litter in the soil. Related results were described by Lamarche et al. (2011) after studying the impact of endochitinase expressing white spruce on soil fungal communities. This study was carried out under greenhouse conditions using two different natural soils. The work reported the absence of changes between transgenic and control tree rhizospheres, with the same fungal community structure in the soils from transgenic or control plants after 8 months. Thus, the authors concluded that the type of soil and the developmental state of the seedlings had a much more significant impact on fungal community than the insertion and expression of the transgene in the transgenic white spruce lines.

The analysis of ectomycorrhizal associations was also studied in transgenic American chestnut (*Castanea dentata*) and *Eucalyptus camaldulensis* overexpressing genes with a potential effect on soil microbiota. Oxalic acid is a fungal pathogenicity factor secreted in plant-pathogen interaction and typically accumulates during infection. Calcium-oxalate crystals are typically produced during infection removing calcium ions and affecting different enzymes and plant processes. Thus, the expression of oxalate oxidase to remove oxalic acid has been used to protect crops from pathogen infections (Walz et al. 2008). Therefore, the expression of oxalate oxidase in transgenics has the potential to influence associated microbial communities. D'Amico et al. (2011) reported the effect of expression of an oxalic oxidase gene on ectomycorrhizal associations in transgenic American chestnut. The study was carried out in greenhouse for one year. The transgene expression was observed in the root of the plant, but no differences in ectomycorrhizal fungal abundance were observed when compared the GMTs with control plants. Authors inferred from this work that the transgenic plants were able to form mycorrhizal associations in a similar manner to the wild type. A related final conclusion was presented by Lelmen et al. (2010) after analysis of the mycorrhizal colonization of transgenic *Eucalyptus camaldulensis*. This GMTs overexpressed allene oxide cyclase, a enzyme involved in the biosynthesis of jasmonate. Interestingly, the transgenic lines presented enhanced tolerance to saline conditions compared to control plants. The mycorrhizal colonization of roots protected the plant from common environmental stress as salt stress. Besides the enhanced salt tolerance of the transgenics, no differences in the colonization were observed under normal or saline conditions.

3.5 Potential Impact of GMTs for Bioremediation

Bioremediation is the technology to decontaminate soils from inorganic or organic pollutants by utilizing the natural metabolic potentials of soil microorganisms and/or plant roots. Phytoremediation can be defined as the use of plants to treat contaminated soil and water resources, whereas rhizoremediation specifically involves the removal of pollutants from contaminated sites by the rhizosphere (the interface constituted by the mutual interaction of plant roots and associate plant microorganisms), and it is considered as the most evolved process of phytoremediation (Pajuelo et al. 2014).

Phytoremediation was first applied for the removal of inorganic pollutants, i.e. heavy metals, from soil, by using naturally hyperaccumulating plants. Plant species are considered as hyperaccumulators if not only tolerate heavy metal concentrations that are highly toxic to the other plants, but also actively take up large amount of the heavy metals from the soil and translocate them to the shoots, accumulating them at concentrations 100–1000-fold higher than those found in non-hyperaccumulating species (Rascio and Navari-Izzo 2011). However, the phytoremediation capacity of natural hyperaccumulators is limited by their small size, slow growth rates, and restricted growth habitat (Meagher and Heaton 2005). Therefore, significant removal of the pollutants should be achieved if genes involved in metabolism, uptake, or transport of heavy metals, and also organic pollutants could be transferred to tree species such as poplar and willow, with their high biomass and extensive root systems (Eapen et al. 2007; Doty 2008; Maestri and Marmiroli

2011). Since willow transformation protocols have not yet been published, the focus on these studies has been on poplar (Doty 2008) using genes derived from plant microbes and animals (Abhilash et al. 2009).

Transgenic poplar with enhanced uptake and metabolism of a variety of toxic compounds as heavy metals, explosives, chlorate organic solvents, herbicides, etc., were developed (Rugh et al. 1998; Bittsanszky et al. 2005; Choi et al. 2007; Doty et al. 2007; Lyyra et al. 2007) and promising results regarding their phytoextraction ability were found in greenhouse (Gullner et al. 2001; Koprivova et al. 2002; Che et al. 2003; Van Dillewijn et al. 2008) and field tests (Hur et al. 2011, 2012; Shim et al. 2013).

Within the approaches followed to detoxify heavy metals, several authors reported the transformation of plants with bacterial genes which encode for enzymes with a capacity to reduce the toxic forms of mercury (Hg^{2+}) in molecular mercury (Hg^{0}) that volatilizes into the atmosphere. Rugh et al. (1998) obtained transgenic trees resistant to toxic concentrations of mercury by inserting a modified mercuric reductase (*mer*A) gene construct into proembryogenic masses of yellow poplar (*Liriodendron tulipifera*). This mercuric ion reductase was also used by Che et al. (2003) to transform Eastern cottonwood (*Populus deltoides*). Choi et al. (2007) and Lyyra et al. (2007) used combinations of *mer*A with an organomercurial lyase gene (*mer*B), which liberates Hg^{2+} from organic compounds, to increase mercury resistance of a non-flowering mutant hybrid poplar (*Populus alba* × *P. tremula*) and *Populus deltoides*, respectively.

Other approaches used glutathione-mediated detoxification systems. Glutathione (GSH), the tripeptide γ-Glu–Cys–Gly, is the metabolic precursor of metal-binding phytochelatins, and GSH and the glutathione S-transferase isoenzyme family play crucial roles in the degradation of several herbicides (Gullner et al. 2001). *Populus tremula* × *P. alba* materials were engineered for resistance to cadmium (Koprivova et al. 2002), zinc (Bittsanszky et al. 2005), and the chloroacetanilide herbicides acetochlor and metolachlor (Gullner et al. 2001) by overexpression of bacterial genes encoding γ-glutamylcysteine synthetase, which increased the content of glutathione, whereas Shim et al. (2013), used a *Saccharomyces cerevisiae* Cadmium factor1 (ScYCF1), a yeast transporter for Cd detoxification, that pumps glutathione-conjugated Cd into the vacuole, resulting in poplar trees with an improved Cd tolerance and uptake activity.

Trees were also engineered to tolerate and accumulate 2,4,6-trinitrotoluene (TNT). Hybrid aspen (*Populus tremula* × *P. tremuloides*) was transformed by van Dillewijn et al. (2008) to express the bacterial nitroreductase gene, *pnrA* (which reduces TNT to 4-hydroxylamino-2,6-dinitrotoluene), and led to increased resistance to TNT and its removal from contaminated waters and soil.

Mammalian genes, as cytochrome P450 2E1, have also been used in phytoremediation. Doty et al. (2007) obtained transgenic (*Populus tremula* × *P.alba*) plants which overexpressed this enzyme, that oxidizes a wide range of small, volatile hydrocarbons, including trichloroethylene, perchloroethen, chloroform, vinyl chloride or benzene. Increased rates of removal of these pollutants were obtained in hydroponic conditions, but field studies are required to confirm that these potential activities actually accelerate the degradation of the respective pollutants.

Regarding the use of recombinant microorganisms of the rhizosphere for their use in rhizoremediation, several successful studies were performed with herbaceous plants (Pajuelo et al. 2014). In the case of trees, *Salix* plants were inoculated with transgenic *Pseudomonas fluorescens* strains engineered for bioremediations by introducing a biphenyl operon which conferred an increased growth rate in soils contaminated with plychlorinated biphenyls (Aguirre de Cárcer et al. 2007b; Power et al. 2011).

Besides their potential usefulness to recover polluted soils, the environmental safety of all of these trees needs to be established before their release can be approved.

One of the specific risks posed by this kind of GMO has been associated specifically to those trees engineered for mercury volatilization. It has been claimed that after remaining in the atmosphere for up 2 years, elemental mercury will finally precipitate with rain and snow, and converted again to ionic and organic mercury. These depositions may lead to dispersion and redistribution of this metal from contaminated soils sites to other less contaminated areas, although other authors consider that pollution through bioremediation would be negligible in comparison to fossil fuel burning and incineration (Mertens 2008). Other risks have been related to the putative ability of trees engineered for phytoremediation to displace other plant species that can live in these habitats, as rare metal tolerant flora (EFSA 2009). Also, and in contrast with GMTs engineered for other purposes, tree engineered for phytoremediation intentionally change physical and chemical properties of their target soils, resulting in a modified environment of the rhizosphere and may be of the bulk soil. These environmental alterations, in turn, could affect the structure of indigenous microbial and fungal communities (Mertens 2008), although it may be difficult to assign to these structural changes a positive or a negative connotation. Indeed, a detailed understanding of the effect of trees (transgenic or not) on the activities of rhizosphere and bulk soil microorganism could help to optimize phytoremediation systems and enhance their use.

As mentioned in Sect. 2, research focused on soil microbial communities of engineered herbaceous plants has been conducted (Dunfield and Germida 2004). However, only a limited number of studies have focused on impacts of transgenic trees engineered for phytoremediation on native microbial communities (Hur et al. 2011, 2012) or the effect of transgenic microbes living in tree roots and engineered for rhizoremediation on other organisms in the rhizosphere (Aguirre de Cárcer et al. 2007b).

Hur et al. (2011) developed a comparative analysis of the bacterial and archaeal communities found in the rhizosphere of wild-type poplars versus poplar genetically modified for mercury volatilization, all of them cultivated on soils contaminated with heavy metals. The analyses were performed at various growth stages, using next-generation sequencing (Schuster 2008). The rate of changes in the structure of the microbial community was slower in wild-type poplars than in transgenic poplars, although it was dependent on the stages of poplar growth. After 3 year of culture, the microbial community of the non-transgenic was similar to that of the transgenic poplars, indicating the stabilization of microbial community. In a similar study, Hur et al. (2012) showed that the overall structure of the rhizosphere fungal community was not significantly influenced by transgenic poplars, although minor changes in fungal diversity were detected in association with the genetic modification of trees, such as the presence of specific taxa in these trees and the faster rate of community change during poplar growth. In a different approach, the rhizosphere of *Salix* sp. plants growing in a soil contaminated with polychlorinated biphenyls (PCBs) were inoculated with transgenic or non-transgenic *Pseudomonas fluorescens* strains engineered for bioremediation (Aguirre de Cárcer et al. 2007b). Physiological profiles were used to detect differences in bacterial community functions in bulk soil and rhizosphere samples. The introduced transgenes had no effect on the function and structure of the bacterial community in bulk soil, although they enhanced biodegradation of PCBs. However, the transgenes affected the development of functionally and genetically distinct bacterial communities in the rhizosphere.

Although these studies provide new insights about the potential effects of GM organisms on soil communities, it is necessary to keep in mind the diversity of results obtained with non-transgenic plants used for phytoremediation (Siciliano and Germida 1999; Gremion et al. 2004; Epelde et al. 2008, 2009; Xiong et al. 2010), including the effect of trees as pine and poplar (Palmroth et al. 2007), or willow (Aguirre de Cárcer et al. 2007a; Yergeau et al. 2014). In addition to other factors as the use of fertilizers and soil amendments, the presence of these trees contributed to the development of bacterial communities distinct from the original communities at functional and structural levels. In the study of Yergeau et al. (2014), non-transgenic willows were planted in contaminated and non-contaminated soils in a greenhouse, and the active microbial communities and the expression of functional genes in the rhizosphere and bulk soil were compared. The combined selective pressure of contaminants and rhizosphere resulted in higher expression of genes related to competition (antibiotic resistance and biofilm formation) in the contaminated rhizosphere. Distinct communities were also observed in the rhizosphere of trees planted in different (non-contaminated) soil types (Gottel et al. 2011), indicating the difficulty of assigning beneficial or unfavourable connotations to the changes found in contaminated soils with transgenic trees.

As a resume, no effect on the environment was reported by the studies performed so far regarding GMTs used for phytoremediation. Besides it should be considered that trees used in phytoremediation would be planted in marginal soils previously altered by man action, and that if no intervention is performed, already represent a major environmental risk (Peuke and Rennenberg 2005; Pajuelo et al. 2014).

4 Conclusions

The risk assessment of a GM requires a case-by-case approach to determine a potential effect on the receiving environment and human health. In the case of forest GMTs, a number of available studies may disclose a general view about the ecological impacts that forest GMT plantations can produce on soil features. To date, there are published greenhouse or field studies of different species of GM poplar, silver birch, white spruce, *Eucalyptus* and American chestnut plants. These species were engineered to alter gene expression affecting different traits of the tree (summarized in Table 1). Several indicators have been used to determine potential impacts of GMTs on soil ecosystems and function. Studies of microbial communities are considered in most of them, especially for studying effects on mycorrhizal fungi because of their relevance in the acquisition of nutrients from soil and potential direct effect on other indicators. Changes in the rate of degradation of organic compounds in soil, evaluation of microbial respiration or the study of nematode population are also recommended indicators in crop plants that have been only used in a very few studies in GMTs (Table 1). Considering that transfer of the transgene from decaying plant material to existing microorganisms is unlikely to occur under natural field conditions (Hay et al. 2002), most of the studies have focused on the alteration in mycorrhizal fungal community. The changes reported in most cases are not significant. Only in a few reports, a limited effect of GMTs on fungal community has been observed, but it has been considered not to have ecological relevance (Table 1). The general lack of studies about how normal agricultural practices affect soil ecosystem indicators creates problem for establishing base lines and for the evaluation of the changes observed in relation to GM plants. More information on how soil microbial communities and other indicators may differ in different soils and environments are required for studies to describe the potential impact of GMT plantations. Similarly, the lack of studies on how traditional silviculture practices affect soil indicators make it difficult to assess the beneficial or negative impact of introduced GMTs in soil ecosystems.

Forestry plantations and cultivated land are human-modified ecosystems and plant type and growth stage, soil type and normal cultivation practices have shown to have more impact on soil indicators than the introduction of a transgenic plant (Dunfield and Germida 2004; Griffiths et al. 2007). The same conclusions can be drawn from the few studies available on polluted soils where different microbial communities are intimately associated with different soils (Epelde et al. 2009; Gottel et al. 2011) and the effect of GMTs on microbial indicators is not biologically significant (Hur et al. 2011, 2012).

Table 1 Indicators used in the evaluation of GMTs on soil ecosystem and observed effects

Type of study[a]/GMT	Indicator used	Effect on soil ecosystem or indicator compared to control plants	References
GMTs with altered lignin biosynthesis			
F/*Populus tremula* × *P. alba*	Soil microbial communities	No effect observed	Pilate et al. (2002)
F/*P. tremula* × *P. alba*	CO_2 production Microbial biomass	No significant differences	Tilston et al. (2004)
F/*P. tremula x P. alba*	Microbial pathogens Leaf-feeding insects Soil organisms Decomposition of roots	Short-term decomposition of roots was slightly enhanced No ecological impacts observed	Halpin et al. (2007)
G/*P. tremuloides*/	Microbial community	No potential ecological impact	Bradley et al. (2007)
F/*Betula pendula*/	Decomposition rate of leaf litter Formation of ectomycorrhizas	No differences were observed	Seppänen et al. (2007)
GMTs expressing rolC from Agrobacterium rhizogenes			
F/*P tremula* × *P. tremuloides*	Mycorrhizal diversity	In general, no significant differences were found One line showed a specific pattern of ectomycorrhizal fungi	Kaldorf et al. (2002)
F/*P tremula* × *P. tremuloides*	Mycorrhizal diversity	In general no relevant differences observed Severe phenotypic changes in GMT had a significant impact on ectomycorrhizal fungi	Kaldorf et al. (2004) and unpublished
Other GMTs expressing genes with a potential effect on soil microbiota			
F/*B. pendula* overexpressing chitinase	Litter mass loss Fungal and total microbial biomass Basal respiration Nematode populations	No relevant changes in litter quality and decomposition were observed	Vauramo et al. (2006)
G/*Picea glauca* overexpressing chitinase	Fungal community biomass and structure	No significant effect on fungal community	Lamarche et al. (2011)

(continued)

Table 1 (continued)

Type of study[a]/GMT	Indicator used	Effect on soil ecosystem or indicator compared to control plants	References
G/*Castanea dentata* overexpressing oxalate oxidase	Mycorrhizal fungi	No significant effect on fungal colonization	D'Amico et al. (2011)
G/*Eucalyptus camaldulensis* salt tolerant	Mycorrhizal fungi	No significant effect on fungal colonization	Lelmen et al. (2010)
Bioremediation assays			
F/*P. alba* × *P. tremula* expressing mercury metabolizing genes	Microbial community	No significant differences observed	Hur et al. (2011)
F/*P. alba* × *P. tremula* expressing mercury metabolizing genes	Fungal community	Minor changes observed in fungal diversity No significant effects on the overall structure of fungal community	Hur et al. (2012)
G/non transformed *Salix viminalis* × *S. schwerinii* inoculated with GM bacteria engineered for rhizomediation	Microbial community	Effects observed in the rhizosphere but not in bulk soil	Aguirre de Cárcer et al. (2007b)

[a]*F* field study; *G* greenhouse or laboratory study

References

Abhilash PC, Jamil S, Singh N (2009) Transgenic plants for enhanced biodegradation and phytoremediation of organic xenobiotics. Biotech Adv 27:474–488

Agerer R (1991) Characterization of ectomycorrhiza. In: Norris JR, Read DJ, Varma AK (eds) Methods in microbiology, vol 23. Academic Press, London, pp 25–73

Aguirre de Cárcer D, Martín M, Karlson U, Rivilla R (2007a) Changes in bacterial populations and in biphenyl dioxygenase gene diversity in a polychlorinated biphenyl-polluted soil after introduction of willow trees for rhizoremediation. Appl Environ Microbiol 73:6224–6232

Aguirre de Cárcer D, Martín M, Mackova M, Macek T, Karlson U, Rivilla R (2007b) The introduction of genetically modified microorganisms designed for rhizoremediation induces changes on native bacteria in the rhizosphere but not in the surrounding soil. ISME J 1:215–223

Bittsanszky A, Komives T, Gullner G, Gyulai G, Kiss J, Heszky L, Radimszky L, Rennenberg H (2005) Ability of transgenic poplars with elevated glutathione content to tolerate zinc($^{2+}$) stress. Environ Int 31:251–254

Bradley KL, Hancock JE, Giardina CP, Pregitzer KS (2007) Soil microbial community responses to altered lignin biosynthesis in *Populus tremuloides* vary among three distinct soils. Plant Soil 294:185–201

Che D, Meagher RB, Heaton ACP, Lima A, Rugh CL, Merkle SA (2003) Expression of mercuric ion reductase in Eastern cottonwood (*Populus deltoides*) confers mercuric ion reduction and resistance. Plant Biotech J 1:311–319

Choi YI, Noh EW, Lee HS, Han MS, Lee JS, Choi KS (2007) Mercury-tolerant transgenic poplars expressing two bacterial mercury-metabolizing genes. J Plant Biol 50:658–662

D'Amico K, Horton T, Maynard C, Powell W (2011) Assessing ectomycorrhizal associations and transgene expression in transgenic *Castanea dentata*. BMC Proceed 5:O54
Doty SL (2008) Enhancing phytoremediation through the use of transgenic plants and entophytes. New Phytol 179:318–333
Doty SL, James CA, Moore AL, Vajzovic A, Singleton GL, Ma C, Khan Z, Xin G, Kang JW, Park JY, Meilan R, Strauss SH, Wilkerson J, Farin F, Strand SE (2007) Enhanced phytoremediation of volatile environmental pollutants with transgenic trees. Proc Natl Acad Sci USA 104:16816–16821
Dunfield KE, Germida JJ (2004) Impact of genetically modified crops on soil- and plant-associated microbial communities. J Environ Qual 33:806–815
Eapen S, Singh S, D'Souza SF (2007) Advances in development of transgenic plants for remediation of xenobiotic pollutants. Biotech Adv 25:442–451
EFSA Scientific Panel on Genetically Modified Organnism (GMO) (2009) Scientific opinion on guidance for the risk assessment of genetically modified plants used for non-food or non-feed purposes, on request of EFSA. EFSA J 1164, 42 pp
Epelde L, Becerril JM, Hernández-Allica J, Barrutia O, Garbisu C (2008) Functional diversity as indicator of the recovery of soil health derived from *Thlaspi caerulescens* growth and metal phytoextraction. Appl Soil Ecol 39:299–310
Epelde L, Mijangos I, Becerril JM, Garbisu C (2009) Soil microbial community as bioindicator of the recovery of soil functioning derived from metal phytoextraction with sorghum. Soil Biol Biochem 41:1788–1794
Fladung M (1990) Transformation of diploid and tetraploid potato clones with the rolC gene of *Agrobacterium rhizogenes* and characterization of transgenic plants. Plant Breed 104:295–304
Fladung M (1996) Transgenic potato plants can be used to evaluate the stability of foreign genes: reversions and chimeras in multiple copies of *rolC* harboring clones. Agronomie 16:113–121
Fladung M, Ballvora A (1992) Further characterization of rolC transgenic tetraploid potato clones, and influence of daylength and level of rolC expression on yield parameters. Plant Breeding 109:18–27
Fladung M, Muhs HJ (2000) Field release with *Populus tremula* (*rolC*-gene) in Großhansdorf. Release of transgenic trees—present achievements, problems, future prospects. Umweltbundesamt, Humboldt-Universität, Berlin, pp 40–45
Fladung M, Kumar S, Ahuja MR (1997) Genetic transformation of *Populus* genotypes with different chimeric gene constructs: Transformation efficiency and molecular analysis. Trans Res 6:111–121
Gottel NR, Castro HF, Kerley M, Yang Z, Pelletier DA, Podar M, Karpinets T, Uberbacher E, Tuskan GA, Vilgalys R, Doktycz MJ, Schadt CW (2011) Distinct microbial communities within the endosphere and rhizosphere of *Populus deltoides* roots across contrasting soil types. Appl Environ Microbiol 77:5934–5944
Gremion F, Chatzinotas A, Kaufmann K, Von Sigler W, Harms H (2004) Impacts of heavy metal contamination and phytoremediation on a microbial community during a twelve-month microcosm experiment. FEMS Microbiol Ecol 48:273–283
Griffiths BS, Caul S, Thompson J, Birch ANE, Cortet J, Andersen MN, Krogh PH (2007) Microbial and microfauna community structure in cropping systems with genetically modified plants. Pedobiol 51:195–206
Gullner G, Komives T, Rennenberg H (2001) Enhanced tolerance of transgenic poplar plants overexpressing gamma-glutamylcysteine synthetase towards chloroacetanilide herbicides. J Exp Bot 52:971–979
Häggman H, Raybould A, Borem A, Fox T, Handley L, Hertzberg M, Lu MZ, Macdonald P, Oguchi T, Pasquali G, Pearson L, Peter G, Quemada H, Seguin A, Tattersall K, Ulian E, Walter C, McLean M (2013) Genetically engineered trees for plantation forests: key considerations for environmental risk assessment. Plant Biotech J 11:785–798
Halpin C, Thain SC, Tilson EL, Guiney E, Lapierre C, Hopkins DW (2007) Ecological impacts of trees with modified lignin. Tree Genet Genomes 3:101–110

Hay ID, Morency M-J, Séguin A (2002) Assessing the persistence of DNA in decomposing leaves of genetically modified poplar trees. Can J For Res 32:977–982

Hur M, Kim Y, Song HR, Kim JM, Choi YI, Yi H (2011) Effect of genetically modified poplars on soil microbial communities during the phytoremediation of waste mine tailings. Appl Environ Microbiol 77:7611–7619

Hur M, Lim YW, Yu JJ, Cheon SU, Choi YI, Yoon SH, Park SC, Kim DI, Yi H (2012) Fungal community associated with genetically modified poplar during metal phytoremediation. J Microbiol 50:910–915

Icoz I, Stotzky G (2008) Fate and effects of insect resistant Bt crops in soil ecosystems. Soil Biol Biochem 40:559–586

Kaldorf M, Fladung M, Muhs HJ, Buscot F (2002) Mycorrhizal colonization of transgenic aspen in a field trial. Planta 214:653–660

Kaldorf M, Renker C, Fladung M, Buscot F (2004) Characterization and spatial distribution of ectomycorrhizas colonizing aspen clones released in an experimental field. Mycorrhiza 14:295–306

Koprivova A, Kopriva S, Jäger D, Will B, Jouanin L, Rennenberg H (2002) Evaluation of transgenic poplar overexpressing enzymes of glutathione synthesis for phytoremediation of cadmium. Plant Biol 4:664–670

Lamarche J, Stefani FO, Seguin A, Hamelin RC (2011) Impact of endochitinase-transformed white spruce on soil fungal communities under greenhouse conditions. FEMS Microbiol Ecol 76:199–208

Lelmen KE, Yu X, Kikuchi A, Shimazaki T, Mimura M, Watanabe KN (2010) Mycorrhizal colonization of transgenic *Eucalyptus camaldulensis* carrying the mangrin gene for salt tolerance. Plant Biotech 27:339–344

Lyyra S, Meagher RB, Kim T, Heaton A, Montello P, Balish RS, Merkle SA (2007) Coupling two mercury resistance genes in Eastern cottonwood enhances the processing of organomercury. Plant Biotech J 5:254–262

Maestri E, Marmiroli N (2011) Transgenic plants for phytoremediation. Inter J Phytorem 13 (sup1):264–279

Meagher RB, Heaton ACP (2005) Strategies for the engineered phytoremediation of toxic element pollution: mercury and arsenic. J Ind Microbiol Biotech 32:502–513

Mertens M (2008) Assessment of environmental impacts of genetically modified plants: implementation of the biosafety protocol development of assessment bases fkz 20167430/07. BfB-Skripten 217, 234 pp

Pajuelo E, Rodríguez-Llorente ID, Lafuente A, Pérez Palacios P, Doukkali B, Caviedes MA (2014) Engineering the rhizosphere for the purpose of bioremediation: an overview. CAB Rev 9(001)

Palmroth MRT, Koskinen PEP, Kaksonen AH, Münster U, Pichtel J, Puhakka JA (2007) Metabolic and phylogenetic analysis of microbial communities during phytoremediation of soil contaminated with weathered hydrocarbons and heavy metals. Biodegradation 18:769–782

Peuke AD, Rennenberg H (2005) Phytoremediation: molecular biology, requirements for application, environmental protection, public attention and feasibility. EMBO Rep 6:497–500

Pilate G, Guiney E, Holt K, Petit-Conil M, Lapierre C, Leple JC, Pollet B, Mila I, Webster EA, Marstorp HG, Hopkins DW, Jouanin L, Boerjan W, Schuch W, Cornu D, Halpin C (2002) Field and pulping performances of transgenic trees with altered lignification. Nature Biotechnol 20:607–612

Power B, Liu X, Germaine KJ, Ryan D, Brazil D, Dowling DN (2011) Alginate beads as a storage, delivery and containment system for genetically modified PCB degrader and PCB biosensor derivatives of *Pseudomonas fluorescens* F113. J App Microbiol 110:1351–1358

Pritsch K, Munch JC, Buscot F (2000) Identification and differentiation of mycorrhizal isolates of black alder by sequence analysis of the ITS region. Mycorrhiza 10:87–93

Rascio N, Navari-Izzo F (2011) Heavy metal hyperaccumulating plants: how and why do they do it? And what makes them so interesting? Plant Sci 180:169–181

Rugh CL, Senecoff JF, Meagher RB, Merkle SA (1998) Development of transgenic yellow poplar for mercury phytoremediation. Nature Biotechnol 16:925–928
Schmülling T, Schell J, Spena A (1988) Single genes from *Agrobacterium rhizogenes* influence plant development. EMBO J 9:2621–2639
Schmülling T, Fladung M, Großmann K, Schell J (1993) Hormonal content and sensitivity of transgenic tobacco and potato plants expressing single rol genes of *Agrobacterium rhizogenes* T-DNA. Plant J 3:371–382
Schuster SC (2008) Next-generation sequencing transforms today's biology. Nat Methods 5:16–18
Seppänen SK, Pasonen H-L, Vauramo S, Vahala J, Toikka M, Kilpeläinen I, Setälä H, Teeri TH, Timonen S, Pappinen A (2007) Decomposition of the leaf litter and mycorrhiza forming ability of silver birch with a genetically modified lignin biosynthesis pathway. Appl Soil Ecol 36:100–106
Shim D, Kim S, Choi Y-I, Song W-Y, Park J, Youk ES, Jeong S-C, Martinoia E, Noh E-W, Lee Y (2013) Transgenic poplar trees expressing yeast cadmium factor 1 exhibit the characteristics necessary for the phytoremediation of mine tailing soil. Chemosphere 90:1478–1496
Siciliano SD, Germida JJ (1999) Enhanced phytoremediation of chlorobenzoates in rhizosphere soil. Soil Biol Biochem 31:299–305
Smith E, Glandorf DCM, Bergmans JEN (2010) General surveillance for effects of GM crops on the soil ecosystem. RIVM Report. National Institute for Public Health and the Environment, The Netherlands
Tilston EL, Halpin C, Hopkins DW (2004) Genetic modification to lignin biosynthesis in field-grown poplar trees have inconsistent effects on the rate of woody trunk decomposition. Soil Biol Biochem 36:1903–1906
Van Dillewijn P, Couselo JL, Corredoira E, Delgado E, Wittich RM, Ballester A, Ramos JL (2008) Bioremediation of 2, 4, 6-trinitrotoluene by bacterial nitroreductase expressing transgenic aspen. Environ Sci Technol 42:7405–7410
Vauramo S, Pasonen H-L, Pappinen A, Setälä H (2006) Decomposition of leaf litter from chitinase transgenic silver birch (*Betula pendula*) and effects on decomposer populations in a field trial. Appl Soil Ecol 32:338–349
Walz A, Zingen-Sell M, Loeffler M, Sauer M (2008) Expression of an oxalate oxidase gene in tomato and severity of disease caused by *Botrytis cinerea* and *Scierotinia sclerotiorum*. Plant Pathol 57:453–458
Xiong JB, Wu L, Tu S, Van Nostrand JD, He Z, Zhou J, Wang G (2010) Microbial communities and functional genes associated with soil arsenic contamination and the rhizosphere of the arsenic hyperaccumulating plant *Pteris vittata* L. Appl Environ Microbiol 76:7277–7284
Yergeau E, Sanschagrin S, Maynard C, St-Arnaud M, Greer CW (2014) Microbial expression profiles in the rhizosphere of willows depend on soil contamination. ISME J 8:344–358

Potential Environmental Impact of Insect-Resistant Transgenic Trees

Marcel Robischon

Abstract Resistance to phytophagous insects is a trait that is highly desirable in commercially grown forest trees. A considerable body of literature exists that describes approaches of genetic engineering to render a tree impalatable or toxic to a wider or narrower range of insects. Such a biotechnological interference with natural ecosystem processes naturally raises concerns about unwanted outcomes and ecological biosafety. There are a growing number of studies focusing on non-target or wider ecosystem effects of such trees in field trials. In this paper, recent work on transgenic trees with enhanced resistance to insect herbivory and potential environmental effects is reviewed. As yet no coherent overall picture has emerged, yet a few instances of unexpected outcomes of the exposure of transgenic insect resistant trees to biocenoses have emerged. With ongoing research, and a longer history of transgenic trees in the field, further results, observations and insights can be expected.

1 Introduction—Insect Resistance as a Commercially Important Trait

The insect fauna is a key factor in the success or failure in achieving material and non-material aims in the management of forests or tree plantations. Even small shifts in the balance of phytophagous versus predatory organisms and in the biological systems of multiple multi-tropic levels and highly branched food chains can alter the forest ecosystem and the economic output dramatically. In particular, large-scale monocultures are susceptible to major insect calamities that can spread and devastate stands rapidly.

M. Robischon (✉)
Lebenswissenschaftliche Fakultät, Humboldt-Universität zu Berlin,
Unter den Linden 6, 10099 Berlin, Germany
e-mail: marcel.robischon@hu-berlin.de

C. Vettori et al. (eds.), *Biosafety of Forest Transgenic Trees*,
Forestry Sciences 82, DOI 10.1007/978-94-017-7531-1_9

Beyond classical approaches in insect pest control by the choice of appropriate silvicultural techniques, use of naturally less susceptible plant material, and the maintenance of a sound ecological balance in the stands, or, alternatively, the application of pesticides, the use of transgenic methods has been discussed and tested experimentally since decades.

Since the first transgenic insect resistant trees were generated (McCown et al. 1991), plant biotechnology in general and in work on trees in particular has advanced at a rapid pace. Agrobacterium-based transformation has been developed in recent years as a standard method even for plants that are normally not a host for virulent wild-type strains of *Agrobacterium tumefaciens* as a plant pathogen, such as typically gymnosperms (Alvarez and Ordás 2013, and references to earlier work therein). Increasingly, effective high-throughput methods allowing the generation of larger and larger numbers of transgenic lines in a short time have been designed. The combination of transgenic and classical breeding methods has made significant progress with the use of transgenic lines expressing the *Populus* Flowering Locus T (PtFT), allowing a shortening of generation times and faster crossings between transgenic and elite non-transgenic lines (Zhang et al. 2010).

Nevertheless, large-scale commercial use of transgenic forest trees is in most places still a scenario of the future, with the only examples of extensive commercially used transgenic plantations so far being found in China (Zhang et al. 2011b). The recent example of the transgenic plum pox virus-resistant *Prunus domestica* line "Honey sweet" being the first tree to have fully gone through the USA regulatory process and having received an unconditional registration in 2011 (Scorza et al. 2013) may, by setting an example, mark the beginning of commercial plantations of transgenic woody plants in the United States. Also in 2010, large-scale field trials using transgenic eucalyptus hybrids with enhanced cold tolerance in the southern United States were approved (USDA APHIS 2012). In Brazil, a potential commercial release of genetically modified eucalyptus is in evaluation (Ledford 2014). In the light of regulatory environments being altered to become more friendly to transgenic plants and transgenic plants being altered to become more amenable to the natural environment, planting of insect-resistant transgenic lines of eucalyptus, poplar, pines, spruce or other trees suitable for plantation forestry at a commercial scale appears to be only a matter of time. The issue of wider environmental effects of insect-resistant transgenic forest trees is however still unresolved. Effects on non-target organisms, downstream effects and unexpected feedback loops may unintentionally be activated, and still only a comparatively small number of studies allows some insights into the potential risks that may arise from the use of transgenic forest trees genetically improved for insect resistance.

1.1 Bt Transgenes—A Classic Approach

To be well prepared for changes and risks a field of (bio)technology may bring with it, it is important to evaluate the latest data on technical developments as to insights concerning effects of both novel systems emerging as well as longer established techniques. One of these biotechnical approaches tested by 'time' at a human, though not necessarily at an evolutionary scale, is the use of Bt transgenes.

The expression of CRYSTAL proteins (Cry), expressed from genes cloned from the bacterial insect pathogen *Bacillus thuringiensis*, and hence referred to as Bt toxins is without doubt the most commonly used transgenic strategy aiming for insect resistance in plants. Upon oral uptake by an insect and proteolysis in the gut, these toxins induce the disruption of cell membranes in the epithelium, causing the death of the insect. Bt spray as a biological pest control and the expression of Bt genes are well established in non-woody agricultural crops. BtCRY proteins have the noteworthy advantage of a high specificity of individual types of these proteins against defined groups of target pests. In general, Cryl, Cry2, and Cry9 work against Lepidoptera; Cry3, Cry7 and Cry8 target coleopterans, and Cry4, Cry10 and Cry11 Diptera, with some CRY proteins also targeting nematodes. Not surprisingly, Bt toxins were the first peptides targeted against insects to be expressed in woody species, and not surprisingly either, the first tree-systems were poplar (McCown et al. 1991). *Populus* species and hybrids, being central woody model organisms in research and biotechnical development and at the same time widely used in fast growing plantations in temperate and boreal regions not only stood at the very beginning, but still remain a central object of research on insect resistant transgenic tree lines (Wullschleger et al. 2002; Génissel et al. 2003a, b). Field trials with transgenic poplars have been carried out since the 1990s (reviewed by Robischon 2006a).

In 2002, about a decade after work on the first Bt poplars was published, Bt poplars were released in China and in 2003 commercial plantations were established (Valenzuela et al. 2006). At a much smaller scale, non-commercial field trials with Bt poplars have been carried out in other parts of the world, in particular in North America (Klocko et al. 2014) and in Europe for example in France (Génissel et al. 2003a).

In subtropical and tropical regions, instead of poplars often *Eucalyptus* species fulfil the role of a fast growing producer of woody biomass. The performance of eucalypts can be equally impacted by insect pests and research on transgenic insect resistant lines has been published. Harcourt et al. (2000), for example, report the generation of a Cry3A Bt transgenic River Red gum *Eucalyptus camaldulensis*, a species that is in many parts of the world of supreme importance in large-scale monoculture forest plantation, often as a biomass-based energy source, or for lignocellulosic fibre for the pulp and paper industry.

Conifers are less likely to be used in classical fast growing biomass-for-energy plantations. However, they are important in the production of structural timber as well as a source of fibre for pulp and paper. Researchers in biotechnology therefore early on studied this group of tree species and their potential for improvement by transgenic methods. The feasibility of achieving insect resistance in Bt-transgenic conifers has been tested in loblolly pine *Pinus taeda* (Tang and Tian 2003), a tree of paramount significance in forest plantations in particular in the southern United States. Another candidate pine to be modified for insect resistance using Bt genes is the Monterey pine *Pinus radiata* (Grace et al. 2005), a species that 'radiated' in the hands of foresters from being a highly endemic local species limited to a small section of the Californian coast into the world's most widely planted pine, covering several million ha in particular in the subtropical and temperate parts of the southern hemisphere (Lavery and Mead 2000). No less than 56 insect species from 44 genera have been found feeding on *P. radiata*, of which however only a small handful, mostly coleopteran, cause damage at a level that would be of major economic concern (Furniss and Carolin 1977). Work on the expression of Bt genes in Chinese white pine *Pinus armandii*, a species which is in some parts of China used in plantation forestry, has been published by Liu et al. (2010).

Pinus is not the only conifer genus in which an enhanced insect resistance would be economically desirable. White spruce *Picea glauca*, a tree whose native distribution spans the boreal forests of North America, is of high economic importance as a source of timber and for the pulp and paper industry. Lachance et al. (2007) generated Bt cry1Ab transgenic *Picea glauca* enhancing its resistance to against the Eastern Spruce Budworm *Choristoneura fumiferana* (Lepidoptera: Tortricidae). Given the great economic importance of conifers one may expect for the future more work on those coniferous species that were already studied in the context of Bt gene expression technologies, as well as an expansion of the work on further conifers with relevance to plantation forestry.

Bt approaches in forest tree biotechnology are well covered by numerous earlier reviews (Robischon 2006a; Ye et al. 2011; Häggman et al. 2013). Therefore, this paper's focus is on most recent work, and on work that specifically addresses the question of potential unwanted environmental impacts, rather than looking at the efficiency in terms of pest control.

1.2 Alternative Approaches to Target Insect Digestive Systems

The midgut appears to be in insects a kind of Achilles heel, as the majority of transgenic approaches to insect resistance published so far focus on this target in one way or other. This includes even most of the work that does not make use of Bt genes. **Chitinases** for example, are a group of chitinolytic enzymes that are involved in plant defense mechanisms by cleaving and hydrolyzing chitins in the filaments of fungi and in insect peritrophic membranes, thus causing an—eventually lethal—abrasion of the insects' gut lining. Genes coding for chitinases have been cloned and expressed in plants mainly with the intention to enhance their resistance against pathogenic fungi. Noël et al. (2005) expressed a fungal chitinase in black spruce and hybrid poplar and Pasonen et al. (2004) did so in silver birch to enhance resistance against fungal pathogens. Moreover, in view of the importance of chitins in an insects body, Yang et al. (2013a) made transgenic *Populus cathayana* expressing a *Manduca sexta* (Lepidoptera: Sphingidae) chitinase together with a *Buthus martensii* scorpion toxin (BmIT).

Avidin and **Streptavidin** are biotin-binding proteins, the former from birds' eggs, and the latter from Streptomyces bacteria. They have been shown to have antimicrobial activity, and expression of these proteins has been tested as a means to elevate plants' resistance to microbial infections. It however has also been found to additionally enhance resistance against insects, which, upon ingestion of this protein, develop biotin deficiency and eventually perish from malnutrition. The application of avidin in pest control has been reviewed by Christeller et al. (2010, and references therein). Data published by Marwick et al. (2003) suggest that avidin and streptavidin expression in apple trees conferred a high level of resistance against larvae of the light brown apple moth (*Epiphyas postvittana*, Lepidoptera: Tortricidae) with mortality being significantly higher, and the larvae's weight gain being significantly lower. While it appears questionable whether a fruit potentially causing a vitamin B deficiency in the consumer would find a market easily, the example is still of considerable interest for being a tree-model. Testing the applicability in a non-food tree crop Burgess et al. (2005) produced avidin-expressing rose gum *Eucalyptus grandis* plants.

Another target in insect metabolism is the functionality of their digestive enzymes. **α-Amylase inhibitors** are proteins that affect hydrolysis in the gut of insects, rendering them unable to digest starch. α-Amylase inhibitors are therefore attractive candidates for reducing plant tissues digestibility to herbivore insects.

Analogously, **proteinase inhibitors** bind digestive proteases rendering them unable to cleave peptide bonds and therefore nonfunctional. Ingestion of such proteins may thus lead to malnourishment and eventually the starvation of the targeted insect.

Cysteine proteinase inhibitors in poplars have been shown to convey resistance to the leaf beetles *Chrysomela tremula* and *Chrysomela populi* (Coleoptera: Chrysomelidae), respectively, to a varying degree (Leplé et al. 1995; Delledonne et al. 2001). The cysteine protease inhibitor oryzacystin I was expressed together with a Bt gene in hybrid poplar *Populus alba* × *glandulosa* by Zhang et al. (2011a, b), who found in feeding experiments a significantly higher mortality in the coleopteran target pest. Less obvious are the effects of trypsin proteinase inhibitor overexpression published so far. Confalonieri et al. (1998), upon expressing a Kunitz trypsin proteinase inhibitor in black poplar *Populus nigra*, found no significant effect on larval mortality, growth or pupal weight in *Lymantria dispar* (Lepidoptera, Lymantriidae) and *Clostera anastomosis* (Lepidoptera: Notodontidae) feeding on this material. Zhang et al. (2002) however found an elevated insect resistance in poplar lines expressing a cowpea trypsin inhibitor. Also, a hybrid triploid poplar [(*P. tomentosa* × *P. bolleana*) × *P. tomentosa*] genetically engineered with a cowpea trypsin inhibitor (CpTI) grown in the field for 2 years was shown by Zhang et al. (2005) to have efficient resistance against the forest tent capterpillar *Malacosoma disstria* (Lepidoptera: Lasiocampidae), the gypsy moth, *Lymantria dispar* (Lepidoptera, Lymantriidae) and the willow moth *Stilpnotia candida* (Lepidoptera: Lymantridae). Maheshwaran et al. (2007) expressed in apple trees a precursor protein that, in the course of processing, gives rise to chymotrypsin and trypsin inhibitors.

They fed the resulting transgenic material to the light brown apple moth and found as a result a reduced body weight in larvae and morphological deformities.

A further method to target insect digestive systems is provided by **agglutinins** such as lectins, a group of carbohydrate binding proteins for which genes have been cloned from numerous sources, including various plant species. Lectins as for instance the *Galanthus nivalis* snowdrop lectin (GNA) work insecticidal by binding to epithelial cells in the midgut tract and causing disruption of their membranes. In contrast to Bt toxins, lectins can be used to target sap sucking insects such as aphids. The gene coding for *Galanthus nivalis* agglutinin has been tested in heterologous expression systems for its potential to enhance tolerance of plants to insects as well as other pests and diseases. *Galanthus nivalis* GNA genes have been expressed in various crop plants to enhance resistance against invertebrate herbivores. McCafferty et al. (2008) used it in Papaya (*Carica papaya*), a plant that is for its single-stemmed habitus, a height of several metres and a lifetime of several years to some extent tree-like, and that is grown in monoculture plantations, and hence lends itself for a comparison with trees used in short-rotation plantation forestry. As for truly lignified woody species, as far as is evident from the current body of literature, heterologous expression of plant lectins has been experimentally performed mainly in species of ornamental or horticultural value. Yang et al. (2000) for example engineered grapefruit (*Citrus paradisi*) with a *Galanthus nivalis* agglutinin (GNA) gene and Shang et al. (2008) report such work on an ornamental cultivar of the Japanese Spindle Tree *Euonymus japonicus*. No comparable work however appears to have been published on any 'true' forest trees.

Anionic peroxidases catalyze the oxidation of aromatic compounds and are involved in polymerizing phenolic monomers. The fact that they are induced as a consequence to wounding of a plant is in accordance with the assumption that the respective oxidation products may be more toxic to the herbivore causing the wounding, or to microbes infecting the wound than the original substrate. Thus the expression of these enzymes in a crop plant can enhance resistance to insects. In trees the effect of expression of a tobacco peroxidase in American sweetgum (*Liquidambar styraciflua*) on insect herbivory has been tested early on by Dowd et al. (1998). Although sweetgum is a highly interesting experimental system, it remained rather the exception. More work has been published on phenol oxidases in the poplar model. One example is the work by Wang and Constabel (2004), who expressed a *Populus trichocarpa* × *P. deltoides* polyphenol oxidase in *Populus tremula* × *P. alba*. Forest Tent Caterpillars fed with such polyphenol oxidase overexpressing poplar leaf material showed reduced growth rates in first instar larvae in autumn and in fourth instar larvae in winter (Wang and Constabel 2004; Barbehenn et al. 2007). Such experiments on the lymantriid caterpillar *Orgyia leucostigma* (Lepidoptera: Lymantriidae) found in one case decreased growth rates, but not in another (Barbehenn et al. 2007), leading the authors to the conclusion that this system may on its own be insufficient to achieve a satisfactory protection of a plant against lepidopteran caterpillars.

1.3 Neurotoxic Trees

'Getting on the insects nerves' rather than delivering a toxic or enzymatic 'punch into the gut' is another approach to protecting a plant from insect herbivory. Engineering a plant to produce **neuroactive substances** may thus be a way to impact herbivorous insects also in a plantation forestry context. One such substance whose content can be enhanced in a transgenic plant is tryptamine, namely by expression of tryptophan carboxylase genes. Some tryptamines such as serotonin have a function as neurotransmitters. Others are known as psychoactive drugs, underlining their strong effect on neuronal functioning. In 2003, Gill et al. published the results of bioassays on hybrid aspen (*Populus tremula* × *P. alba*) overexpressing a tryptophan carboxylase gene cloned from the 'happy tree' *Camptotheca acuminata*. They found larval growth and leaf consumption being reduced in forest tent caterpillars feeding on the transgenic plants and these parameters actually being inversely correlated with tryptamine levels in the transgenic tissue.

Yet tryptamines are among the more benign compounds affecting neuronal systems. Increasingly, work is published that makes use of neurotoxic proteins of a highly fascinating origin: scorpion venoms. The infamously lethal scorpion *Androctonus australis* (Buthidae) for example, whose natural geographic range stretches from northern Africa to India, boasts one of the most potent scorpion toxins known to science, making it popular in the pet trade. Some of the proteins isolated from its venom show an insect-specific toxicity, which promises a potential for biotechnological applications, including work on insect resistance in crop plants. Expression of a gene called ***A**ndroctonus **a**ustralis* **I**nsect **T**oxin (AaIT) cloned from this scorpion species in a hybrid poplar (*P. deltoides* × *P. simonii*) was reported in 2000 by Wu et al. According to this work a wide resistance was achieved. More specifically, the gypsy moth (*Lymantria dispar*, Lepidoptera: Lymantriidae) showed upon feeding on this transgenic material a lower weight gain and a higher larval mortality (Wu et al. 2000).

A relative of Androctonus is *Mesobuthus martensii* (Buthidae), the Asian Golden Scorpion, which equally holds a store of potentially useful polypeptides in its venom and is a stock of traditional Chinese medicine. From this venom highly insectotoxic peptides referred to as ***B**uthus **m**artensii* **I**nsect **T**oxins (BmIT) have been purified (Escoubas et al. 2000) and the respective genes cloned. Expression in poplar has been reported by Yang et al. (2010, 2013a).

Venoms quite *en par* in toxicity to the scorpion ones are found in several spiders such as *Atrax robustus*, Australia's notorious Sydney funnel-web spider. From this venom ω-atracotoxins have been found to selectively affect insects by blocking voltage activated channels (Chong et al. 2007). These peptides have been expressed in crop plants, including woody species. Zhan et al. (2001) report expression of a spider insecticidal peptide gene in birch (*Betula platyphylla*). Cao et al. (2010) describe expression of a fused gene containing an ω-atracotoxin and the Bt CryIA (b) gene C terminal in *Populus simonii* × *P. nigra* and finding a strong effect against Gypsy moths larvae. This paper gives an example of work in which two different resistance genes, Bt and a non-Bt resistance gene, several of which are discussed below, have been combined. Similarly, Tian et al. (2000) who transformed a hybrid poplar with two genes coding for insecticidal proteins, the Bt Cry1Ac and an arrowhead (*Sagittaria sagittifolia*) proteinase inhibiting API gene, found an increased mortality with insects feeding on the leaves of these plants. Interestingly, additive effects could be shown in the second generation of a lepidopteran target pest, *Clostera anachoreta* (Lepidoptera: Notodontidae) being grown on a transgenic poplar that combined expression of both, an arrowhead proteinase inhibitor and Cry1AC (Yuan et al. 2011). The combination of different genes that are thought to enhance insect resistance however does not always lead to an added effect, as illustrated in work by Zhan et al. (2001).

2 Transgenic Insect-Resistant Trees and Their Potential Ecological Impact

The currently available literature suggests a future trend of expansion of areas planted with insect-resistant transgenic plants, and a further extension of the use of such technologies into the forest sector. The long discussed question of potential risks is therefore as relevant as ever.

Proponents point out that application of transgenic methods will reduce the use of pesticides and suggest that the technique may further contribute to conservation aims and environmental targets by enhancing a timber and fibre source that spares natural forests. It also has been argued further that, with exotic insect pests continuing to reach new ecosystems, transgenic approaches can be useful to preserve native forests and to rescue endangered native trees (Adams et al. 2002; Merkle et al. 2007).

Opponents argue that it involves incalculable risks, such as negative effects on non-target organisms, including other herbivores, or predatory organisms that may be affected via the food chain. Also discussed are effects on decomposition of leaf biomass and thus nutrient cycles, which may also have an impact on aquatic ecosystems (Gatehouse et al. 2011; Carstens et al. 2012).

2.1 Unwanted Effects of Bt—Hitting, Yet Missing the Target

In parallel to the largest number of studies on transgenic trees designed to have enhanced resistance to insect making use of Bt genes, also the largest number of studies on biosafety aspects of transgenic insect-resistant trees are taking Bt transgenic lines into the focus, and again it is transgenic poplar lines on which most work has been published so far.

Hu et al. (2007) in their work on a Bt transgenic *Populus nigra* plantation in China report that "*the variety, number and parasitic ratio of the natural enemies of insects in the transgenic poplar plantation… were more than those in the non-transgenic poplar plantation*". Zhang et al. (2011b) found that their Bt-Cry3A transgenic poplar line had no effects on non-target insects, i.e. non-coleopterans. This however does not answer questions concerning other herbivorous beetles, which certainly would be expected to be susceptible to this particular Bt too.

Few studies appear to look at on-target species of the same order as of the target insects, i.e. which non-target coleopterans are affected by Cry3Aa, which non-target lepidoptera are affected by Cry1Ab and so forth. There may be an underlying thinking of any leaf-eating insect being by nature a pest and hence intended to be a target. However, it ought to be kept in mind that often only a very small segment of the insect biocenosis feeding on a plant cause in fact significant damage, as shown for example for radiata pine by Furniss and Carolin (1977).

With the high specificity in individual Bt toxins being already well known for a long time it does not come as a surprise that coleopterans are not affected by a specific anti-*lepidopteran* toxin. However, in some cases Bt toxins may fail to have an effect even on their particular target group. In a study by Axelson et al. (2011) on leaf litter in a stream it was found that aquatic detritus feeding beetles were not affected by dead leaves of a transgenic poplar expressing Cry3Aa, i.e. the 'beetle Bt'.

2.2 Insect Communities—Impacts on a Delicate Balance

Even if non-target insects may not be directly impacted it is plausible that due to a competitive advantage being created for non-susceptible taxa a shift in species composition may occur. This may lead to the original target being indeed reduced in its impact, but secondary pests moving up the scale of importance, resulting potentially in a similar economic damage as in conventional crops.

Axelsson et al. (2012) found that a high expression Cry3Aa-Bt line of a transgenic hybrid poplar was indeed subject to lower herbivory, but at the same time there was no positive effect of this in terms of biomass accumulation. There was no effect on hymenopterans or lepidopterans found, suggesting no non-target effect. Also there was a target-species, the beetle *Byctiscus populi* (Coleoptera, Attelabidae) that was seemingly unaffected by the beetle-specific Bt toxin, quite similar to the above mentioned case reported by Axelsson et al. (2011).

Zhang et al. (2011b) found that the density of sucking pests on the Bt transgenic hybrid poplars increased in comparison to the wild type. However, also other non-targeted insects, including two lepidopterans, decreased in their density. One explanation for this given by Zhang et al. (2011b) is linked to an increase in two predatory coccinelid beetles on these Bt-Cry3A (i.e. coleopteran-targeted) lines. Their presence in larger numbers than on the wild type plants may be due to shifts towards a higher density of sucking insects that are expected to be an attractive prey for these beetles.

2.3 The Food Chains—Handing Down the Toxins?

Hu et al. (2007) carried out experiments with inoculating pupae of *Apocheimia cinerarius* (Lepidoptera: Geometridae) and *Orthosia incerta* (Lepidoptera: Noctuidae) collected from Bt transgenic *Populus nigra* plantation in China with the wasp *Chouioia cunea* (Hymenoptera: Eulophidae), and found that there was no effect of the host pupae coming from the transgenic plantation on wasp hatching rate. Barraclough et al. (2009) however found a reduced survival of the endoparasitic braconid *Meteorus pulchricornis* (Hymenoptera Braconidae) on the New Zealand native pine looper *Pseudocoremia suavis* (Lepidoptera: Geometridae) feeding on Bt transgenic *Pinus radiata*. Lei et al. (2011) studied the development of the parasitoid *Microplitis pallidipes* (Hymenoptera: Braconidae) on larvae of the oriental armyworm *Mythimna separata* (Lepidoptera: Noctuidae) feeding on an artificial diet containing different levels of Cry1Ab toxin and found that the toxin caused a slower development of the larvae leading to a decrease weight of cocoons and the imago of *Microplitis pallidipes* (Hymenoptera: Braconidae). This result suggests a downstream effect of the Bt toxin in the food chain, potentially harming parasitoids.

3 Leaf Litter Decomposition and Potential Effects in Terrestrial and Aquatic Ecosystems

The potential impact of transgenic plant material onto the environment does not end with the death of the transgenic tissue. Decomposition and litter breakdown are central processes in terrestrial and aquatic ecosystems and are dependent on detritus-feeding organisms, in particular arthropods and microbes.

Results on whether Bt leaks into the soil from roots of Bt transgenic plants are ambiguous. Saxena and Stotzky (2000) showed that the Bt toxin was released from roots of Bt corn into the soil, while Head et al. (2002) could not find the toxin in soil on which Bt cotton had been grown for several years.

LeBlanc et al. (2007) studied rhizosphere soil samples taken from the one Canadian field trial of *Picea glauca* expressing Cry1Ab and indeed discovered a significantly different composition of microbial communities in the rhizosphere of the transgenic trees. Differences could not only be shown between the Bt lines and the wild type, but also between Bt lines, wild type and transgenic lines expressing only GUS and the selectable marker NPTII. However, when Lamarche and Hamlin (2007) studied nitrogen fixing microbial communities in the soil of the same field trial, they found no difference in comparison to soil collected in plantations of spruce expressing only GUS and the marker gene or such consisting only of wild type plants.

In decomposition experiments in a natural stream in Sweden, Axelsson et al. (2011) found a strongly altered insect abundance on transgenic Cry3Aa expressing poplar leaf litter submerged in mesh bags, as compared to wild type leaves subject to the same treatment. Quite unexpectedly, the insect abundance on the transgenic material was much higher, with the composition of the insect community being virtually unchanged, and beetles, against which Cry3Aa is targeted forming only a small part of the sample and appearing unaffected too.

There are several publications (Pasonen et al. 2004; Noël et al. 2005; Stefani and Hamelin 2010) on expression of chitinases and other genes that are expected to enhance resistance to fungal pathogens in trees. Lamarche et al. (2011) found in greenhouse experiments no difference between fungal communities in the soil in the rhizosphere of *Picea glauca* expressing a *Trichoderma harzianum* chitinase, a line created in the first place to render it resistant against fungal pathogens.

As addressed above, chitinases can have an insecticidal effect suggesting that a potential non-target effect on insects cannot be excluded. However, as yet there are no studies on this question available.

4 The challenge of Insect Resistance Management

As mentioned above, insect resistance management is a hot and controversial topic, given the as yet uncertain effects on resistance development. Critics quote evidence from non-woody plants that transgenic work will lead eventually to insect resistances—which in turn is denied by supporters. Questions concerning potential problems in insect resistance management thus are still unanswered.

A line in Cao et al. (2010) saying that "*a recent study demonstrated transgenic plants expressing Bt genes have developed insect resistance*" nicely captures the situation for it remains in the greater picture unclear whether it is the plants or the insects that are rendered resistant in the end.

Large-scale plantations may potentially turn out as effective tools to involuntarily select for insect strains with high resistance against the only widely applied and applicable environmentally friendly biocontrol, the Bt toxin. This concern is supported by the history of resistance development in crop plants. In 2003 still no cases of a resistance in insects having developed as a consequence to exposure to Bt crops in the field had been reported. "*Resistance to Bt toxin surprisingly absent from pests*" is the title of a news piece in *Nature Biotechnology* of that year (Fox 2003). What this headline states however is, strictly speaking, false, since at that time a resistance against conventional Bt spray had already been found in the diamondback moth *Plutella xylostella* (Lepidoptera: Plutellidae). Bt resistant corn borers and other insect had already been generated and grown in laboratory experiments, proving already then that the emergence of a resistance against the Bt toxin was certainly within the realm of the possible. Not surprisingly it did eventually happen. In 2007, Van Rensburg reported a field-evolved resistance to Bt-transgenic corn in the stem borer, *Busseola fusca* (Lepidoptera: Noctuidae) in South Africa. Soon field-evolved resistance to a Bt transgenic crop and an increased frequency of resistance alleles was also confirmed in the corn earworm *Helicoverpa zea* (Lepidoptera: Noctuidae) (Tabashnik et al. 2008). Bt transgenic corn is grown on a much larger area and also for a longer time than any Bt-transgenic tree. Therefore it is plausible that, depending on the management strategies, Bt-resistances in insects that are relevant to forest trees will emerge at a later point in the future.

Hu et al. (2007) found the insect resistance remaining stable between 1997 and 2001 in Bt transgenic *Populus nigra* in plantations in China. Zhang et al. (2011a) take the lower numbers of coleopteran pests in a plantation of transgenic hybrid poplars expressing Bt together with a proteinase inhibitor in comparison to wild type plantations, assessed in a 3-year period as an indication for the transgenic poplar retaining their resistance against the coleopteran targets. Zhang et al. (2011b) also, having performed an assessment over three years, argue that, as in Bt transgenic hybrid poplar plantations fewer of the targeted insects were observed, this is an indication of the plants retaining their insect resistance.

One way to address the problem of resistances arising is to introduce multiple resistance genes. However, it is a question of whether this just buys time rather than preventing the arising of resistances efficiently and long term.

5 Non-Bt Approaches and biosafety

The body of literature concerning the safety aspects of insect-resistant transgenic trees in the environment is largely focusing on Bt transgenic lines. Although non-Bt approaches are on the rise there are few studies looking at transgenic trees expressing other proteins for enhanced insect resistance.

5.1 Chitinases

Birch is a source of high-value timber and is important as a forest tree, in particular in the northern countries, and a host plant to a multitude of herbivorous insects. Work to develop approaches to reduce herbivory on commercially grown silver birch (*Betula pendula*) includes expression of chitinase. Vihervuori et al. (2008) studied insect communities on lines of transgenic birch (*Betula pendula*) that were expressing a sugar beet chitinase gene and were originally made to test the potential for enhancing fungal resistance. They found indeed a shift in insect communities on transgenic trees compared to the wild type ones, with the surprising result being, given the anti-insect effect of chitinases shown elsewhere, that the transgenic lines in fact turned out to be generally *more* attractive to the main group of insect present on birch here, the aphids. The insect diversity appeared to be unaffected by the level of transgene expression.

5.2 Proteinases

Protease inhibitors act in a highly unspecific manner and affect a range of proteolytic enzymes. Given that proteases do not only play an important role in the insect gut, but in many other compartments of many other organisms, it does not seem unreasonable that overproducing these proteins in a plant system may lead to non-target effects. As reviewed by Schlüter et al. (2010, and references therein) results on non-target effects of protease-inhibitor expressing transgenic plants are ambiguous. With positive, negative and insignificant effects being observed on different non-target organisms, it cannot be excluded that non-desirable effects on ecosystems may arise upon a larger scale use of such plants.

Until now, no transgenic trees that overexpress solely protease inhibitor genes have been tested in the field. A change in protease activity and a resultant resistance of the insects can be observed already in greenhouse experiments and in work in crop plants (e.g. Dunse et al. 2010).

5.3 Lectins

As yet, no studies as to the non-target effects of lectin expression are reported from transgenic trees. However, there is indication from crop plants that there may be downstream effects in the food chain. In an example of work on agricultural crops Hogervorst et al. (2009) studied imagines of parasitoid wasps feeding on honeydew excreted by aphids on an artificial diet containing GNA. They found that the life span of these wasps was significantly reduced. However, this observation could not be repeated with wasps fed on honeydew excreted by aphids feeding on a GNA transgenic wheat line. Given that GNA expression was generally very low in these plants this finding does not exclude an effect being present in other cases or in the field. For trees as yet only a very limited amount of work with GNA expressing lines has been published. However, a potential combination of an aphid developing resistance and, in addition, the parasitoid effectiveness in aphid control being weakened due to the toxin being delivered via the honeydew would surely give reason for concern. Hogervorst et al. (2009) point out a need to study honeydew as a potential route of exposure of non-target organisms to transgenic products, such as GNA and other toxins. Honeydew producing insects are found on a large number of forest trees, as illustrated by the commercial production of honey from their sugary excretions in many parts of the world.

5.4 Avidin

Christeller et al. (2006) found that avidin readily binds to the surface of soil particles and further maintained its insecticidal activity, as tested in bioassays using the light brown apple moth (*Epiphyas postvittana*, Lepidoptera: Tortricidae). In work reviewed and summarized by Christeller et al. (2010), effects of avidin-transgenic tobacco on non-target organisms were absent or non-significant. The authors (Christeller et al. 2010) however, in view of the potential lower nutritional value suggest that the technique should be applied preferably in non-food crops, explicitly naming pine, eucalyptus, poplar and willow.

5.5 Tryptamine

Tryptamines are neuroactive substances that may affect a wide range of organisms, provided they do possess neurons. Gill et al. (2003) point out that "*Elevated tryptamine levels can be expected to have positive and/or negative impacts on other organisms that depend on their interaction with poplar to successfully complete their life cycle*". However, there is altogether an insufficient amount of literature to draw further conclusions.

5.6 Scorpion and Spider Toxins

It is not unlikely that biomass of transgenic trees would in a plantation situation not only be a potential source of feed to insects and other invertebrates, but also to other herbivores. Yang et al. (2013a) present a study that found no effect in rabbits that were fed green fodder from transgenic *Populus cathayana* Rehd. leaves expressing a *Manduca sexta* chitinase and a *Buthus martensii* scorpion toxin (BmIT). The same authors also performed a study in which the transgenic material was fed to the animals in the form of dried leaf pellets, and, as before neither found an effect on weight or any pathological symptoms nor morphological changes in the rabbits (Yang et al. 2013b).

The risks of bringing recombinant spider neurotoxins into the environment are likely to be comparable to those associated to the use of scorpion derived genes. However, no studies on this topic have yet been published.

5.7 Effects of Non-anti-insect Transgenic Trees on Insects

While in the previous sections work on transgenic trees specifically engineered to have phenotypes that enhance insect resistance, it is worthwhile to consider what effects other transgenic lines of forest trees, carrying genes that may at times be combined with Bt or other insect resistance enhancing genes, may have on the insect fauna. Other transgenic lines that do not target insect may equally have effects on invertebrates, as "*Changes in herbivore preference or performance on genetically modified plants could result from transgene expression, either though the primary… or secondary (pleiotropic) functions … of the novel protein*" (Post and Parry 2011, and references therein).

Again there is little data to build on. As yet no studies are available on the effects of transgenic material expressing antifungal polypeptides further down the food chain. It is imaginable that effects onto fungal symbionts may affect social insects and thus have wider implications onto ecosystems—a potential risk that is to the author's best knowledge as yet completely unstudied (Robischon 2006b).

A study by Post and Parry (2011) found that fitness in gypsy moth fed on genetically modified American chestnut that expressed a wheat oxalate oxidase gene (a gene which, if expressed in *Zea mays* resulted in a reduction of feeding by European corn borer (*Ostrinia nubialis*, Lepidoptera: Pyralidae; Ramputh et al. 2002) was *positively* affected by this diet.

There is a general concern in parts of the public that marker genes that are involved in most transformation events, such as NPTII may have unpredicted environmental or health effects. Schnitzler et al. (2010) studied invertebrates in a field trial consisting of transgenic *Pinus radiata* lines expressing said marker gene as well as different genes studied for their role in development, such as *CONSTANS*, *LEAFY* and *APETALA*. Given that NPTII is not known to be entomotoxic it does not come as a surprise that no significant difference in invertebrate communities collected from the transgenic trees as compared to the wild type trees could be found.

Hjältén et al. (2007) studied transgenic hybrid aspen clones (*P. tremula* × *P. tremuloides*) that were transformed for what was believed to be a non-insect resistance trait, the overexpression of a maize sucrose phosphate synthase. The original aim of this work was to enhance biomass production. However, it turned out that by introducing this gene also levels of endogenous phenolics and nitrogen were altered. In a line that stood out by having high levels of salicin, tremuloidin, condensed tannins and nitrogen, but at the same time lower concentrations of coumaric acid and several flavonoids a lower level feeding by a leaf beetle (*Phratora vitellinae, Coleoptera: Chrysomelidae*) than in the wild type was observed. Kosonen et al. (2012), working on a transgenic poplar engineered to have high levels of condensed tannins by expression of a MYB134 transcription factor leading to a high expression of the corresponding MYB134 gene, found that the brassy willow beetle *Phratora vitellinae* (*Coleoptera: Chrysomelidae*) preferred the transgenic line over the wild type, but appeared to be particular attracted by salicylates which are feeding stimulants to this beetle.

High tannin levels have been observed to be correlated with reduced herbivory in forest trees, and therefore high tannin content has been at times considered to be part of defense systems, with the empirical evidence however being inconsistent. A study by Boeckler et al. (2014) found that forest tent caterpillars (*Malacosoma disstria,* Lepidoptera: Lasiocampidae) and gypsy moth (*Lymantria dispar*, Lepidoptera: Lymantriidae) show an enhanced preference for transgenic aspen material that expressed a tannin regulatory gene and had as a result significantly elevated foliar condensed tannin levels. Given that the insects also showed an increased weight gain, faster pupation and other indicators of fitness this system does not appear as a suitable way to enhance resistance.

Hjältén et al. (2013) also compared feeding by *Phratora vitellinae* in two hybrid aspen clones (*P. tremula* × *P. tremuloides* and *P. tremula* × *P. alba*) that were both transformed either with a cry3Aa Bt gene or had an altered lignin composition due to overexpression of caffeic acid 3-*O*-methyltransferase (COMT) or Cinnamyl Alcohol Dehydrogenase (CAD). While in the Bt lines a negative effect on the feeding could be found, this was not the case in the COMT or CAD lines. Tiimonen et al. (2005) fed leaves of a transgenic silver birch (*Betula pendula*) expressing a COMT (caffeate/5-hydroxyferulate *O*-methyltransferase) gene and displaying a phenotype of altered lignin composition to a range of, but found no significant difference in feeding preference of growth performance between wild type control and transgenic line.

Depending on the future development and future expansion of transgenic plantations it can however not be excluded that there may be effects on insect populations within a suite of larger scale ecological effects. For example, it seems natural that large-scale *Eucalyptus* plantations in the southern USA, made possible solely by the development of transgenic freeze-resistant lines, would harbour drastically altered insect communities than the original, native forest cover. This is demonstrated already by eucalyptus plantations worldwide. Work on freezing resistance in transgenic poplar has been reported by Zhou et al. (2010).

6 Conclusions

Effects that were so far scientifically proven do not suggest all too drastic consequences of the use of transgenic insect-resistant trees, as far as the direct impact of the respective transgene is concerned. It would however be premature and in fact imprudent to give all-clear signals. The number of studies on trees is very small. The number of studies on agricultural crops is much larger, but the scope for comparisons is limited due to the entirely different structure of production systems, and any conclusions drawn from observations in agriculture are only of limited meaning to forestry.

Finally, transgenic methods make it increasingly possible to grow plant species not only outside their original and natural geographic range, but also in completely different climatic zones, as for example in the case of the cold-resistant *Eucalyptus* trees. It can also be expected that transgenes combining several aspects, such as insect resistance, cold resistance or salt tolerance will greatly enlarge the area in which a particular clone can be planted, hence expanding the effects of the large-scale monocultures, and also including more naturally occurring insects into the 'unplanned' exposure trials. The emergence of insects that are resistant against their specific Bt-toxin in Bt-transgenic trees is therefore only a matter of time, and, according to current knowledge, can only be delayed by appropriate management strategies.

It is, in spite of past research, difficult or impossible to estimate what further 'unwanted effects' may potentially emerge in the wake of the use of insect resistant transgenic forest trees.

Kosonen et al. (2012) emphasize that "*transgenic plants are powerful tools, but that enhancing one secondary pathway may lead to unexpected effects on other pathways, and thus impact characteristics such as plant resistance against herbivores, especially in changing climate conditions*". Post and Parry (2011) add on that matter that "*the imprecise nature of transgenesis may inadvertently alter tree phenotype, thus potentially impacting ecologically dependent organisms*". With an increasing number of transgenic lines expressing multiple foreign genes in combination there will be an increased complexity of environmental effects and impacts on entomocenoses and zoocenoses in general, making an assessment of potential non-target effects increasingly difficult.

The problem of unwantedly affecting non-target organisms in tree biotechnology may however change with the advent of new technologies, which allow a shift away from the overexpression of transgenes under foreign promoters to modulating the expression of the plants own, native genes. A new tool of tree transgenesis is emerging with the development of the clustered regulatory interspaced short palindromic repeats (CRISPR) associated nuclease (Cas) system (CRISPR/Cas). Tools to design CRISPR/Cas9 constructs for targeted genome editing are now available for *Populus trichocarpa* (Heigwer et al. 2014).

Whichever technology however is applied to generate genetically altered trees, it has to be considered in risk assessment that classical mechanistic thinking is unsuitable to study living systems that are prone to yield phenomena that are unexpected and that may at first even appear paradox.

At a time when the long-term effects even of "simple" transgenes expressing only a Bt gene are hardly satisfactorily studied, we are looking at much more complex constructs with proteins interacting in unknown ways and having potentially very different effects on target and non-target organisms. Therefore, a continuous monitoring at a high level is needed to ensure phytophagous insects in non-natural stands with a non-natural genotype will not become a factor unpredictable to future foresters.

References

Adams JM, Piovesan G, Strauss S, Brow S (2002) The case for genetic engineering of native and landscape trees against introduced pests and diseases. Conserv Biol 16(4):874–879

Alvarez JM, Ordás RJ (2013) Stable agrobacterium-mediated transformation of maritime pine based on kanamycin selection. Sci World J. http://dx.doi.org/10.1155/2013/681792

Axelsson EP, Hjältén J, LeRoy CJ, Whitham TG, Julkunen-Tiitto R, Wennström A (2011) Leaf litter from insect-resistant transgenic trees causes changes in aquatic insect community composition. J Appl Ecol 48(6):1472–1479. doi:10.1111/j.1365-2664.2011.02046.x

Axelsson EP, Hjältén J, LeRoy CJ (2012) Performance of insect-resistant *Bacillus thuringiensis* (Bt)-expressing aspens under semi-natural field conditions including natural herbivory in Sweden. Forest Ecol Manage 264:167–171

Barbehenn RV, Jones CP, Yip L, Tran L, Constabel CP (2007) Does the induction of polyphenol oxidase defend trees against caterpillars? Assessing defenses one at a time with transgenic poplar. Oecologia 154(1):129–140

Barraclough EI, Burgess EPJ, Philip BA, Wohlers MW, Malone LA (2009) Tritrophic impacts of Bt-expressing transgenic pine on the parasitoid *Meteorus pulchricornis* (Hymenoptera: Braconidae) via its host *Pseudocoremia suavis* (Lepidoptera: Geometridae). Biol Control 49 (2):192–199

Boeckler GA, Towns M, Unsicker SB, Mellway RD, Yip L, Hilke I, Gershenzon J, Constabel CP (2014) Transgenic upregulation of the condensed tannin pathway in poplar leads to a dramatic shift in leaf palatability for two tree-feeding Lepidoptera. J Chem Ecol 40(2):150–158

Burgess EPJ, Todd JH, Philip BA (2005) Insecticidal efficacy of transgenic avidin-expressing *Eucalyptus grandis* against a leaf-chewing lepidopteran pest. HortResearch client report 13836/2005. HortResearch, Auckland, New Zealand, pp 12

Cao C-W, Liu G-F, Wang Z-Y, Yan S-C, Ma L, Yang C-P (2010) Response of the gypsy moth, *Lymantria dispar* to transgenic poplar, *Populus simonii* × *P. nigra*, expressing fusion protein gene of the spider insecticidal peptide and Bt-toxin C-peptide. J Insect Sci 2010(10):200. doi:10.1673/031.010.20001

Carstens K, Anderson J, Bachman P, De Schrijver A, Dively G, Federici B, Hamer M, Gielkens M, Jensen P, Lamp W, Rauschen S, Ridley G, Romeis J, Waggoner A (2012) Genetically modified crops and aquatic ecosystems: considerations for environmental risk assessment and non-target organism testing. Transgenic Res 21(4):813–842. doi:10.1007/s11248-011-9569-8

Chong Y, Hayes JL, Sollod B, Wen S, Wilson DT, Hains PG, Hodgson WC, Broady KW, King GF, Nicholson GM (2007) The ω-atracotoxins: selective blockers of insect M-LVA and HVA calcium channels. Biochem Pharmacol 74(4):623–638

Christeller JT, Markwick NP, Poulton J, O'Callaghan M (2006) Binding of an insecticidal transgene product to soil: biological activity of soil-bound avidin and the effects of time and microbial activity. Soil Biol Biochem 38(8):2043–2052

Christeller JT, Markwick NP, Burgess EPJ, Malone LA (2010) The use of biotin-binding proteins for insect control. J Econ Entomol 103(2):497–508. doi:10.1603/EC09149

Confalonieri M, Allegro G, Balestrazzi A, Fogher C, Delledonne M (1998) Regeneration of *Populus nigra* transgenic plants expressing a Kunitz proteinase inhibitor (Kti3) gene. Mol Breed 4(2):137–145

Delledonne M, Allegro G, Belenghi B, Balestrazzi A, Picco F, Levine A, Zelasco S, Calligari P, Confalonieri M (2001) Transformation of white poplar (*Populus alba* L.) with a novel *Arabidopsis thaliana* cysteine proteinase inhibitor and analysis of insect pest resistance. Mol Breeding 7(1):35–42

Dowd PF, Lagrimini LM, Herms DA (1998) Differential leaf resistance to insects of transgenic sweetgum (*Liquidambar styraciflua*) expressing tobacco anionic peroxidase. Cell Mol Life Sci 54(7):712–720

Dunse KM, Stevens JA, Lay FT, Gaspar YM, Heath RL, Anderson MA (2010) Coexpression of a potato type I and II proteinase inhibitor gives cotton plants protection against insect damage in the field. Proc Natl Acad Sci USA. doi:10.1073/pnas.1009327107

Escoubas P, Stankiewicz M, Takaoka T, Pelhate M, Romi-Lebrun R, Wu FQ, Nakajima T (2000) Sequence and electrophysiological characterization of two insect-selective excitatory toxins from the venom of the Chinese scorpion *Buthus martensi*. FEBS Lett 483(2–3):175–180

Fox J (2003) Resistance to Bt toxin surprisingly absent from pests. Nat Biotechnol 21(9):958–959

Furniss RL, Carolin VM (1977) Western forest insects. U.S. Department of Agriculture, Miscellaneous Publication 1339. Washington, DC, 654 p

Gatehouse AMR, Ferry N, Edwards MG, Bell HA (2011) Insect resistant biotech crops and their impacts on beneficial arthropods. Philos Trans R Soc B Biol sci 366(1569):1438–1452

Génissel A, Leplé J-C, Millet N, Augustin S, Jouanin L, Pilate G (2003a) High tolerance against *Chrysomela tremulae* of transgenic poplar plants expressing a synthetic cry3A gene from *Bacillus thuringiensis* ssp tenebrionis. Mol Breeding 11:103–110

Génissel A, Augustin S, Courtin C, Pilate G, Lorme P, Bourguet D (2003b) High level of resistance to Bt plants in the absence of man-made changes. Proc R Soc Lond B 270:791–797
Gill RI, Ellis BE, Isman MB (2003) Tryptamine-induced resistance in tryptophan decarboxylase transgenic poplar and tobacco plants against their specific herbivores. J Chem Ecol 29(4):779–793
Grace LJ, Charity JA, Gresham B, Kay N, Walter C (2005) Insect resistant transgenic *Pinus radiata*. Plant Cell Rep 24(2):103–111
Guynup S (2010) Blending old and new in reforesting. Sci Am 82–83
Häggman H, Raybould A, Borem A, Fox T, Handley L. Hertzberg M, Lu MZ, Macdonald P, Oguchi T, Pasquali G, Pearson L, Peter G, Quemada H, Séguin A, Tattersall K, Ulian E, Walter C, McLean M (2013) Genetically engineered trees for plantation forests: key considerations for environmental risk assessment. Plant Biotechnol J 11(7):785–798
Harcourt RL, Kyozuka J, Floyd RB, Bateman KS, Tanaka H, Decroocq V, Llewellyn DJ, Zhu X, Peacock WJ, Dennis ES (2000) Insect- and herbicide-resistant transgenic eucalypts. Mol Breeding 6(3):307–315
Head G, Surber JB, Watson JA, Martin JW, Duan JJ (2002) No detection of Cry1Ac protein in soil after multiple years of transgenic Bt cotton (Bollgard) use. Environ Entomol 31(1):30–36
Heigwer F, Kerr G, Boutros M (2014) E-CRISP: fast CRISPR target site identification. Nat Methods 11:122–123
Hjältén J, Lindau A, Wennström A, Blomberg P, Witzell J, Hurry V, Ericson L (2007) Unintentional changes of defence traits in GM trees can influence plant herbivore interactions. Basic Appl Ecol 8(3):434–443
Hjältén J, Axelsson EP, Julkunen-Tiitto R, Wennström A, Pilate G (2013) Innate and introduced resistance traits in genetically modified aspen trees and their effect on leaf beetle feeding. PLoS ONE 8(9):e73819. doi:10.1371/journal.pone.0073819
Hogervorst PAM, Wäckers FL, Woodring J, Romeis J (2009) Snowdrop lectin (*Galanthus nivalis* agglutinin) in aphid honeydew negatively affects survival of a honeydew-consuming parasitoid. Agric For Entomol 11(2):161–173
Hu JJ, Li SM, Lu MZ (2007) Stability of insect-resistance of Bt transformed *Populus nigra* plantation and its effects on the natural enemies of insects. For Res Chin Acad For 20:656–659 (in Chinese, English abstract)
Klocko A, Viswanath V, Ma C, James RR, Cardineau G, Skinner J, Oppert BS, Payne P, Miller L, Meilan R (2014) Bt-Cry3Aa expression reduces insect damage and improves growth in field-grown hybrid poplar. Can J For Res 44(1):28–35
Kosonen M, Keski-Saari S, Ruuhola T, Constabel CP, Julkunen-Tiitto R (2012) Effects of overproduction of condensed tannins and elevated temperature on chemical and ecological traits of genetically modified hybrid aspens (*Populus tremula* × *P. tremuloides*). J Chem Ecol 38(10):1235–1246
Lachance DA, Hamel LP, Pelletier F, Valéro JR, Bernier-Cardou M, Chapman K, van Frankenhuyzen K, Séguin A (2007) Expression of a *Bacillus thuringiensis* cry1Ab gene in transgenic white spruce and its efficacy against the spruce budworm (*Choristoneura fumiferana*). Tree Genet Genomes 3(2):153–167
Lamarche J, Hamelin RC (2007) No evidence of an impact on the Rhizosphere Diazotroph community by the expression of *Bacillus thuringiensis* Cry1Ab toxin by Bt white spruce. Appl Environ Microbiol 73(20):6577–6583
Lamarche J, Stefani FO, Séguin A, Hamelin RC (2011) Impact of endochitinase-transformed white spruce on soil fungal communities under greenhouse conditions. FEMS Microbiol Ecol 76 (2):199–208
Lavery PB, Mead DJ (2000) *Pinus radiata*: a narrow endemic from North America takes on the world. In: Richardson DM (ed) Ecology and biogeography of pinus. Cambridge University Press, Cambridge, pp 433–441
LeBlanc PM, Hamelin RC, Filion M (2007) Alteration of soil rhizosphere communities following genetic transformation of white spruce. Appl Environ Microbiol 73(13):4128–4134
Ledford H (2014) Brazil considers transgenic trees. Genetically modified eucalyptus could be a global test case. Nature 512:357. doi:10.1038/512357a

Lei Z, Shanjun J, Xingfu J, Haixia Y, Lizhi L (2011) The influence of Cry1Ab toxin on the growth and development of *Microplitis pallidipes*. Plant Prot Beijing 37(6):107–111 (in Chinese, English abstract)

Leplé JC, Bonade-Bottino M, Augustin S, Pilate G, Le Tan VD, Delplanque A, Cornu D, Jouanin L (1995) Toxicity to *Chrysomela tremulae* (Coleoptera: Chrysomelidae) of transgenic poplars expressing a cysteine proteinase inhibitor. Mol Breeding 1(4):319–328

Liu XZ, Liu Z, Yang YM, Zhang HY (2010) Production of transgenic *Pinus armandii* plants harbouring btCryIII(A) gene. Biol Plant 54(4):711. doi:10.1007/s10535-010-0126-8

Maheswaran G, Pridmore L, Franz P, Anderson MA (2007) A proteinase inhibitor from *Nicotiana alata* inhibits the normal development of light-brown apple moth, *Epiphyas postvittana* in transgenic apple plants. Plant Cell Rep 26(6):773–782

Markwick NP, Docherty LC, Phung MM, Lester MT, Murray C, Yao JL, Mitra DS, Cohen D, Beuning LL, Kutty-Amma S, Christeller JT (2003) Transgenic tobacco and apple plants expressing biotin-binding proteins are resistant to two cosmopolitan insect pests, potato tuber moth and lightbrown apple moth, respectively. Transgenic Res 12(6):671–681

McCafferty HRK, Moore PH, Zhu YJ (2008) Papaya transformed with the *Galanthus nivalis* GNA gene produces a biologically active lectin with spider mite control activity. Plant Sci 175 (3):385–393

McCown BH, McCabe DE, Russell DR, Robinson DJ, Barton KA, Raffa KF (1991) Stable transformation of Populus and incorporation of pest resistance by electric discharge particle acceleration. Plant Cell Rep 9(10):590–594

Merkle SA, Andrade GM, Nairn CJ, Powell WA, Maynard CA (2007) Restoration of threatened species: a noble cause for transgenic trees. Tree Genet Genomes 3(2):111–118

Noël A, Levasseur C, Le VQ, Séguin A (2005) Enhanced resistance to fungal pathogens in forest trees by genetic transformation of black spruce and hybrid poplar with a *Trichoderma harzianum* endochitinase gene. Physiol Mol Plant Pathol 67(2):92–99

Pasonen HL, Seppanen SK, Degefu Y, Rytkonen A, von Weissenberg K, Pappinen A (2004) Field performance of chitinase transgenic silver birches (*Betula pendula*): resistance to fungal diseases. Theor Appl Genet 109(3):562–570

Post KH, Parry D (2011) Non-target effects of transgenic blight-resistant American chestnut (Fagales: Fagaceae) on insect herbivores. Environ Entomol 40(4):955–963

Ramputh AI, Arnason JT, Cass L, Simmonds JA (2002) Reduced herbivory of the European corn borer (*Ostrinia nubilalis*) on corn transformed with germin, a wheat oxalate oxidase gene. Plant Sci 162(3):431–440

Robischon M (2006a) Field trials with transgenic trees—state of the art and developments. In: Fladung M, Ewald D (eds) Tree transgenesis; Recent developments. Springer, Berlin

Robischon M (2006b) Gift und Gentechnik im Garten der Ameisen Hymenoptera Formicidae. Entomol Z 116(2):61–65 (in German, English abstract)

Saxena D, Stotzky G (2000) Insecticidal toxin from *Bacillus thuringiensis* is released from roots of transgenic Bt corn in vitro and in situ. FEMS Microbiol Ecol 33(1):35–39

Schlüter U, Benchabane M, Munger A, Kiggundu A, Vorster J, Goulet MC, Cloutier C, Michaud D (2010) Recombinant protease inhibitors for herbivore pest control: a multitrophic perspective. J Exp Bot 61(15):4169–4183

Schnitzler F-R, Burgess EPJ, Kean AM, Philip BA, Barraclough EI, Malone LA, Walter C (2010) No unintended impacts of transgenic pine (*Pinus radiata*) trees on above ground invertebrate communities. Environ Entomol 39(4):1359–1368

Scorza R, Callahan A, Dardick C, Ravelonandro M, Polak J, Malinowski T, Zagrai J, Cambra M, Kamenova I (2013) Genetic engineering of Plum pox virus resistance: 'HoneySweet' plum—from concept to product. Plant Cell Tissue Organ Cult 115(1):1–12

Shang A-Q, Tain C-W, Zhao L-J, Tian Y-C (2008) Establishment of genetic transformation system of *Euonymus japonicus* 'Cu Zhi' mediated by *Agrobacterium tumefaciens*. Acta Hortic Sinica 35(3):409–414

Stefani FOP, Hamelin RC (2010) Current state of genetically modified plant impact on target and non-target funi. Environ Rev 18:441–475

Tabashnik BE, Gassmann AJ, Crowder DW, Carrière Y (2008) Insect resistance to *Bt* crops: evidence versus theory. Nat Biotechnol 26(2):199–202

Tang W, Tian Y (2003) Transgenic loblolly pine (*Pinus taeda* L.) plants expressing a modified delta-endotoxin gene of *Bacillus thuringiensis* with enhanced resistance to *Dendrolimus punctatus* Walker and *Crypyothelea formosicola* Staud. J Exp Bot 54(383):835–844

Tian YC, Zheng JB, Yu HM, Liang HY, Li CQ, Wang JM (2000) Studies of transgenic hybrid poplar 741 carrying two insect-resistant genes. Acta Bot Sinica 42:263–268

Tiimonen H, Aronen T, Laakso T, Saranpää P, Chiang V, Ylioja T, Roininen H, Häggman H (2005) Does lignin modification affect feeding preference or growth performance of insect herbivores in transgenic silver birch (*Betula pendula* Roth)? Planta 222(4):699–708

USDA Aphis (2012) Permit application 11-052-101rm received from ArborGen. Field testing of genetically engineered *Eucalyptus grandis* × *Eucalyptus urophylla*. Final Environmental Assessment, Apr 2012

Valenzuela S, Balocchi C, Rodríguez J (2006) Transgenic trees and forestry biosafety. Electr J Biotechnol 9(3) (Special Issue)

Van Rensburg JBJ (2007) First report of field resistance by the stem borer, *Busseola fusca* (Fuller) to Bt-transgenic maize. S Afr J Plant Soil 24(3):147–151

Vihervuori L, Pasonen HL, Lyytikäinen-Saarenmaa P (2008) Density and composition of an insect population in a field trial of chitinase transgenic and wild-type silver birch (*Betula pendula*) clones. Environ Entomol 37(6):1582–1591

Wang JH, Constabel CP (2004) Polyphenol oxidase overexpression in transgenic Populus enhances resistance to herbivory by forest tent caterpillar (*Malacosoma disstria*). Planta 220 (1):87–96

Wu NF, Sun Q, Yao B, Fan YL, Rao HY, Huang MR, Wang MX (2000) Insect-resistant transgenic poplar expressing AaIT gene. Chin J Biotechnol Sheng Wu Gong Cheng Xue Bao 16:129–33 (in Chinese, English abstract)

Wullschleger SD, Jansson S, Taylor G (2002) Genomics and forest biology: Populus emerges as the perennial favorite. Plant Cell 14(11):2651–2655

Yang ZN, Ingelbrecht I, Louzada E, Skaria M, Mirkov TE (2000) Agrobacterium-mediated transformation of the commercially important grapefruit cultivar Rio Red (*Citrus paradisi* Macf.). Plant Cell Rep 119(12):1203–1211

Yang L, Sun Y, Xie L, Liang A (2010) A novel approach for in situ bud transformation of Populus by Agrobacterium. Scand J For Res 25(1):2010. doi:10.1080/02827580903228870

Yang LY, Sun Y, Hao YS, Wang YX (2013a) Effects of transgenic poplar leaves with binary insect-resistance genes used as feed for rabbits. World Rabbit Sci 21(4):257–261. doi:10.4995/wrs.2013.1188

Yang L, Sun Y, Wang Y, Hao Y (2013b) Effects of dietary transgenic poplar leaf pellets on performance and tissues in rabbits. J Sci Food Agric 11. doi:10.1002/jsfa.6388

Ye X, Busov V, Zhao N, Meilan R, McDonnel LM, Coleman HD, Mansfield SD, Chen F, Li Y, Cheng Z-M (2011) Transgenic Populus trees for forest products, bioenergy, and functional genomics. Crit Rev Plant Sci 30:415–434

Yuan S, Yang M, Gao B (2011) Additive insect-resistant effects of transgenic triploid Chinese white poplar against *Clostera anachoreta*. Front Agric China 5(2):237–240. doi:10.1007/s11703-011-1076-4

Zhan YG, Liu ZH, Wang YC, Wang ZY, Yang CP, Liu GF (2001) Transformation of insect resistant gene into birch. J Northeast For Univ 29:4–6

Zhang Q, Lin SZ, Zhang ZY, Lin YZ (2002) Test of insect-resistance of transgenic poplar with CpTI gene. For Stud China 4:27–32

Zhang Q, Zhang ZY, Lin SZ, Lin YZ (2005) Resistance of transgenic hybrid triploids in *Populus tomentosa* Carr. against 3 species of Lepidopterans following two winter dormancies conferred by high level expression of cowpea trypsin inhibitor gene. Silvae Genetica 54(3):108–116

Zhang H, Harry DE, Ma C, Yuceer C, Hsu CY, Vikram V, Shevchenko O, Etherington E, Strauss SH (2010) Precocious flowering in trees: the FLOWERING LOCUS T gene as a research and breeding tool in Populus. J Exp Bot 61(10):2549–2560. doi:10.1093/jxb/erq092 Epub 2010 Apr 20

Zhang B, Chen M, Zhang X, Luan H, Diao S, Tian Y, Su X (2011a) Laboratory and field evaluation of the transgenic Populus alba × Populus glandulosa expressing double coleopteran-resistance genes. Tree Physiol 31(5):567–573

Zhang B, Chen M, Zhang X, Luan H, Tian Y, Su X (2011b) Expression of Bt-Cry3A in transgenic *Populus alba* × *P. glandulosa* and its effects on target and non-target pests and the arthropod community. Transgenic Res 20(3):523–532

Zhou Z, Wang MJ, Zhao ST, Hu JJ, Lu MZ (2010) Changes in freezing tolerance in hybrid poplar caused by up- and down-regulation of PtFAD2 gene expression. Transgenic Res 19 (4):647–654

Potential Impacts of GM Trees on the Environment and on Plant "Omics": Questionnaire-Based Responses

Stefano Biricolti, Detlef Bartsch, Wout Boerjan, Matthias Fladung, Debora C.M. Glandorf, Jeremy Sweet and Fernando Gallardo

Abstract A questionnaire has been formulated among the participants of the COST action FP0905 "Biosafety of forest transgenic trees" and other scientists in order to collect comments and personal opinions on some general aspects of the impact of the introduction of transgenic forest trees and on the use of "omics" strategies for environmental risk assessment (ERA). Beyond the personal opinions in perceiving the complexity of the topic, some interesting hints have emerged. Almost all responders agree that important biosafety issues can only be addressed by conducting field releases of transgenic trees. Despite the recent publication of numerous "omics" studies in relation to GM crop assessment, large-scale methods that can be internationally certified and accepted are presently not available.

S. Biricolti (✉)
Dipartimento di Scienze delle Produzioni Agroalimentari e dell'Ambiente, Università degli Studi di Firenze, Viale delle Idee 30, 50019 Sesto Fiorentino, Italy
e-mail: stefano.biricolti@unifi.it

D. Bartsch
Federal Office of Consumer Protection and Food Safety (BVL), Mauerstrasse 39-41, 10117 Berlin, Germany

W. Boerjan
VIB Department of Plant Systems Biology and UGent Department of Plant Biotechnology and Bioinformatics, Technologiepark 927, 9052 Ghent, Belgium

M. Fladung
Thünen Institute (TI) of Forest Genetics, Sieker Landstr. 2, 22927 Grosshansdorf, Germany

D.C.M. Glandorf
GMO Office, National Institute of Public Health and the Environment, P.O. Box 1, 3720 BA Bilthoven, The Netherlands

J. Sweet
Sweet Environmental Consultants, 6 Green Street, Cambridge CB24 5JA, UK

F. Gallardo
Departamento de Biología Molecular Y Bioquímica, Facultad de Ciencias e Instituto Andaluz de Biotecnología, Unidad Asociada UMA-CSIC, Universidad de Málaga, Málaga, Spain

C. Vettori et al. (eds.), *Biosafety of Forest Transgenic Trees*, Forestry Sciences 82, DOI 10.1007/978-94-017-7531-1_10

1 Introduction

Transformation or genetic modification is a modern plant biotechnology technique for the transfer of genes from one organism into almost any other one, producing what we call a genetically modified (GM) or genetically engineered (GE) organism. However, concerns have been raised that this breakdown of species barriers could lead to novel and unpredictable rearrangements of genetic information in the genome of the resulting GM organisms. In the past 20 years, plant breeders have taken advantage of these new techniques to insert new characteristics that are either unattainable by traditional agricultural methods or require long time periods along the whole breeding process.

The benefits of GM plants for plant breeders as well as farmers have been considerable, e.g. by providing novel forms of plant protection that can replace the use of agrochemicals. A large proportion of maize, soybean and cotton produced worldwide are GM. However, there is a current debate whether GM plants might contain unexpected, potentially undesirable changes in overall metabolite composition, as a direct consequence of the introduction of (a) new gene(s) into the genome. The concept of unintended effects in GM plants encompasses all effects that go beyond those intended by the original modification and that might impact primarily on health or the environment. Predictable and unpredictable unintended effects may or may not prove to have relevance in terms of safety, but must be taken into account when assessing risk parameters. However, appropriate analytical technology and acceptable metrics for full compositional similarity may require technology and tool development.

The targeted approach for analysis of single compounds with special focus on important nutrients and critical toxicants has been widely accepted by international bodies as part of the concept of "substantial equivalence". It has been successfully applied to the safety assessment of the first generation of GM crops. While the selection of compounds for such analysis is the first step in the targeted approach, there are no generally accepted and harmonised guidelines that define the full extent of analyses required to fulfil statutory risk assessment (RA) procedures.

The major plant genome sequencing projects (*Arabidopsis*, poplar, rice, etc.) can be considered as huge reference books and are important sources of information. Microarrays, next generation sequencing (NGS; RNASeq), and proteomics, however, measure components that are subject to constant change. These changes may be cyclical, developmental, or responses to changes in the environment. The approaches provide comprehensive 'snapshots' of the cell, tissue or organ at the level of messenger RNA (the expressed genes or transcriptome) and protein (proteome), respectively, and will eventually reveal how the components function and interact as working parts of the cell machinery. The entire collection of metabolites in the cell is called the metabolome and the science of measuring it (metabolomics) forms the third of the new "omics" technologies. It uses established methods of analytical chemistry such as flow injection electrospray ionisation mass spectrometry (FIE-MS), Fourier transform ion cyclotron resonance mass spectrometry

(FT-ICR-MS), Fourier transform infrared (FTIR) spectroscopy and nuclear magnetic resonance (NMR), by seeking to measure "everything at once" rather than isolating single compounds of interest. The results again reflect a changeable composition measured at a specific moment in time. The development of these three technologies allows us to think in terms of whole biological systems, leading to understanding and prediction of how a change in one part of the system will affect other parts and ultimately the whole system. Without doubt, all these techniques are capable of generating volumes of data that are far greater than anything that biologists have faced before.

In applying these techniques to plants, it is essential that multiple reference samples are studied in order to assess inherent environmental variability, thus enabling the experimental samples to be evaluated in the context of the baseline situation as a whole. Variable patterns in transcriptome, proteome or metabolome depend on growth conditions, geography, season or variety. It is worth noting that comparisons of GM trees with "omics" platforms are expected to be less complicated than comparisons of annual GM crops. In fact, annual GM crops have been so far selected by a process based not only on the suitable expression of a new trait but they have also undergone a number of crosses and backcrosses to introgress the new trait into elite lines and to obtain homozygosity of the transgene. With forest trees, self-fertilisation is not a viable approach for obtaining homozygous single-copy clones (the genotypes usually selected for commercialisation) due to long juvenile period, high level of heterozygosity leading to inbreeding depression and to self-incompatibility and sterility phenomena. As a consequence, in GM trees, the transgene is harboured by only one of the two homologous chromosomes, and is expected to remain unchanged during the whole transgenic clone's lifetime. Also, backcrossing is not normally used to produce large quantities of transgenic woody plants due to the long time needed to produce and test large numbers of transgenic progenies and because backcrossing breaks up the genetic constitution of the elite variety that was transformed. Therefore, to maintain the traits of the original untransformed cultivar and the additional features conferred by the transgene, GM trees must be vegetatively propagated on a large scale for field plantations. With plants like forest trees that propagate vegetatively, comparative analyses would include the parental variety used to produce the transgenic lines. This is different from crops that reproduce sexually and where the comparator is a near-isogenic line that is genetically different from the GM variety at many loci throughout the genome. In the "omics" approach, the choice of the appropriate comparator is of fundamental importance to disaggregate the natural genetic variability of the conventional plant from that induced by the GM.

2 Questionnaire on Opinions

In considering these differences between trees and crop plants, the opinion of some participants involved in research on transgenic organisms was sought on factors which could affect the procedures for testing and RA (risk assessment) of GM trees.

Specifically, they were asked what they consider as the most relevant environmental impacts of GM trees and what information they believe can be obtained from "omics" analyses that could be useful in the ERA. The resulting information would then help in contributing to the debate on potential impacts of GM trees and to inform policy makers.

A total of 20 participants responded, most of them having direct experience with GM plants, RA and/or regulation of GM plants.

First of all, some responders, before answering the specific questions, made some remarks which should be pointed out, even if the comments were not necessarily shared by all the participants.

- There are now hundreds of studies (the European Community spent 300 million Euros already) showing that transgenic plants are not inherently more risky for the environment than classically bred hybrids, just because they are GM. Therefore, a case-by-case approach should be followed.
- Exotic species, and new varieties obtained by classical breeding, might pose risks for the environment and for human health (e.g. allergies) that could be far greater than those caused by GM plants, because the differences in the genome of new hybrids compared to their parents are typically greater, and therefore more unpredictable, than those of GM plants that usually have only a few extra genes introduced.
- Transgenic plants have already been cultivated for decades, and they appear to be safe and predictable. Every year, 150 million ha of transgenic plants are cultivated, and none of the alarming scenarios have come to reality.
- For the above reported reasons (i) the safety measures that have been taken, starting from early days of cultivating GM crop and increasingly so up to today, need to be reconsidered, (ii) the fact that GM plants are treated differently from classically bred hybrids is scientifically not defendable.
- If no safety measures are taken for non-GM hybrids, then the extreme safety measures for GM plants have to be reconsidered.
- It would be worth deregulating particular classes of GM plants for which the safety and environmental impact is negligible, e.g. promoter–reporter constructs, genome editing, knock-outs of endogenous genes, etc. One can then spend time and money on GM plants, and classically bred plants, which cultivation may interfere with the environment or human health (e.g.: when cultivating a new potentially toxic plant variety engineered for non-food uses close to other stands of the same species that are used for human food and animal feed).

Taking into consideration the above mentioned premises the answers of the participants on the following questions are given below:

Which of the potential adverse effects of GMO (section C2, EU Directive 2001/18/EC) are most relevant to consider in the case of GM trees?

Several participants pointed to the difficulty in answering this question without taking into consideration the inserted trait. To this aim, it should be underlined that this survey has been specifically targeted at the evaluation of "omics" strategies as a tool for environmental risk assessment. Therefore, comments and suggestions should be focussed on a general point of view, but additional information on a "case-by-case" basis would be considered when there is an indication of a potential hazard or impact on the environment or when a risk to the environment had been identified in the ERA.

Generally speaking, considering the fact that woody plants (and thus GM trees) survive for many years on the same spot, most concerns expressed by the participants were related to the putative modifications of the dynamics of populations in the receiving environment. In this respect, the presence of wild relatives and the ability of the GM tree to outcross with them must also be taken into consideration when evaluating the invasiveness or persistence of the GM tree in a specific environment. Many responders outline that such characteristics should also be assessed when breeders bring any new varieties onto the market. New varieties (GM or not) are often selected for higher resistance/tolerance to biotic and abiotic stresses, and these resistance alleles can also affect allele frequencies of natural stands of related species in the neighbourhood.

Other aspects to be taken into consideration as a potential adverse effect are: an altered susceptibility to pathogens (possibly due to the onset of hyper resistance) and the impacts on soil biogeochemistry due to the residuals of the GM trees in the soil.

With a root system exploring deeply into the soil and persistent for many years, woody plants (and thus GM trees) could modify the carbon and nitrogen biological cycles, modifications which cannot be easily monitored. Less concern has been raised by the responders about the putative toxicity or allergenicity to man and animals and risks related to the potential spread of antibiotic resistance marker genes from trees to soil microorganisms.

Which of the following mechanisms do you consider most probable in potential adverse effects of GM trees?

- *The spread of the GMO(s) in the environment*
- *The transfer of the inserted genetic material to other organisms, or the same organism whether GM or not*
- *Phenotypic and genetic instability*
- *Interactions with other organisms*
- *Changes in management, including, where applicable, in agricultural practices*

Most answers show that horizontal and vertical gene transfer are exposure pathways to be taken into consideration, followed by the issue that containment strategies will not be able to limit the establishment and spread of the GM trees in the environment.

Also the interactions of GM trees with other organisms possibly modifying microbial, plant and animal communities constitute a concern for some of the responders.

Only a few participants stressed the occurrence of phenotypic and genetic instability in the GM trees related to the long life span of these organisms and even fewer raised problems about the changes in management and cultivation practices.

Questions about relevance of plant "omics" in ERA
According to the *Directive 2001/18/EC (annex II, D.2.)* information has to be provided on immediate and/or delayed environmental impact resulting from direct and indirect interactions of the GM plant with the environment. One of the questions debated is whether measurable indicators correlated to the requested information can be drawn from an "omics" approach.

(a) *To determinate a potential adverse effect, do you think that "omics" variations between a GM tree and its non-GM parents are relevant to/can be used in the ERA of GM trees?*

Metabolomic approaches reveal new metabolites being synthesised in both GM and new hybrid plants in comparison to their parents, some of which might be toxic to humans or animals. For example, each interspecific hybrid will make numerous metabolites of which the abundance is different from that in the parents, or which are not at all present in either of the parents. Plants contain an estimated number of 5000–10,000 metabolites. For 10 % of these metabolites, the structure is known; however, from the majority of the metabolites, the structure is unknown. For almost none of these known metabolites we have sufficient knowledge on what effects they have on humans and animals, and if they do, whether it is positive or negative and in which doses, and if these supposed activities depend on the type of human/animal gut flora or not. Also, the occurrence of these new metabolites will depend on the organ analysed, its developmental stage (juvenile, adult) and its interaction with the environment. Thus, while in some cases it might be good to do a targeted metabolite profiling, an "omics" approach is not advisable.

Studies in Arabidopsis on the effects of knocking out individual genes of the lignin biosynthesis pathway have shown that many new metabolites, that are absent or below the detection limit in wild type, are being made in mutants, which are not necessarily transgenic, but could be caused by a spontaneous mutation or a mutagen. Also dozens to hundreds of genes are typically differentially expressed in mutants (e.g. Vanholme et al. 2012).

Therefore, even if "omics" studies in some cases can assist in the identification of changes and new traits in GM trees (Yanfang et al. 2011; Xu 2014) most participants did not recommend an "omics" approach for ERA of GMOs.

The main difficulty with the "omics" approach is that, in the current state of knowledge, it is very difficult to identify the biological relevance of the observed changes. It is crucial to have substantial information on the natural variation (baseline data) in wild-type or "traditional" plants in order to be able to interpret "omics" data correctly.

Furthermore, most changes at the "omics" level do not penetrate to the phenotypic level. In other words, whether a plant induces or represses the expression of dozens of genes, this does not necessarily have any consequences on the development of these plants. This is called "phenotypic buffering" (Fu et al. 2009; Boerjan and Vuylsteke 2009).

(b) *Do you think that a change in plant "omics" can be correlated to indicators given above for ERA?*
Most participants pointed out that "omics" technologies represent promising tools for the detection of unintended effects, but these methods must be validated and harmonised and, at present, no "omics" output can be correlated to indicators for potential adverse effects on the environment (though this opinion was not shared by some participants). Several studies have shown that the extent of natural variation of gene expression is larger than the variation due to the insertion and expression of the transgenes. Furthermore, blindly looking for changes in molecular profiles can give misleading answers.

Please note that according to the Directive 2001/18/EC (Annex II, D.2.) information and indicators have to be provided on intended and unintended changes to the GM plant in order to determine the potential impacts of genetically modified plants (GMP) and such information was considered for the ERA of crop plants. Do you think that additional information or indicator(s) have to be considered for a potential environmental effect of GM trees?
For most participants, the indicators listed in the *Directive 2001/18/EC (annex II, D.2.)* are valid for GM crops and also fully applicable to GM trees because these indicators are broadly defined. To conduct an ERA for GM trees, the indicators may be tailored to specific biological characteristics of trees, in particular those related to the life span of trees: the long cohabitation in the same environment for several decades can affect the impact on the environment (ecosystem) and biodiversity (most concerns were related to the stability of the transgene and to the long-term effects on the fauna and soil microbial communities).

It should be added that concerns were expressed about the fact that the genetic diversity in next generation forests will be eroded if the local varieties are driven out of the market and superseded by GE trees and their products as only a limited number of commercial varieties will be grown. In this context, the beneficial functions of genetic diversity in populations ought to be considered. Of course, the same holds for classically bred (non-GM) varieties that are deployed at large scale.

What are your general recommendations for post-release monitoring of potential environmental impacts of GM trees?
About this question, different opinions were voiced: (i) if no adverse effect is revealed from the comparison of the GM trees with their non-GM comparator, there is no need for any post-release monitoring for these GM trees as is the case for non-GM tree plantations. (ii) if post-release monitoring is done for GM trees, it should also be extended to new conventionally bred varieties as well. Very few participants believed that GM trees should not be released on the market.

Most respondents made some recommendations and suggested specific topics to be monitored: first of all, monitoring will be affected by the longevity of the trees. Monitoring platforms in this case diverge from GM crops. Looking at the following recommendations, it should be taken into account that monitoring should be proportionate in time and space, and consider already identified risks on a case-by-case basis:

- It was suggested to monitor the soil communities and/or their functions underneath GM trees to detect whether there are any differences compared with plantations of non-GM trees.
- Behaviour of insects and other animals (survival, feeding patterns and reproduction) associated with GM trees should also be monitored.
- Initially, monitoring studies of GM trees can be performed in contained environments (e.g. glasshouses). Subsequently long-term, controlled field testing of GM trees compared with their parental lines and related wild types should be conducted at several sites. Traceability of GM trees is another additional concern.
- If risks or uncertainties are identified in the ERA, the effects of climate change on these risks of GM and conventional forest reproductive material should be considered both during the ERA and during monitoring.

What are the main research topics that the EU has to support to cope with potential adverse effects of GM trees on the receiving environment?
Almost all responders agree on the importance of research to address concerns related to biosafety and that the issues that are central to safe deployment can only be addressed by conducting field release of transgenic trees.

Many participants outline the need for a deeper knowledge of the biology of transgenic higher plants, including GM trees and have suggested several topics on which research should be focused:

Improvement of RA technologies

- Developing of appropriate tools to assist collecting data which is relevant to determining changes and impacts of trees cultivated at large scale, be it GM or non-GM. This could include "omics".
- Collecting baseline data on how trees interact with their receiving environment and other organisms.
- Establishing base line data on rates of gene flow and gene introgression from cultivated into wild type trees.

Gene flow and fitness assessment

- Seed and pollen dissemination, impact on non-targeted species and environment, be it for GM or non-GM trees.

- Potential invasiveness of trees cultivated at large scale, be it GM or non-GM trees.
- Development of genetic containment strategies for suppressing transgene dissemination (inducible germination, sterility, control of transition between juvenile and reproductive phase).

Biotic Interactions

- Field trials and silvicultural management of GM trees, ecological impact and biodiversity conservation evaluation
- The assessment of the possible impacts of GM plants on soil microbial communities and functions in comparison to those induced by non-GM trees.

Human Health

- Monitoring of literature originating from clinical research regarding possible allergenic effects on human health.

Legislation Policy

- To support policy level in EU countries regarding/reviewing of legislation/laws for GMOs, with the aim of providing safe scientific atmosphere for GM tree research.

Improving genetic transformation methods

- To avoid pleiotropic adverse effects and reduce risks in the receiving environment, multigene co-transformation, targeting of regulator genes and plastid or mitochondrial transformation should be investigated, in addition to the use of gene-editing techniques such as CRISPR/Cas9 system.

Participants were also invited to make any additional comments that they considered relevant to the topic of this survey.

In addition to the comments reported at the start of the questionnaire, only one participant made two main comments

- More opportunities should be created to finance educational projects with the aim of making people more aware of the need for objective and critical thinking regarding GMOs (i.e. to explain the differences and similarities between traditional breeding and biotechnology)
- A recommendation for national and EU funders to support research initiatives for establishing ethical-based platforms and transparent committees for transgenic biotechnology and research on national levels.

3 Conclusions

Despite the recent publication of numerous "omics" studies in relation to GM crop assessment, large-scale methods that can be internationally certified and accepted are presently not available. Using metabolomics as data generation instrument (measuring more metabolites but with less precision) for GM forest trees assessment would provide little or no added value, since it does not yet substitute the currently used analytical methods.

Basic research should be carried out to improve analytical methods and evaluate the reliability of the results. A weight-of-evidence approach for a better determination of the consistency of the observed differences, and determination of their non-transient nature and of their biological relevance, are all recommended. Modelling is needed to analyse observed differences in various metabolic pathways. Subsequently, a tiered approach to the potential use of "omics" could be proposed, which would follow a decision tree incorporating parameters from traditional safety assessments and establish, on a case-by-case basis, whether "omics" use is helpful or not at a later stage.

Therefore, it seems unlikely that a new GM plant would exhibit extensively altered gene expression, protein or metabolite profiles if the key metabolome as well as phenotypic characteristics like growth, flowering, fruit development, or seed production are equivalent to non-GM counterparts. Nevertheless, biological variability is a common phenomenon from a systems biology point of view. Differences attributed to transgenesis have already been reported in the published "omics" studies. However, when a larger set of references was included in the comparative assessments (i.e. beyond the pairwise comparison of a GM line and its near isogenic line), the most pronounced differences were consistently found between the various conventional varieties, a trend basically linked to the crop diversity maintained or created by plant breeders. This should be put in perspective, taking into account that conventional breeding is generally regarded safe, despite the fact that the nature of the changes in new conventional cultivars is usually unknown.

References

Boerjan W, Vuylsteke M (2009) Integrative genetical genomics in Arabidopsis. Nat Genet 41 (2):144–145

Fu J, Keurentjes JJB, Bouwmeester H, America T, Verstappen FWA, Ward JL, Beale MH, de Vos RCH, Dijkstra M, Scheltema RA, Johannes F, Koornneef M, Vreugdenhil D, Breitling R, Jansen C (2009) System-wide molecular evidence for phenotypic buffering in Arabidopsis. Nat Genet 41(2):166–167

Vanholme R, Storme V, Vanholme B, Sundin L, Christensen JH, Goeminne G, Halpin C, Rohde A, Morreel K, Boerjan Wout (2012) A systems biology view of responses to lignin biosynthesis perturbations in Arabidopsis. Plant Cell 24:3506–3529

Xu L (2014) Overexpression of a new cellulose synthase gene (PuCesA6) from Ussuri poplar (Populus ussuriensis) exhibited a dwarf phenotype in transgenic tobacco. Plant Omics J 7(1):54–62
Yanfang Y, Qi T, Liu H, Qiu D (2011) Tree omics and biotechnology in China. Plant Omics J 4(6):288–294

Part III
Socio-economic Considerations for GM Forest Trees

Approaches and Tools for a Socio-economic Assessment of GM Forest Tree Crops: Factors for Consideration in Cost–Benefit Analyses

Vassiliki Kazana, Lambros Tsourgiannis, Valasia Iakovoglou, Christos Stamatiou, Angelos Kazaklis, Paraskevi Koutsona, Dimitrios Raptis, Anastasia Boutsimea, Mirjana Šijačić-Nikolić, Cristina Vettori and Matthias Fladung

Abstract Decisions related to the use of genetically modified (GM) forest trees could be more rational if they would take into account socio-economic considerations in addition to environmental risk assessment. This chapter presents an overview of available socio-economic approaches and tools for assessment of GM

V. Kazana (✉) · V. Iakovoglou · C. Stamatiou · A. Kazaklis · P. Koutsona · D. Raptis · A. Boutsimea
Department of Forestry and Natural Environment Management, Eastern Macedonia and Thrace Institute of Technology, 1st km Drama-Mikrohori, 66100 Drama, Greece
e-mail: vkazana@gmail.com

V. Iakovoglou
e-mail: viakovoglou@yahoo.gr

C. Stamatiou
e-mail: stamatioy@gmail.com

A. Kazaklis
e-mail: kazaklis@otenet.gr

P. Koutsona
e-mail: ekoutsona@yahoo.gr

D. Raptis
e-mail: d_rapt@yahoo.gr

A. Boutsimea
e-mail: aboutsimea@gmail.com

L. Tsourgiannis
Region of Eastern Macedonia and Thrace, 67100 Xanthi, Greece
e-mail: ltsourgiannis@gmail.com

M. Šijačić-Nikolić
Faculty of Forestry, University of Belgrade, Kneza Višeslava 1, 11030 Belgrade, Serbia
e-mail: sijacic68@mts.rs

C. Vettori et al. (eds.), *Biosafety of Forest Transgenic Trees*,
Forestry Sciences 82, DOI 10.1007/978-94-017-7531-1_11

forest crops and presents options for their implementation. In particular, it explores the suitability of Cost–Benefit Analysis (CBA), a well-known method in the field of economics, for aiding the decision-making process that regulates the experimental and/or commercial release of GM forest crops. A generic catalogue of potential positive and negative externalities that can reasonably be expected as a result of commercial application of GM forest trees and which are specifically connected to modified traits was compiled to form the basis for CBA. Cost and benefit variables were grouped according to two criteria (i) the sustainability type of variables, namely environmental, economic and social variables and (ii) the affected party. The latter is particularly useful as it is related to the distributional equity of costs and benefits of GM forest trees. Finally, results from a focus group study that was organized as part of COST Action FP0905 in order to identify the most important positive and negative externalities of GM forest plantations in connection to modified traits is also presented. CBA can make a significant contribution to a more rational decision-making process towards the potential release of GM forest trees, as it would add a measure of potential contributions to social welfare. However, further research is required to provide more information on the range of potential positive and negative externalities, their quantification, and predictions at different spatial and temporal scales.

1 Introduction

The potential use of GM forest trees in commercial plantations has been predicted by a number of authors to contribute to increased forest productivity, improved pulp for paper, biofuel production, reduced need for or more efficient use of energy, pesticides and fertilizers, climate change mitigation and preservation of biodiversity (Sedjo 2006; Chapotin and Wolt 2007; FAO 2008, 2010; Hinchee et al. 2009; Flachowsky et al. 2009; Harfouche et al. 2011). However, to date no commercial GM forest crops are grown anywhere in the world with the exception of China. This lag in the commercial utilization of GM technology in forestry has been largely blamed on current biosafety regulations (Van Frankenhuyzen and Beardmore 2004; Valenzuela et al. 2006; Harfouche et al. 2011; Aguilera et al. 2013; Häggman et al. 2013, www.cost-action-fp0905.eu).

C. Vettori
National Research Council (CNR), Institute of Biosciences and Bioresources (IBBR)—UOS Florence, Via Madonna del Piano 10, 50019 Sesto Fiorentino, FI, Italy
e-mail: cristina.vettori@cnr.it

M. Fladung
Thünen-Institute of Forest Genetics, Sieker Landstr. 2, 22927 Grosshansdorf, Germany
e-mail: matthias.fladung@ti.bund.de

In the European Union (EU), regulating decisions in relation to GM forest crops, similarly to GM agricultural crops are based on environmental risk assessments. Clearly, potential risks of GM forest trees, particularly those related to biodiversity and the environment as well as to human and animal health are essential to be taken into account in making biosafety decisions. However, consideration solely based on such criteria may not lead to rational public decisions regarding the approval of GM forest crops for experimental or commercial plantings. Current restrictive legislation runs the risk of unintentionally hampering innovation as evidenced through non-approved cases where the risks might be low and the expected damages small, or of closing the door on developments where the benefits associated with cultivation of GM forest trees might be well worth accepting potential risks.

In order to improve the decision-making process in relation to biosafety of GM forest trees both potential benefits of their cultivation and public acceptance should be considered through socio-economic assessments in addition to risk assessments (Greiter et al. 2011; Spök 2011; Wesseler et al. 2011; Du 2012; Horna et al. 2013; Jorge et al. 2014).

Such socio-economic assessments are not new to the EU where the European GMO Socio-Economics Bureau (ESEB) was established with the aim to organize the frame of socio-economic assessments of GM crops within the EU territory. At the international level, concerns for socio-economic considerations of living modified organisms are reflected in Article 26 of the Cartagena Protocol on biosafety (Falck-Zepeda 2009).

This chapter presents an overview of approaches and tools for *ex-ante* socio-economic assessments and how these can be utilized for decisions about the experimental and/or commercial release of GM forest trees with particular emphasis on Cost–Benefit Analysis (CBA). To assist with such CBAs, a list of positive and negative externalities that can be reasonably expected as a result of GM forest tree cultivation is presented. Cost and benefit variables are grouped according to two criteria, (i) the type of variable, that is, economic, environmental or social, and (ii) the affected party. Information concerning the affected parties is particularly useful when performing CBAs, because it is related to distributional equity issues of benefits and costs.

Finally, we present results from a focus group study that was conducted as part of COST Action FP0905 to identify the most important positive and negative externalities related to GM forest tree plantations and specific modified traits.

2 Approaches and Tools for Socio-economic Assessment of GM Forest Tree Biosafety

There are two main approaches that can be used for socio-economic assessments of GM forest tree biosafety, “Cost–Benefit Analysis” (CBA) and “Multiple Criteria Analysis” (MCA). Both approaches have been successfully used for many years for assessing non-GM forest tree plantations. However, their potential use for socio-economic assessments of GM forest trees in support of decisions about biosafety requires special insight with regard to scope and potential methodological advantages and limitations.

Both CBA and MCA are tight to a portfolio of alternative projects or policy scenarios and are appropriate to be used in cases where social benefits and social costs should be considered in a specific decision-making environment (Hall and Moran 2003; Flannery et al. 2004; Charlier and Valceschini 2008). Formulation of alternative projects may relate to regulation options, that is, to grant or not to grant approval for a specific investment in a plantation, or various investments that may differ depending on a specific GM trait, timescale, and GM plantation area or management type. Alternative policy scenarios can also be formulated on the basis of a combination of different issues of interest, particularly regulation procedures and controls, consumer acceptance of specific GM forest products, timescale, GM forest coverage, forest management and technology type.

In assessing GM forest tree biosafety, CBA analysts should first decide whose benefits and costs of GM forest tree plantations are most important, and then record potential impacts and select measurement indicators. Predictions on potential quantitative impacts need to be made over the life of each alternative project or policy scenario, and money values need to be attached to each impact. Finally, in order to determine present values, a discount rate needs to be included when adding up benefits and costs and a sensitivity analysis needs to be conducted. The social discount rate is the rate at which the CBA analysts should discount the benefits and costs accruing at different times over the life of each alternative project or policy scenario. However, different theories suggest different values for the social discount rate. Therefore, sensitivity analysis is most often required for the value of the social discount rate. The alternative with the largest net social benefits would be deemed to be the "best GM forest tree biosafety decision" and should be recommended to the appropriate authority.

In contrast, a decision based on MCA would not necessarily determine "the best decision" as this would take into account trade-off information to help decision makers arriving at a decision (Kazana et al. 2003). MCA requires decision makers to develop such trade-off information, which necessitates more active participation of all affected parties in the decision-making process. MCA algorithmic procedures are more complex than those used for CBA and often the computational burden of input values is much more demanding. However, when using MCA no monetization of the perceived benefits and costs is required.

Both approaches can be used for socio-economic assessments of GM forest tree biosafety particularly at *ex-ante* level, depending on the specific decision-making environment. However, since CBA has been a requirement for investment planning and analysis under the EU cohesion policy since 2000 for both the private and public sectors (European Commission 2008) it was used as a basis for our preliminary work within the frame of COST Action FP0905.

Although CBA is conceptually rather simple, its implementation in the context of socio-economic assessments of GM forest tree plantations presents specific difficulties. These difficulties are related to the distributional equity of costs and benefits, that is, the question of who are potential winners or losers under each alternative project or policy scenario. Also, potential impacts and their measurement indicators are difficult to determine, as these are the subjects of ongoing research in particular with regard to GM forest tree cultivation methods and their impacts. Predictions of quantitative impacts over time might also be problematic as forest trees are long-lived organisms and potential impacts of GM forest tree cultivation may extend over very long periods of time.

The most difficult aspects of CBA in the assessment of GM tree biosafety relate to the monetization of all benefits and costs. In particularly, this is the case for determining environmental, health and/or social benefits and costs that have no tangible market value and for which specific methods need to be developed to set such values. One commonly used method uses the concept of "Willingness To Pay" (WTP), that is the value of the impact under concern that is related to what people are willing to pay for a specific product. In this approach people's "revealed preferences" are determined by observing their actual behaviour or through their "stated preferences" by simply asking people what they would be willing to pay for a certain resource. Also the "Contingent Valuation Method" (CVM) is a well-known method, where through a survey, people are asked how much they would be willing to pay for a certain positive externality (gain) (for example of a GM forest tree crop), or how much they would be willing to pay to prevent a certain negative externality (loss) from occurring or how much they would be "Willing To Accept" (WTA) to face a loss. Using CVM studies all types of used values, such as direct, indirect and option values, as well as non-use values, such as bequest, existence and intrinsic values can be estimated (Pearce and Moran 1994).

An overview of methods used to evaluate WTP for GMOs in agriculture has been presented by Marks et al. (2003), distinguishing between three different categories, opinion surveys, choice experiments and experimental auction market methods, with the latter recommended as the most reliable method.

To date, comparable studies have not been undertaken for GM forest tree plantations, as presently no commercial plantations are grown anywhere in the world except China. This needs to be addressed once CBA is implemented in the assessment of GM forest tree biosafety, and to this end experiences and relevant published work from GMOs in agriculture might prove very useful.

3 Identification of Potential Benefits and Costs of GM Forest Tree Crops

One important consideration for CBA is the ranking of potential beneficiaries and/or parties that are expected to bear the costs of a specific project or policy scenario. To address these questions, a catalogue of potential benefits (positive externalities) and costs (negative externalities) of GM forest tree plantations was compiled in relation to specific modified traits. Information in this catalogue was based on scientific literature searches, official reports and online sources from around the world.

Identified benefits and costs were allocated to three sustainability impact categories (environmental, economic and social; Tables 1 and 2). To address issues related to distributional equity of costs and benefits, these were further dissected within each category in relation to affected parties, including potential GM producers, consumers, local communities, industry and the natural environment in general.

In order to rank identified positive and negative externalities according to their relative importance, a focus group study was conducted as part of COST Action

Table 1 Potential benefits of GM forest plantations for CBA assessments

Type of variable	Benefit description/trait	Party affected
Environmental	Less chemicals when isolating cellulose from wood (modified lignin content)	Industry/natural environment
	Less energy when isolating cellulose from wood (modified lignin content)	Industry/natural environment
	Less pulp mills pollution	Industry/natural environment
	Fewer insecticides/pesticides/herbicides resistance in forest plantations (Insecticide/pesticide/herbicide resistance)	GM producers/natural environment
	Avoidance of transgene escape (transferred genes become inactive)	Natural environment
	Less trees harvested for consumption needs (modified lignin content)	Consumers/natural environment
	Less old growth logging (high yield of GM forest plantations)	Natural environment
	Higher weed control efficiency (herbicide resistance)	GM producers
	Better forest tree adaptability to suit different forest management practices	GM producers
	Restoration of highly contaminated soils (increased tree stress tolerance)	Natural environment
Economic	Higher pulping efficiency (modified lignin content)	Industry
	Better timber quality higher value product (modified lignin content)	Consumers
	Higher biofuel production efficiency (modified lignin content)	Industry
	Stronger timber construction materials (modified lignin content)	Industry/consumers
	Lower input costs for tree production (pesticide resistance)	GM producers
	Higher tree productivity (disease resistance)	GM producers
	Higher overall market value of GM forest trees (modified lignin content)	GM producers/consumers
	More production (if GM forest trees could be grown in other locations outside their natural distribution)	GM producers/consumers
Social	Protection and conservation of culturally important tree species that are declining due to diseases (disease resistance)	Natural environment/local communities

Table 2 Potential costs of GM forest plantations for CBA assessments

Type of variable	Cost description/trait	Party affected
Environmental	Trees less fit (modified lignin content)	GM producers/natural environment
	Forest trees more vulnerable to viral diseases (modified lignin content)	GM producers/natural environment
	Higher rates of soil decomposition (modified lignin content)	Natural environment
	More pesticide resistant forest species (insect resistance)	GM producers/natural environment
	Less phytophagus and pollen-feeding resistant species (insect resistance)	GM producers/natural environment
	Toxication of insectivores (insect resistance)	Natural environment
	Increased number of secondary pests (insect resistance)	GM producers/natural environment
	Smaller exposure of non-pest insects to pesticides (specifically targeted insecticides)	GM producers/natural environment
	Loss of biodiversity (genetic flow between transgenic and wild trees)	Natural environment
	Adverse effects on biotrophic processes of host ecosystems (if new genetic traits enter these ecosystems)	Natural environment
	Manifestation of forest tree abnormalities (clonal variation)	GM producers
	More selection pressure for resistant weed biotypes (herbicide resistance)	GM producers/natural environment
	More use of broad spectrum herbicides (herbicide resistance)	GM producers/natural environment
	Higher tree mortality (change of lignin levels may lead to less tree viability)	GM producers
Economic	Loss of socio-economic values of non-modified natural forests (use of high yield plantations instead)	Local economy
	Higher cost of pest outbreak control (if pest species become resistant to currently effective chemical and biological control methods)	GM producers
	Higher economic risk of GM forest plantations due to long periods of obtaining approval	GM producers
	Higher cost of new seeds	GM producers

(continued)

Table 2 (continued)

Type of variable	Cost description/trait	Party affected
Social	Restricted access of poor producers to the new seed type markets (higher cost of new seeds)	GM producers
	Cultural adaptation to changing biodiversity status (transgene escape may change species composition, number of species, etc.)	Local communities
	Less adaptable local systems to GM forest trees	GM producers
	The society becomes dependent on outside inputs because of GM forest trees	Local/national economy

FP0905. This focus group consisted of 30 scientists with different scientific expertise and representing 17 member countries from across the COST Action FP0905 network. Represented member countries included Albania, Argentina, Austria, Belgium, Bulgaria, Czech Republic, Estonia, FYROM, Germany, Greece, Israel, Italy, Romania, Serbia, Slovakia, Slovenia and Spain. Details of the profile of this focus group with regard to the age, years of experience and field of expertise of participants are presented in Table 3.

The majority of focus group participants were highly experienced scientists in their respective fields of expertise.

The focus group members were provided with a questionnaire and were asked to rank in terms of importance both the potential benefits and costs of GM forest plantations using a five-point Likert scale. This is the most often used type of rating scale in surveys (Boone and Boone 2012).

In terms of potential benefits of GM forest tree plantations included in Table 1, eight were rated by the majority (more than 60 %) of focus group participants' as very important or extremely important in their country context for consideration in relevant CBA studies (Fig. 1).

Table 3 Profile of focus group members

Variable		Percentage of focus group
Age	25–35	24
	36–45	14
	46–55	52
	56–65	10
Years of experience	<10	17
	10–19	40
	20–29	37
	>30	7
Field of expertise	Molecular biology/forest tree genetics	70
	Ecology/silviculture	13
	Management/policy/economics	17

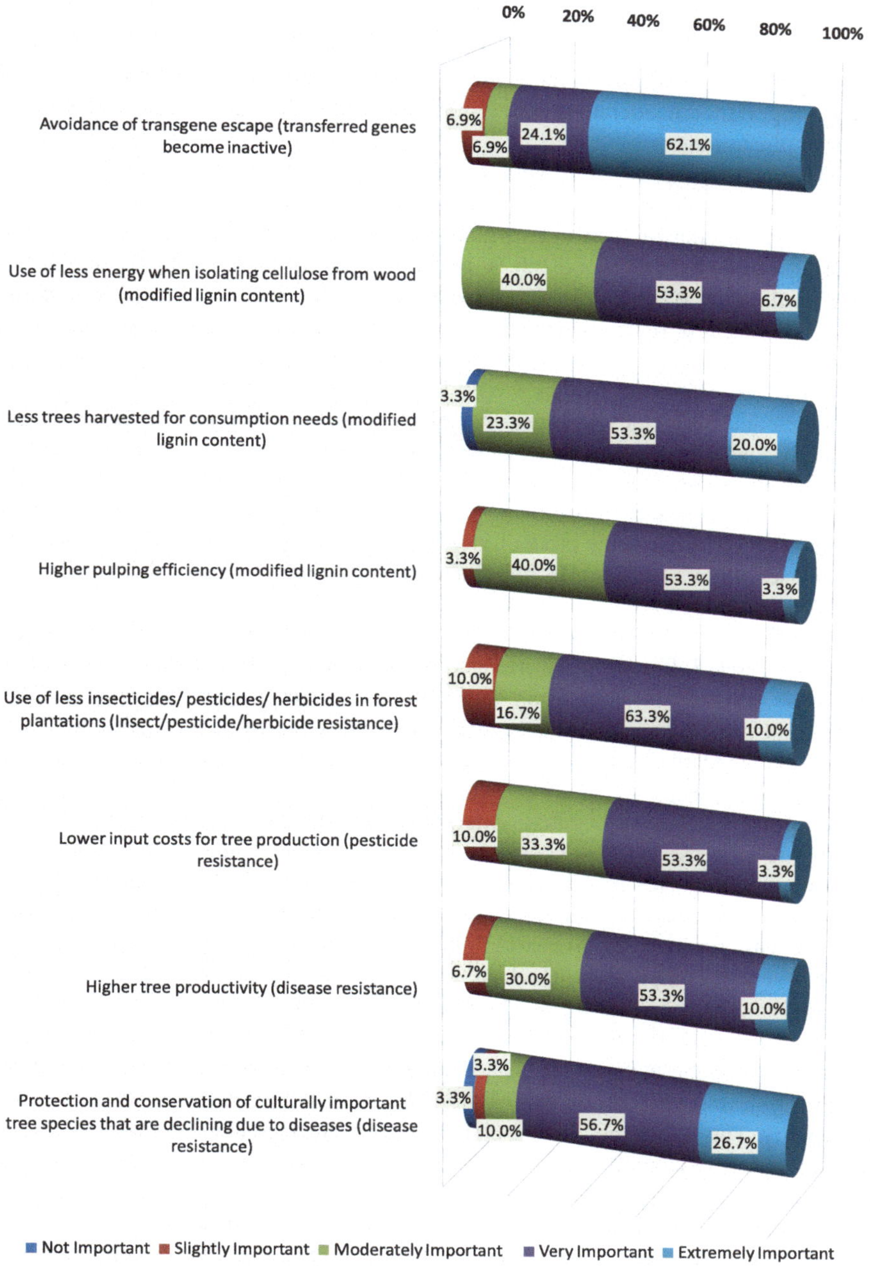

Fig. 1 Response percentage of the COST Action FP0905 focus group indicating the most important benefits of GM forest tree plantations

The avoidance of transgene escape, which is technically achieved by rendering transferred genes inactive through flowering suppression or sterility genes, was rated by the great majority of focus group experts (response rate of almost 86 %) as a very important consideration in decisions about the potential release of GM forest trees in plantations.

GM forest tree plantations with modified lignin content or composition were seen by the majority of focus group participants as very important for inclusion in CBA as processing of such trees would require less energy during the process of isolating cellulose from wood (60 % response rate), because a smaller number of trees would need to be harvested to satisfy consumption requirements (response rate 73 %) and because of an expected higher pulping efficiency (response rate 57 %). Modifications related to insect and pesticide tolerance or herbicide resistance were also rated very important as they would lead to reduction in the use of insecticides, pesticides and herbicides in forest tree plantations (response rate 73 %) and consequently lower input costs for tree production (response rate 57 %) not to mention potential environmental benefits. Finally, GM forest tree plantations with modifications related to disease resistance were rated as very beneficial as these might be characterized by higher tree productivity (response rate 63 %) and aid the protection and conservation of culturally important trees that are declining due to diseases (response rate 83 %).

Potential costs of GM forest tree plantations as identified by the focus group are shown in Table 2. Five issues were rated by the majority of the focus group experts (response rate over 60 %) as very important or extremely important (Fig. 11.2).

Loss of biodiversity due to potential genetic flow between transgenic and wild trees was considered by 80 % of focus group participants as an extremely important negative externality.

A further potentially negative externality includes possible adverse effects on biotrophic processes of host ecosystems that may occur if new genetic traits enter these ecosystems. This was rated as very important by 63 % of focus group experts. Higher tree mortality was also rated as very important by 73 % of the focus group experts. This type of negative externality was seen as a possible result of altered lignin levels or composition which in turn could lead to reduction of tree viability.

Finally, two negative social externalities were identified. First, the need for cultural adaptation to changing biodiversity status which could result following a potential transgene escape. Such an escape was seen to potentially affect species composition and number which 73 % of focus group experts saw as a negative externality. The second negative social externality (with an agreement level of 57 % of the focus group experts) involves the degree of dependency of society on outside inputs because of GM forest trees.

In summary, sustainability categorization of positive and negative externalities related to the GM forest tree plantations (Tables 1 and 2) by COST Action FP0905 focus group experts, identified 4 environmental, 3 economic and 1 social positive externalities as well as 3 environmental and 2 social negative externalities for use in Cost–Benefit Analyses.

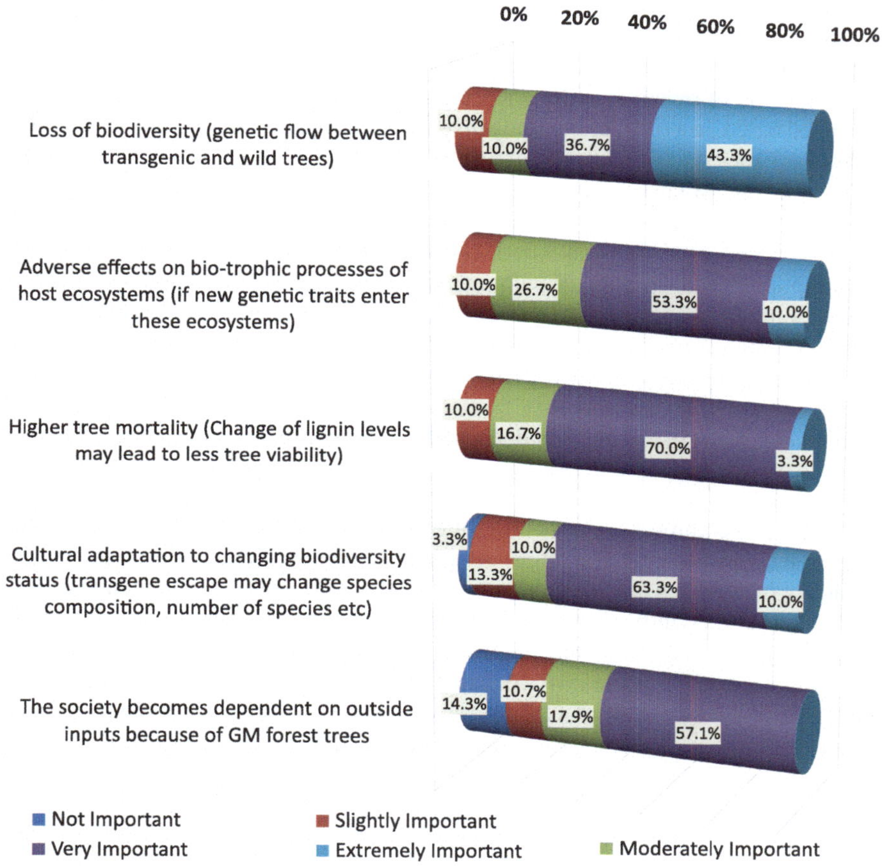

Fig. 2 Response percentage of the COST Action FP0905 focus group indicating the most important costs of GM forest tree plantations

4 Conclusions

The aim of this chapter was to present and discuss potential approaches and tools for a socio-economic assessment of GM forest tree biosafety.

Two general approaches, "Cost–Benefit Analysis" (CBA) and "Multiple Criteria Analysis" (MCA), were discussed, with the main focus on CBA. A preliminary generic catalogue of potential positive and negative externalities that are reasonably expected to occur as a result of commercial application of GM forest trees was compiled to underpin CBA assessments. Grouping of cost and benefit variables was attempted with respect to sustainability variables, including economic, environmental and social variables as well as the affected party. Such information is intended to aid Cost–Benefit Analyses as part of a broader risk assessment of the experimental and/or commercial release of GM forest tree crops with respect to distributional equity of costs and benefits.

A ranking of relative importance of positive and negative externalities was attempted based on a focus group study involving participants of COST Action FP0905 including 30 scientists of different expertise and representing 17 member countries.

The most important positive externalities that were identified included (i) avoidance of transgene escape by rendering transferred genes inactive, (ii) use of less energy when isolating cellulose from wood, (iii) a smaller number of trees harvested for consumption needs, (iv) higher pulping efficiency, (v) use of less insecticides, pesticides or herbicides in forest tree plantations, (vi) lower input costs for tree production, (vii) higher tree productivity and (viii) the protection and conservation of culturally important tree species that are declining due to diseases. The most important negative externalities that should be taken into account in CBA studies were identified as (i) potential loss of biodiversity due to the risk of potential flow between transgenic and wild trees, (ii) adverse effects on biotrophic processes of host ecosystems, because of the risk the new genetic traits to enter these ecosystems, (iii) risk of higher tree mortality, (iv) cultural adaptation to changing biodiversity status due to the risk the transgene escape leading to changes in species composition and number, and (v) dependency of societies on outside inputs because of GM forest tree plantations.

Because it efficiently allows such external factors to be taken into consideration and thereby add an evaluation of potential impacts on social welfare, CBA can make an important informative contribution to a more rational decision-making process in the assessment of GM forest tree biosafety. However, further research should be undertaken to provide more detailed information on the full range of potential positive and negative externalities, their quantification with measurement indicators and predictions at different spatial and temporal scales. Finally, in order to complete the picture, also "Willingness To Pay" (WTP) and/or "Willingness To Accept" (WTA) studies will be required for different types of GM forest tree crops and their potential impacts, taking into consideration public acceptance in the GM biosafety decision-making process.

References

Aguilera J, Nielsen KM, Sweet J (2013). Risk assessment of GM trees in the EU: current regulatory framework and guidance. iForest 6: 127–131

Boone HN, Boone DA (2012) Analyzing Likert data. J Extension, 5092, Article no. 2TOT2. Available at http://www.joe.org/joe/2012april/pdf/JOE_v50_2tt2.pc

Chapotin SM, Wolt JD (2007) Genetically modified crops for the bioeconomy: meeting public and regulatory expectations. Transgenic Res 16(6):675–688

Charlier C, Valceschini E (2008) Importance and limits of the cost-benefit analysis for GMOs regulation. Paper presented at the 110th EAAE seminar "System dynamics and innovation in food networks". Innsbruck-Igls, Austria, 18–22 Feb, pp 373–377

Commission European (2008) Guide to cost-benefit analysis of investment projects. Directorate General Region Policy, Brussels, p 257

Du D (2012) Rethinking risks: should socioeconomic and ethical considerations be incorporated into the regulation of genetically modified crops? Harvard J Law Technol 26(1):376–401

Falck-Zepeda JB (2009) Socio-economic considerations, Article 26.1 of the Cartagena protocol on biosafety. What are the issues and what is at stake? AgBioForum 12(1):90–107

FAO (2008) The potential environmental, cultural and socio-economic impacts of genetically modified trees. UNEP/CBD/SBSTTA/13/INF/6, pp 17

FAO (2010) Forests and genetically modified trees. Rome, Italy, pp 235

Flachowsky H, Hanke M-V, Peil A, Strauss SH, Fladung M (2009) A review on transgenic approaches to accelerate breeding of woody plants. Plant Breed 128:217–226

Flannery ML, Thorne FS, Kelly PW, Mullins E (2004) An economic cost-benefit analysis of GM crop cultivation: an Irish case study. AgBioForum 7(4):149–157

Greiter A, Miklau M, Heissenberger A, Gaugitsch H (2011) Socio-economic aspects in the assessment of GMOS-options for action. Umweltbundesamt Reports, Vienna 52 p

Häggman H, Raybould A, Borem A, Fox T, Handley L, Hertzberg M, Lu M, Macdonald P, Oguchi T, Pasquali G, Pearson L, Peter G, Quemanda H, Seguin A, Tattersall K, Ulian E, Walter C, McLean M (2013) Genetically engineered trees for plantation forests: key considerations for environmental risk assessment. Plant Biotechnol J: 1–14

Hall C, Moran D (2003) Cost-benefit analysis of GM crops in Scotland. Part one: outlining scenarios and categorizing costs and benefits for valuation. Contributed paper 77th annual conference 11th–14th April 2003, Plymouth, pp 19

Harfouche A, Meilan R, Altman A (2011) Tree genetic engineering and applications to sustainable forestry and biomass production. Trends Biotechnol 29(1):11–17

Horna D, Zambrano P, Falck-Zepeda JB (2013) Socioeconomic considerations in biosafety decisionmaking. Methods and implementation. International Food Policy Research Institute, Washington DC, pp 212

Hinchee M, Rottman W, Mullinax L, Zhang C, Chang S, Cunningham M, Pearson L and Nehra N (2009). Short-rotation woody crops for bioenergy and biofuels applications. In Vitro Cellular and Developmental Biology-Plant 45(6): 619–629

Jorge FC, Wechsler S, Livingston M, Mitchell L (2014) Genetically engineered crops in the United States. ERR-162, U.S. Department of Agriculture, Economic Research Service, pp 54

Kazana V, Fawcett R, Mutch WES (2003) A decision support modeling framework for multiple use forest management: The queen Elizabeth forest case study in Scotland. Eur J Oper Res 148(1):102–115

Marks LA, Kalaitzadonakes N, Vickner SS (2003) Evaluating consumer response to GM foods: some, methodological considerations. Current Agriculture, Food and Resource Issues, No 4: 80–94

Pearce DW, Moran D (1994) The economic value of biodiversity. Earthscan and Island Press, London

Sedjo RA (2006) Toward commercialization of genetically engineered forests: economic and social considerations. Resources for the future, pp 46

Spök A (2011) Assessing socio-economic impacts of GMOs, issues to consider for policy development—Final report. Federal Ministry of Health; Federal Ministry for Agriculture, Forestry, environment, and water management. Vienna, pp 127. http://bmg.gv.at/cms/home/attachments/5/0/0/CH1050/CMS1291038713992/assessing_socio-economic_impacts_of_gmos,_band_2_20101.pdf

Valenzuela S, Balocchi C and Rodriguez J (2006). Transgenic trees and forest biosafety, Electronic Journal of Biotechnology Vol. 9, Issue 3, on line at http://www.ejbiotechnology.info/content/vol9/issue3/full/22/

Van Frankenhuyzen K, Beardmore T (2004) Current status and environmental impact of transgenic forest trees. Can J For Res 34:1163–1180

Wesseler J, Scatasta S, Fall EH (2011) Chapter 7 the environmental benefits and costs of genetically modified (GM) crops. In: Carter CA, Moschini G, Sheldon I (eds) Genetically modified food and global welfare. Frontiers of economics and globalization, vol 10. Emerald Group Publishing Limited, pp 173–199

Public Knowledge and Perceptions of Safety Issues Towards the Use of Genetically Modified Forest Trees: A Cross-Country Pilot Survey

Vassiliki Kazana, Lambros Tsourgiannis, Valasia Iakovoglou, Christos Stamatiou, Alexander Alexandrov, Susana Araújo, Saša Bogdan, Gregor Božič, Robert Brus, Gerd Bossinger, Anastasia Boutsimea, Nevenka Ćelepirović, Helena Cvrčková, Matthias Fladung, Mladen Ivanković, Angelos Kazaklis, Paraskevi Koutsona, Zlata Luthar, Pavlina Máchová, Jana Malá, Kostlend Mara, Milan Mataruga, Jana Moravcikova, Donatella Paffetti, Jorge A.P. Paiva, Dimitrios Raptis, Conchi Sanchez, Sandra Sharry, Terezia Salaj, Mirjana Šijačić-Nikolić, Noemi Tel-Zur, Ivaylo Tsvetkov, Cristina Vettori and Nieves Vidal

Abstract Information on public awareness and acceptance issues regarding the use of Genetically Modified (GM) trees in forestry is lacking, although such information is available for GM organisms in agriculture. This is mainly due to the fact that in Europe there is no authorization for commercial planting of GM forest trees. To address this issue and within the frame of a European COST Action on the Biosafety of Transgenic Forest Trees (FP0905), a KAP (Knowledge Attitude

---

V. Kazana (✉) · V. Iakovoglou · C. Stamatiou · A. Boutsimea · A. Kazaklis · P. Koutsona
D. Raptis
Department of Forestry and Natural Environment Management, Eastern Macedonia and Thrace Institute of Technology, 1st km Drama-Mikrohori, 66100 Drama, Greece
e-mail: vkazana@gmail.com

V. Iakovoglou
e-mail: viakovoglou@yahoo.gr

C. Stamatiou
e-mail: stamatioy@gmail.com

A. Boutsimea
e-mail: aboutsimea@gmail.com

A. Kazaklis
e-mail: kazaklis@otenet.gr

P. Koutsona
e-mail: ekoutsona@yahoo.gr

C. Vettori et al. (eds.), *Biosafety of Forest Transgenic Trees*,
Forestry Sciences 82, DOI 10.1007/978-94-017-7531-1_12

Practice) cross-country pilot survey was conducted among university students of different disciplines as sampling subjects. In total, 1920 completed questionnaires from 16 European and non-European countries were evaluated. The results provided novel cross-country insights into the level of public knowledge, particularly of young people and their perceptions on safety issues related to the use of GM forest trees, as well as on their attitude towards the acceptance of GM forest trees

D. Raptis
e-mail: d_rapt@yahoo.gr

L. Tsourgiannis
Directorate of Public Health and Social Care of Regional District of Xanthi, Region of Eastern Macedonia and Thrace, 67100 Xanthi, Greece
e-mail: ltsourgiannis@gmail.com

A. Alexandrov · I. Tsvetkov
Forest Research Institute, Bulgarian Academy of Sciences, Kliment Ohridski Blvd. 132, 1756 Sofia, Bulgaria
e-mail: forestin@bas.bg

I. Tsvetkov
e-mail: tsvet_i@yahoo.com

S. Araújo
Instituto de Tecnologia Química e Biológica, Universidade Nova de Lisboa, Av. da República, 2780-157 Oeiras, Portuga and Instituto de Investigação Científica Tropical (IICT), Biotrop, Rua Da Junqueira, 30, 1349-007 Lisbon, Portugal
e-mail: ssaraujo@gmail.com

S. Bogdan
Faculty of Forestry, Department of Forest, Genetics, Dendrology and Botany, University of Zagreb, Svetošimunska 25, 10000 Zagreb, Croatia
e-mail: sbogdan@sumfak.hr

G. Božič
Slovenian Forestry Institute, Vecna 2, 1000 Ljubljana, Slovenia
e-mail: gregor.bozic@gozdis.si

R. Brus
Department of Forestry and Renewable Forest Resources, Biotechnical Faculty, University of Ljubljana, Vecna pot 83, 1000 Ljubljana, Slovenia
e-mail: robert.brus@bf.uni-lj.si

G. Bossinger
Department of Forest and Ecosystem Science, The University of Melbourne, Creswick, Melbourne, VIC 3363, Australia
e-mail: gerd@unimelb.edu.au

N. Ćelepirović · M. Ivanković
Laboratory of molecular-genetic testing, Division of Genetics, Forest tree breeding and seed science, Croatian Forest Research Institute, Cvjetno Naselje 41, 10450 Jastrebarsko, Croatia
e-mail: nevenkac@sumins.hr

M. Ivanković
e-mail: mladeni@sumins.hr

cultivation. The majority of the respondents, which was more than 60 % in all countries, approved the use of GM forest trees for commercial plantations, excluding natural forests. The majority of respondents also appeared willing to buy products from such plantations, such as wood products, pulp and paper. Over 80 % of the respondents from all countries were in favour of using labelling to identify products of GM origin, while more than 80 % of those would prefer that this labelling be legally mandatory. The top three benefits that were rated as very important in all countries involved the potential lower demand of the GM forest

H. Cvrčková · P. Máchová · J. Malá
Forest and Game Management Research Institute, Strnady 136, 25202 Jíloviště, Czech Republic
e-mail: cvrckova@vulhm.cz

P. Máchová
e-mail: machova@vulhm.cz

J. Malá
e-mail: mala@vulhm.cz

M. Fladung
Thünen-Institute of Forest Genetics, Sieker Landstr. 2, 22927 Grosshansdorf, Germany
e-mail: matthias.fladung@ti.bund.de

Z. Luthar
Agronomy Department, Biotechnical Faculty, University of Ljubljana, Jamnikarjeva 101, 1000 Ljubljana, Slovenia
e-mail: zlata.luthar@bf.uni-lj.si

K. Mara
Agricultural University of Tirana, Tirana, Albania
e-mail: kostlend_mara@yahoo.com

M. Mataruga
University of Banja Luka, Banja Luka, Bosnia and Herzegovina
e-mail: mmataruga@gmail.com

J. Moravcikova · T. Salaj
Institute of Plant Genetics and Biotechnology, Slovak Academy of Sciences, Akademicka 2, PO Box 39A, 95007 Nitra, Slovakia
e-mail: jana.moravcikova@savba.sk

T. Salaj
e-mail: terezia.salaj@savb.sk

D. Paffetti
Department of Agrifood Production and Environmental Science, University of Florence, P. le delle Cascine 28, 50144 Florence, Italy
e-mail: donatella.paffetti@unifi.it

J.A.P. Paiva
Apartado 12, 2780-901 Oeiras, Portugal and Instituto de Investigação Científica Tropical (IICT), iBET, Instituto de Biologia Experimental e Tecnológica, Biotrop, Rua Da Junqueira, 30, 1349-007 Lisbon, Portugal
e-mail: jorge.pinto.paiva@gmail.com

plantations for pesticides, the potential of GM forest trees for restoration of contaminated soils and the potential higher GM forest tree productivity. The top three GM forest tree risks that were perceived as serious hazards in all countries included the potential loss of biodiversity due to gene flow between transgenic and wild trees, the adverse effects of biotrophic processes on host ecosystems and the cultural adaptation to changing biodiversity conditions due to transgene escape. Overall, lack of knowledge regarding the potential benefits and potential risks of the cultivation of GM forest trees was observed in almost all surveyed countries.

1 Introduction

At the current state of biotechnology science, it is feasible to grow Genetically Modified (GM) forest trees on a commercial scale (Van Frankenhuyzen and Beardmore 2004; Williams 2006; Häggman et al. 2013, www.cost-action-fp0905.eu). The most typical genetic modifications include alterations of lignin content and composition, insect resistance and disease resistance, herbicide tolerance, abiotic stress tolerance, growth improvement and reproductive development. It has been argued that the use of GM forest trees in commercial plantations would contribute to increased forest productivity, improved pulp for paper, biofuel production, climate change mitigation, preservation of biodiversity and reduction of energy, pesticides

C. Sanchez · N. Vidal
Instituto de Investigaciones Agrobiológicas de Galicia (CSIC),
Santiago de Compostela, Spain
e-mail: conchi@iiag.csic.es

N. Vidal
e-mail: nieves@iiag.csic.es

S. Sharry
Facultad de Ciencias Agrarias y Forestales, Universidad Nacional de la Plata,
Av.60 y119-c.c. 31 (1900), La Plata, Argentina
e-mail: ssharry@gmail.com

M. Šijačić-Nikolić
Faculty of Forestry, University of Belgrade, Belgrade, Serbia
e-mail: sijacic68@mts.rs

N. Tel-Zur
French Associates Institute for Agriculture and Biotechnology of Drylands,
The Jacob Blaustein Institutes for Desert Research, Ben-Gurion University of the Negar,
Sede-Boqer Campus, 84990 Midreshet Ben-Gurion, Israel
e-mail: telzur@bgu.ac.il

C. Vettori
National Research Council (CNR), Institute of Biosciences and Bioresources
(IBBR)—UOS Florence, Via Madonna del Piano 10, 50019 Sesto Fiorentino, FI, Italy
e-mail: cristina.vettori@cnr.it

and fertilizers utilization (Sedjo 2006; Chapotin and Wolt 2007; FAO 2008, 2010; Hinchee et al. 2009; Flachowsky et al. 2009; Harfouche et al. 2011). Despite these potential advantages, nowadays GM forest trees are not commercially grown anywhere in the world, except China. The main difficulties for the commercialization of GM forest trees are connected to technical limitations, restrictive regulation frameworks, biosafety and socioeconomic considerations (Valenzuela et al. 2006; Farnum et al. 2007; Harfouche et al. 2011; Häggman et al. 2012, 2014).

Technical limitations relate to the required scientific advances in biotechnology, particularly to issues related to gene stability, mass propagation, genetic deployment and ecological impacts. Regulation frameworks impose extensive approval processes on the deliberate release of GM forest trees. These frameworks require comprehensive safety assessments but risk analyses differ between countries. Within the European Union member countries, the rules for the deliberate releases of GM forest trees, mainly implemented through the 2001/18/EC directive, aim at providing high levels of protection for human health and the environment (Aguilera et al. 2013; Häggman et al. 2013). As a result, the approval process for deliberate releases is very time-consuming and expensive (Harfouche et al. 2011).

Socioeconomic considerations are mainly related to potential markets for products deriving from GM forest trees, the costs of patenting and their public acceptance. The latter in particular is influenced by environmental, public health and sociocultural concerns, which have been raised mainly by opinion influencing groups, such as several NGOs. Concerns often focus on potential gene flow between GM and wild trees and consequent implications for the natural environment, increased use of broad spectrum herbicides due to potential increased selection pressure for resistant weed biotypes, potential increase in pesticide resistant forest trees, negative effects on forest tree fitness, potential higher vulnerability of forest trees to viral and other diseases, increased soil decomposition, adverse effects on biotrophic processes in host ecosystems if new genetic traits enter these ecosystems and flowering suppression (El-Lakany 2004; Van Frankenhuyzen and Beardmore 2004; Williams and Davis 2005; Sedjo 2006; Farnum et al. 2007; FAO 2008, 2010). Moreover, concerns focus on the unintentional development of insect and herbicide resistant species as a result of transgene escape that might alter species composition and reduce the number of species present in a given location, thus forcing cultures to adapt to changing biodiversity conditions (Peterson et al. 2000).

The focus of this chapter relates to the public acceptance issue of GM forest tree commercialization. This is due to the fact that information on public attitudes towards the use of GM forest trees in plantations is completely lacking, particularly in Europe, although relevant information is available for the use of GMOs in agriculture (Ferguson et al. 2002; Hossain et al. 2002; Grice et al. 2003; Magnusson 2004; Pereira de Abreu et al. 2006; Shehata and Cox 2007; European Commission 2010; Buah 2011; Pew Research Center 2015). To address this lack of information and to identify the level of public knowledge and perceptions of safety issues, as well as the public acceptance towards the use of GM forest trees in plantations, a cross-country pilot survey was coordinated within the frame of the European COST ACTION FP0905. This COST-ACTION was related to various biosafety aspects, e.g. analyses of the efficiency of existing gene containment strategies to avoid or to minimize gene flow or evaluation of methods to monitor GMTs in the entire production chain (Fladung et al. 2012; Vettori et al. 2014). The pilot survey in particular was focused on young people and it aimed to identify: (i) their level of knowledge about the use of GM forest trees in plantations, (ii) their perceptions about the potential benefits and risks of GM forest tree cultivation and (iii) their attitude towards acceptance of GM forest tree cultivation. Focusing on young people aged 18–35, it was decided, because commercial GM forest plantations are not currently grown and therefore, such a focus group represents the future consumers, developers and/or policy makers.

2 Materials and Methods

A Knowledge Attitude Practice (KAP) pilot survey was conducted during 2012 and 2013 in order to detect novel preliminary cross–country information on the level of public knowledge and perceptions on safety issues related to the cultivation of GM forest trees, as well as the public attitude towards acceptance of GM forest tree cultivation. The KAP survey was focused mainly on the young population. KAP studies are highly focused social surveys usually limited in scope and tailored to a specific problem. They have been used for several types of problems, such as public health problems and water resource and waste management problems (Goutille 2009; Grace et al. 2009; Siwakoti 2009; Vivas et al. 2010). They provide information on what people know about a certain topic (Knowledge), how they feel towards a subject and the ideas they have towards it (Attitude) and the ways they act to demonstrate their knowledge and attitudes (Practice). The KAP pilot survey that is presented in the current article was carried out in 16 countries among University students of three different categories of study fields. These included students of (i) forestry, (ii) other environment-related disciplines, such as botany, biology, agriculture, landscape architecture and environmental science, and (iii) economics including related disciplines, such as accounting, business administration, financial management, management science and marketing. University students were selected as subjects, because in addition to financial survey implementation constraints and ease of access issues, it was considered that they would have a larger potential to advance our understanding of young people's attitudes towards acceptance of GM forest tree commercialization, due to their higher education level. Recent research has shown that the level of education is a much more important factor in determining people's acceptance of GMOs compared to political and/or religious convictions or other demographic factors (Pew Research Center 2015).

Selection of students from different fields of study was decided in order to enhance the data analysis possibilities, particularly in terms of exploring relationships between young people's potential acceptance of GM forest tree cultivation and field of study and/or the year of studies attendance. Furthermore, university students have been widely used as subjects of research also in many other studies, particularly in the fields of social psychology, marketing and consumer research (Peterson 2001; Druckman and Kam 2009).

A questionnaire was designed for this survey with questions organized in four sections (i) socio-demographic questions, (ii) questions of yes/no type related to knowledge about transgenic forest trees, (iii) questions related to the nature of safety concerns about the cultivation of transgenic forest trees and (iv) questions of yes/no type concerning issues relevant to public acceptance of the cultivation of transgenic forest trees.

Two questions of Section 3 included a number of options with a qualitative rating scale of a level of importance in a country context. One of the questions of Section 3 in particular prompted evaluation of different types of potential benefits deriving from the use of transgenic forest crops. For this evaluation, a four-level rating scale was used: "very important", "slightly important", "not important" and "I do not know". Listed potential benefits of GM forest trees included the reduced need for chemicals and energy to process cellulose from wood, the harvesting of a smaller number of trees for consumption, the reduced use of insecticides, pesticides and herbicides, the restoration of contaminated soils, reduction in old growth logging, higher pulping efficiency, better timber quality, more efficient biofuel production, stronger timber for construction and the potential for higher tree productivity.

Another question of this section asked to evaluate options from a list of potential risks related to the release of GM forest trees. Options included the risk of GM forest trees to be less fit, more vulnerable to viral diseases, generate higher rates of soil decomposition, become pesticide resistant, increase the use of broad spectrum pesticides, lead to the loss of biodiversity due to gene flow between transgenic and wild trees, generate adverse effects on biotrophic processes of host ecosystems, increase the costs of controlling pest outbreaks and force cultural adaptation to changing biodiversity conditions due to transgene escape. For this question, a four-level rating scale was used presenting the options "serious hazard, "slight hazard", "no hazard" and "I do not know".

The questionnaires were translated in the national languages of the participating countries and were handed to students, who completed them on-site, following all the necessary clarifications on the questions' technical aspects. A total of 1920 questionnaires were finally collected from across 16 countries located in four continents, Europe (13 countries), South America (1 country), Australia (1 country) and West Asia (1 country) and subjected to statistical analysis. European participants were weighted towards countries in south-eastern, Central and south-west Europe. The number of respondents by country of origin is portrayed in Fig. 1.

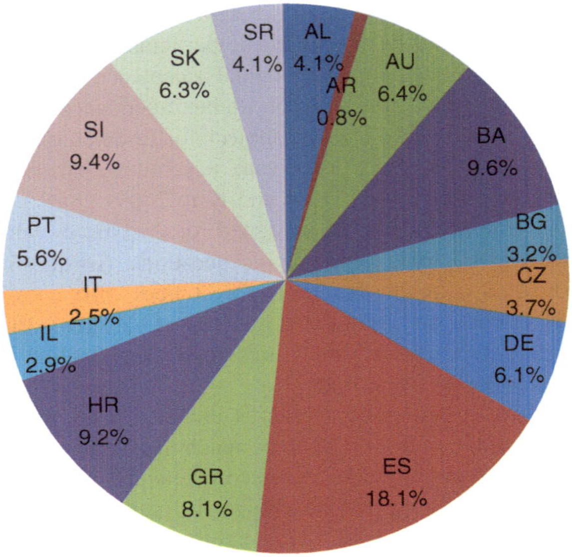

Fig. 1 Distribution of respondents by country of origin. (*AL*) Albania; (*AR*) Argentina; (*AU*) Australia; (*BA*) Bosnia and Herzegovina; (*BG*) Bulgaria; (*HR*) Croatia; (*CZ*) Czech Republic; (*DE*) Germany; (*GR*) Greece; (*IL*) Israel; (*IT*) Italy; (*PT*) Portugal; (*SR*) Serbia; (*SK*) Slovakia; (*SI*) Slovenia; (*ES*) Spain

3 Results and Discussion

3.1 Respondents' Demographic Information

The respondents' demographic characteristics, in particular gender and average age by country of origin are displayed in Table 1. In general, the age of respondents by country ranged from 20 to 27 years of age (all were university students) but was somewhat higher in Argentina (31) and Bulgaria (35), because postgraduate students were included in the sample population for these countries.

3.2 Respondents' Knowledge of GM Forest Trees

The detailed sample results for cross-country comparison of respondents' positive answers about their knowledge of the meaning of GM forest trees, the commercial cultivation of GM forest trees and the market availability of GM forest tree products are portrayed in Fig. 2. More than 60 % of respondents in all countries stated that they knew the meaning of transgenic forest trees, with the exception of Israel, where almost 60 % stated that they did not know what was meant by this term (Fig. 2a).

The highest percentage of respondents, who stated that they knew what transgenic forest trees were, was recorded in Argentina (100 %).

A very high percentage of positive responses (95 %) were also recorded in Portugal. Overall, the high degree of the knowledge to the meaning of GM forest

Table 1 Respondents' demographic profile by country of origin

Countries	Gender and average age		Years
	Male (%)	Female (%)	
Albania	34	66	23
Argentina	47	53	31
Australia	45	55	23
Bosnia and Herzegovina	84	16	22
Bulgaria	44	56	35
Croatia	37	63	20
Czech Republic	39	61	22
Germany	43	57	24
Greece	46	54	22
Israel	55	45	27
Italy	50	50	23
Portugal	29	71	22
Serbia	75	25	20
Slovakia	37	63	22
Slovenia	37	63	21
Spain	43	57	22

trees is attributed to the subjects' high level of education, since they were all university students.

In contrast, more than half of the respondents across most countries stated that they did not know whether GM forest plantations were grown commercially (Fig. 2b). The lowest number of positive responses to this question was recorded in Australia (26 %) and Israel (22 %). In most of the remaining surveyed countries, the number of positive responses was higher ranging from 30 % to around 50 % of the sample population. More than 50 % of the cohort responded positively to this question in Argentina, Portugal, Albania and the Czech Republic (Fig. 2c). Furthermore, about 50 % or more of the respondents in most countries declared that they did not know whether any final products from GM forest plantations, such as wood, biofuel, pulp and paper, were sold on the market. The highest number of positive responses with regard to this question was recorded in Spain and Albania (65 % in both countries), while the lowest level of awareness was recorded in Israel (13 %) and the Czech Republic (29 %).

3.3 Respondents' Attitude About the Importance of the GM Forest Tree Benefits

The response ratings expressed as percentage of the cohort for each of the prompted potential GM forest tree benefits classified as "very important", "slightly important" and "not important" by country of origin are reported in Table 2 under columns A, B and C respectively. In almost 40 % of the surveyed countries about half of the

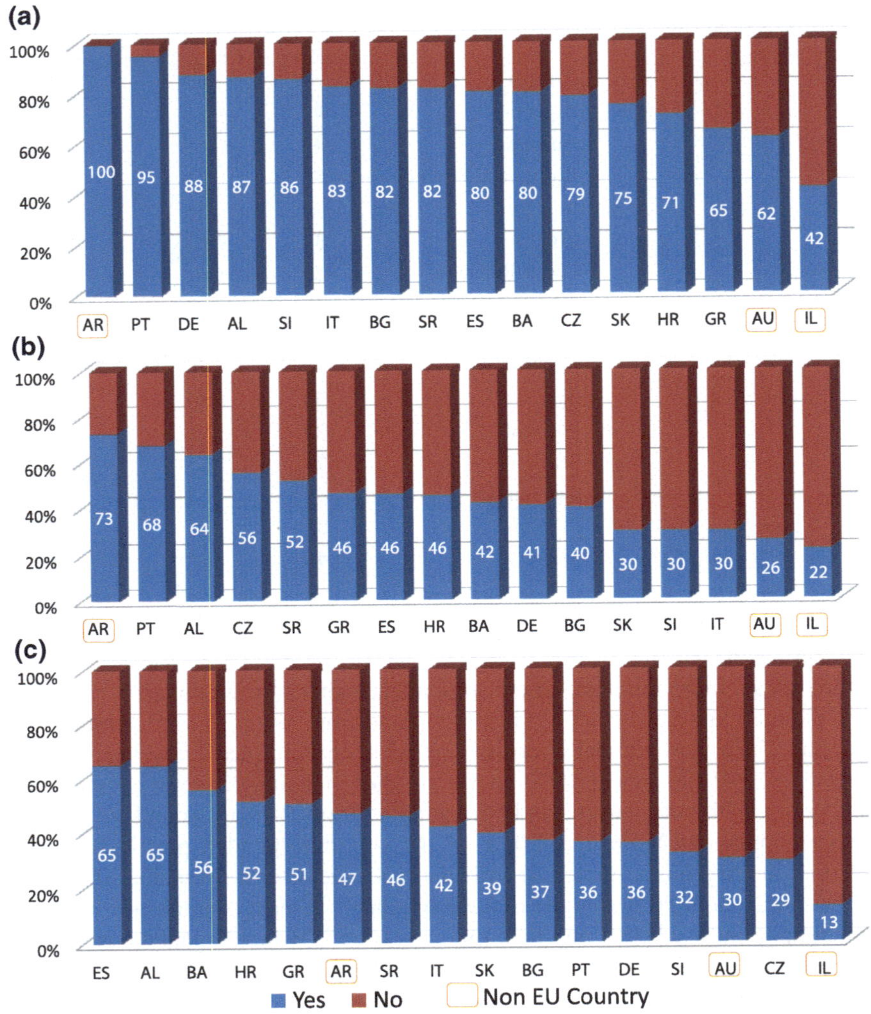

Fig. 2 Respondents' responses on knowledge of GM forest trees **a** meaning **b** commercial cultivation and **c** market availability of products. (*AL*) Albania; (*AR*) Argentina; (*AU*) Australia; (*BA*) Bosnia and Herzegovina; (*BG*) Bulgaria; (*HR*) Croatia; (*CZ*) Czech Republic; (*DE*) Germany; (*GR*) Greece; (*IL*) Israel; (*IT*) Italy; (*PT*) Portugal; (*SR*) Serbia; (*SK*) Slovakia; (*SI*) Slovenia; (*ES*) Spain

respondents—even more in a few cases—stated that they had no knowledge of the potential benefits of GM forest plantations. Therefore, they were not able to rate the importance of the prompted potential benefits in their country context. The detailed distribution of the respondents' "do not know" answers for each potential benefit by country of origin is displayed also in Table 2 under column D for each of the recorded GM forest tree benefits.

Table 2 Distribution of respondents' positive responses about GM forest tree benefits by country of origin (% of cohort)

Country	B1				B2				B3				B4				B5			
	A	B	C	D	A	B	C	D	A	B	C	D	A	B	C	D	A	B	C	D
AL	83	3	5	9	74	23	3	0	78	18	4	0	74	18	4	4	76	9	12	4
AR	13	0	27	60	20	0	27	53	0	0	33	67	7	0	40	53	7	0	13	80
AU	47	39	6	8	57	30	7	5	57	35	3	5	51	36	9	4	65	24	6	6
BA	54	33	4	9	63	21	9	7	68	26	3	3	57	33	6	4	39	39	10	12
BG	59	20	5	15	36	38	7	20	66	20	9	5	62	24	7	7	69	17	3	10
HR	9	3	29	59	3	5	27	65	3	3	23	70	7	4	36	53	8	12	49	32
CZ	10	3	33	54	4	3	30	63	1	3	26	70	4	7	27	61	4	4	27	64
GE	45	46	4	4	56	30	9	5	59	30	7	4	35	47	9	9	48	33	9	11
GR	60	29	5	6	57	32	8	3	70	21	6	3	47	38	5	10	62	28	5	5
IL	11	11	48	30	4	17	37	43	2	13	24	61	4	9	39	48	2	4	22	72
IT	52	38	6	4	49	45	4	2	81	19	0	0	73	25	2	0	84	8	4	4
PT	46	46	2	6	55	32	7	6	45	47	4	4	44	46	6	4	66	25	6	4
SR	67	26	1	6	69	23	4	4	74	17	3	6	67	21	6	6	40	37	14	9
SK	3	2	32	63	1	1	13	85	5	10	32	53	8	8	36	48	6	8	28	59
SL	4	4	37	54	2	13	37	48	4	8	21	67	5	12	31	52	6	6	37	52
ES	67	20	5	8	43	29	13	16	76	16	7	2	32	37	23	8	77	10	6	7

(continued)

Table 2 (continued)

Country	B6				B7				B8				B9				B10				B11			
	A	B	C	D	A	B	C	D	A	B	C	D	A	B	C	D	A	B	C	D	A	B	C	D
AL	60	24	8	8	64	30	6	0	56	36	8	0	49	22	22	8	68	18	15	0	69	17	14	0
AR	7	0	40	53	13	0	20	67	7	7	33	53	13	7	40	40	13	0	33	53	0	0	33	67
AU	48	30	13	9	35	44	8	13	48	37	9	6	48	28	14	10	43	38	12	7	60	31	4	5
BA	62	23	9	6	21	41	15	23	70	19	6	5	53	26	9	12	56	29	7	8	75	18	2	5
BG	0	0	0	0	50	26	12	12	19	42	12	27	62	21	5	12	46	35	5	14	61	19	18	2
HR	9	11	39	40	7	6	34	53	24	16	40	20	7	8	32	52	5	10	36	49	6	4	33	57
CZ	7	6	33	54	23	10	46	21	4	3	31	61	9	21	36	34	6	11	53	30	1	3	30	66
GE	20	35	15	30	26	46	9	19	32	42	13	13	48	32	16	4	25	39	19	17	47	40	9	5
GR	40	42	9	9	26	49	16	9	41	38	18	3	31	39	23	7	31	42	21	6	59	25	13	3
IL	2	6	26	67	2	26	33	39	7	26	39	28	8	25	36	32	6	24	41	30	7	15	41	37
IT	52	30	16	2	21	47	16	16	27	29	40	4	40	41	17	2	37	40	18	5	65	20	6	9
PT	17	42	28	13	25	56	8	11	33	47	12	8	43	38	10	9	19	52	18	11	44	46	5	6
SR	53	30	5	12	29	41	12	18	65	21	6	8	47	36	8	9	55	30	10	5	72	18	2	8
SK	3	7	24	66	2	8	42	49	21	11	50	18	10	15	33	42	7	15	45	33	7	3	38	52
SL	7	20	43	30	2	5	30	64	3	9	52	35	5	17	42	36	2	8	34	56	3	6	41	51
ES	27	36	20	17	44	37	13	7	48	30	12	10	50	25	13	12	43	36	15	5	60	23	11	6

A Very important, *B* Slightly important, *C* Not important, *D* I do not know

B1 Less chemicals and energy to process cellulose; *B2* Harvesting of a smaller number of trees for consumption needs; *B3* Less pesticides in for-est plantations; *B4* Less herbicide treatments of forest plantations, *B5* Restoration of contaminated soil; *B6* Less old growth logging; *B7* Higher pulping efficiency; *B8* Better timber quality/higher value product; *B9* More efficient biofuel production from GM forest trees; *B10* Stronger timber construction materials, *B11* Higher tree productivity

AL Albania; *AR* Argentina; *AU* Australia; *BA* Bosnia and Herzegovina; *BG* Bulgaria; *HR* Croatia; *CZ* Czech Republic; *DE* Germany; *GR* Greece; *IL* Israel; *IT* Italy; *PT* Portugal; *SR* Serbia; *SK* Slovakia; *SI* Slovenia; *ES* Spain

Table 3 Overall importance indices of GM forest tree benefits (all countries)

Benefits	Very important	Slightly important	Not important
B1	1.89	0.65	0.25
B2	1.78	0.68	0.24
B3	2.07	0.57	0.20
B4	1.73	0.73	0.29
B5	1.97	0.52	0.25
B6	1.24	0.68	0.33
B7	1.17	0.94	0.32
B8	1.52	0.83	0.38
B9	1.57	0.80	0.36
B10	1.38	0.86	0.38
B11	1.91	0.58	0.30

The remaining respondents, who rated the prompted potential benefits of GM forest plantations in almost half of the surveyed countries, rated as very important for their countries the potential lower demand of the GM forest plantations for pesticides (B3), the potential higher GM forest tree productivity (B11), the lower demand for chemicals and energy to process cellulose if GM forest trees will be used (B1), the need to harvest a smaller number of trees for consumption needs, if GM forest trees will be used (B2) and the potential of GM forest trees for restoration of contaminated soils (B5). Four of the five potential GM forest tree benefits that were rated as very important by the respondents of the survey were environmental in context and one was economic.

To rank the potential GM forest tree benefits an overall importance index was calculated for each potential benefit and each rating scale for the cohort from all the participating in the survey countries following a weighting normalizing procedure.

The results are displayed in Table 3. It is evident that the overall ratings of the potential GM forest tree benefits rated as very important were much higher than those rated as slightly important. On the other hand, the overall ratings of the potential GM forest tree benefits rated as not important were very low.

The top three potential benefits that were rated as very important in all countries involved the lower demand for pesticides in the GM forest plantations, the potential of GM forest trees for restoration of contaminated soils and the potential high productivity of GM forest trees.

3.4 *Respondents' Attitude About the Importance of the GM Forest Tree Risks*

In contrast with the general pattern recorded in the cohort of students regarding the knowledge of potential benefits of GM forest plantations, in most of the surveyed countries the respondents appeared to be more aware of the potential risks of GM forest plantations and to be able to rate the importance of the prompted potential risks in their country context. This is evident in the detailed distribution of the respondents' "do not know" answers for each potential risk by country of origin displayed in Table 4. The respondents who rated the prompted potential risks of GM forest plantations in almost half of the surveyed countries stated that the loss of biodiversity (R6), the adverse effects of biotrophic processes on host ecosystems (R7), the higher vulnerability of GM forest trees to viral diseases (R2), the increased use of broad spectrum herbicides (R5) and the cultural adaptation to changing biodiversity conditions (R9) would constitute potential serious hazards for their countries. From the five potential GM forest tree risks rated as serious hazards by the respondents of the survey four were environmental and one social in context. The percent cohort response ratings for each of the prompted potential GM forest tree risks rated as "serious hazard", "slight hazard" and "no hazard" by country of origin are reported in Table 4 under columns A, B and C respectively.

The overall importance indices of each of the potential GM forest tree risks and each rating scale were calculated for the total population sample from all the participating countries in the survey following a weighting normalizing procedure in order to rank the potential risks. The results are displayed in Table 5. It is evident that the overall ratings of the potential GM forest tree risks perceived as serious hazards were higher than those perceived as slight hazards, although for some risks the importance ratings are quite close. On the other hand, the overall ratings of the potential GM forest tree risks perceived as no hazards were very low compared with the ratings of the other two categories. The top three GM forest tree risks that were perceived as serious hazards, in all countries, involved the loss of biodiversity, the adverse effect of biotrophic processes on host ecosystems and the cultural adaptation to changing biodiversity conditions.

3.5 Respondents' Attitudes Towards Acceptance of GM Forest Tree Cultivation

The distribution of positive responses by country of origin, with respect to the four questions concerning: (a) agreement to the approval of GM forest tree commercial cultivation, (b) willingness to purchase forest products of GM tree origin, (c) preferences towards GM forest tree labelling policy and (d) agreement with the mandatory labelling of forest products of GM origin are provided in Figs. 3, 4 and 5. The respondents appeared very positive about the approval of GM forest trees being planted in plantations, excluding natural forests for all the countries that were involved in this survey. Even in the countries with the lowest numbers of positive responses to the relevant question, such as Bulgaria, Croatia, Greece and Serbia, almost 60 % of the respondents would agree with permission to be granted for commercial GM forest tree plantations.

Table 4 Distribution of respondents' positive responses about GM forest tree risks by country of origin (% of cohort)

Country	B1				B2				B3				B4				B5			
	A	B	C	D	A	B	C	D	A	B	C	D	A	B	C	D	A	B	C	D
AL	28	56	10	5	65	29	7	0	54	31	4	12	26	39	31	4	60	27	9	4
AR	40	33	20	7	13	20	33	33	40	13	13	33	13	13	40	33	13	20	13	53
AU	24	53	8	15	57	30	5	8	45	37	6	12	42	32	15	11	51	28	9	12
BA	37	52	5	6	59	30	4	7	17	46	20	17	33	35	20	12	51	32	8	9
BG	45	27	5	23	48	27	7	19	30	40	12	18	29	29	24	18	46	34	5	15
HR	7	3	55	36	6	1	32	61	9	15	46	31	8	16	49	27	10	2	33	55
CZ	1	17	57	24	6	13	50	31	9	56	30	6	4	28	45	23	13	9	39	40
GE	18	49	22	11	24	55	15	6	11	55	21	14	23	50	21	6	33	35	18	14
GR	61	30	5	4	52	35	5	8	41	38	11	10	35	45	16	4	49	30	12	9
IL	11	43	30	17	7	30	41	22	11	24	32	33	13	53	25	9	13	30	41	17
IT	46	36	9	9	62	25	5	8	33	38	16	13	27	48	17	8	59	27	4	10
PT	65	14	10	11	31	62	6	2	29	55	8	8	35	48	14	3	23	70	4	3
SR	45	42	5	8	58	33	1	8	31	39	23	7	41	31	14	14	59	26	11	4
SK	8	13	56	24	10	8	42	41	18	13	43	26	12	8	46	34	13	10	33	45
SL	5	11	54	30	2	11	47	41	9	31	42	18	3	19	41	37	6	8	34	52
ES	13	53	17	17	51	37	6	5	58	31	3	8	44	39	10	8	64	24	4	9

(continued)

Table 4 (continued)

Country	B6				B7				B8				B9			
	A	B	C	D	A	B	C	D	A	B	C	D	A	B	C	D
AL	80	17	4	0	59	31	6	4	58	35	8	0	53	35	6	6
AR	13	13	20	53	33	40	27	0	27	27	47	0	33	20	13	33
AU	65	29	2	4	56	28	3	13	39	45	7	10	32	43	12	13
BA	72	17	4	7	49	31	8	12	40	37	9	14	33	38	10	19
BG	71	20	2	7	63	22	0	15	41	35	7	17	44	31	7	19
HR	2	2	14	82	9	1	27	63	12	9	43	37	15	8	34	43
CZ	3	11	20	66	17	11	37	34	7	21	53	19	14	19	37	30
GE	51	31	16	3	38	42	7	13	28	42	19	10	26	43	21	10
GR	65	25	5	5	54	27	7	12	45	35	10	10	42	38	10	10
IL	6	11	35	48	7	11	37	44	13	32	35	20	20	32	33	15
IT	90	10	0	0	77	17		6	37	48	13	2	88	12	0	0
PT	10	88	1	1	28	69	1	2	44	46	7	4	40	50	7	3
SR	76	14	7	3	57	30	5	8	42	32	17	9	45	35	9	11
SK	5	4	26	64	8	8	35	48	5	14	45	36	15	14	37	34
SL	2	9	17	72	10	13	31	46	6	16	53	25	11	27	38	24
ES	86	9	1	3	55	29	4	12	35	43	14	9	80	13	3	4

A Serious hazard, *B* Slight hazard, *C* No hazard, *D* I do not know

R1 Forest trees less fit, *R2* Forest trees more vulnerable to viral diseases, *R3* Higher rates of soil decomposition, *R4* More pesticide resistant forest species, *R5* More use of broad spectrum herbicides, *R6* Loss of biodiversity, *R7* Adverse effects on bio-trophic processes of host ecosystems, *R8* Increased cost of controlling pest outbreaks, *R9* Cultural adaptation to changing biodiversity conditions

AL Albania; *AR* Argentina; *AU* Australia; *BA* Bosnia and Herzegovina; *BG* Bulgaria; *HR* Croatia; *CZ* Czech Republic; *DE* Germany; *GR* Greece; *IL* Israel; *IT* Italy; *PT* Portugal; *SR* Serbia; *SK* Slovakia; *SI* Slovenia; *ES* Spain

Table 5 Overall importance indices of GM forest tree risks (all countries)

Risks	Serious hazard	Slight hazard	No hazard
R1	1.36	1.06	0.37
R2	1.65	0.89	0.31
R3	1.33	1.12	0.33
R4	1.17	1.06	0.43
R5	1.69	0.82	0.28
R6	2.09	0.62	0.17
R7	1.86	0.82	0.23
R8	1.44	1.09	0.39
R9	1.77	0.91	0.28

R1 Forest trees less fit, *R2* Forest trees more vulnerable to viral diseases, *R3* Higher rates of soil decomposition, *R4* More pesticide resistant forest species, *R5* More use of broad spectrum herbicides, *R6* Loss of biodiversity, *R7* Adverse effects on biotrophic processes of host ecosystems, *R8* Increased cost of controlling pest outbreaks, *R9* Cultural adaptation to changing biodiversity conditions

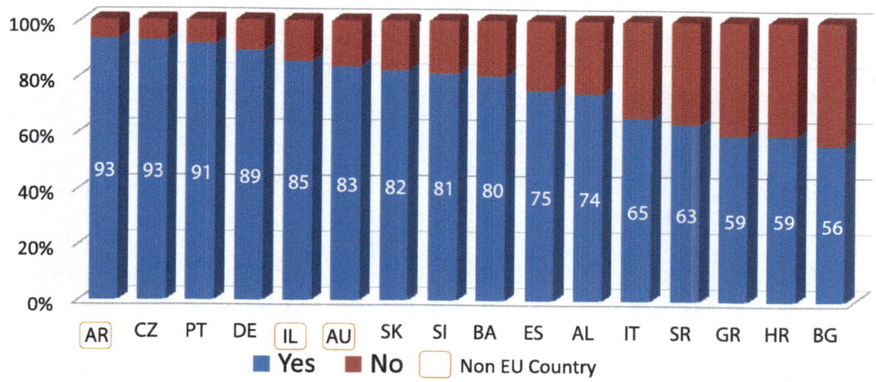

Fig. 3 Distribution of respondents' positive responses by country of origin regarding their agreement with the approval of commercial cultivation of GM forest trees. (*AL*) Albania; (*AR*) Argentina; (*AU*) Australia; (*BA*) Bosnia and Herzegovina; (*BG*) Bulgaria; (*HR*) Croatia; (*CZ*) Czech Republic; (*DE*) Germany; (*GR*) Greece; (*IL*) Israel; (*IT*) Italy; (*PT*) Portugal; (*SR*) Serbia; (*SK*) Slovakia; (*SI*) Slovenia; (*ES*) Spain

The majority of respondents also appeared willing to buy products from such plantations, such as wood products or pulp and paper. These findings were comparable with the results of a recent study, which attempted to explore consumers' potential buying behaviour towards transgenic forest products in Greece (Tsourgiannis et al. 2015).

Over 80 % of respondents from all countries were in favour of using labelling to identify products of GM origin. The great majority of those respondents, also over 80 % in all countries, argued that this labelling should be legally mandatory.

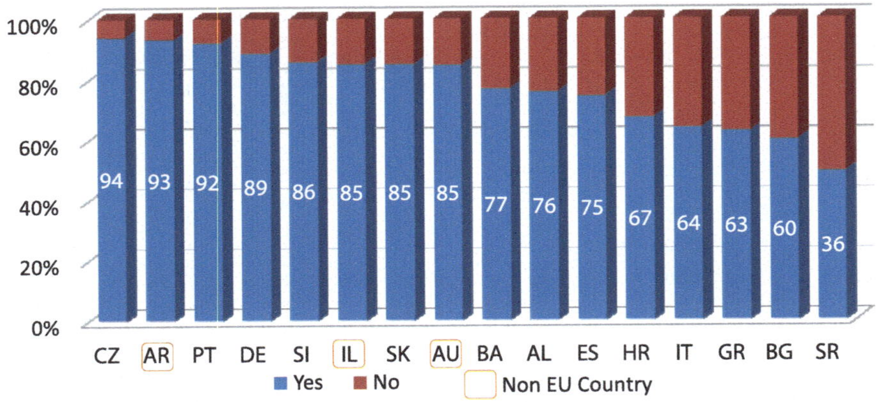

Fig. 4 Distribution of respondents' positive responses regarding their willingness to purchase GM forest tree products by country of origin. (*AL*) Albania; (*AR*) Argentina; (*AU*) Australia; (*BA*) Bosnia and Herzegovina; (*BG*) Bulgaria; (*HR*) Croatia; (*CZ*) Czech Republic; (*DE*) Germany; (*GR*) Greece; (*IL*) Israel; (*IT*) Italy; (*PT*) Portugal; (*SR*) Serbia; (*SK*) Slovakia; (*SI*) Slovenia; (*ES*) Spain

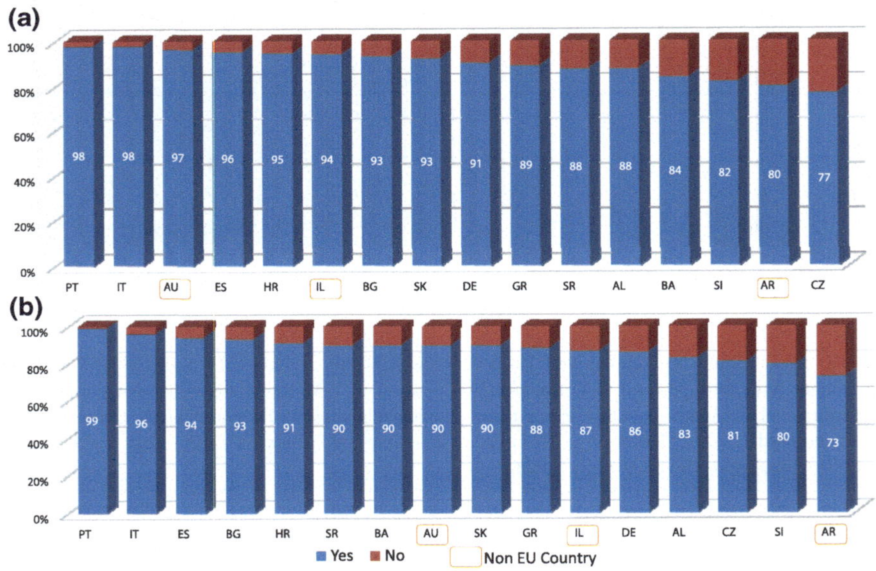

Fig. 5 Distribution of respondents' positive responses by country of origin regarding their agreement with A) the labelling policy and B) the mandatory labelling of forest products of GM origin. (*AL*) Albania; (*AR*) Argentina; (*AU*) Australia; (*BA*) Bosnia and Herzegovina; (*BG*) Bulgaria; (*HR*) Croatia; (*CZ*) Czech Republic; (*DE*) Germany; (*GR*) Greece; (*IL*) Israel; (*IT*) Italy; (*PT*) Portugal; (*SR*) Serbia; (*SK*) Slovakia; (*SI*) Slovenia; (*ES*) Spain

4 Conclusions

In this article the preliminary results of a KAP cross-country pilot study on the public knowledge, perceptions and attitudes towards the potential risks and benefits related to the GM forest tree cultivation are presented. The results were based on responses from young, educated people aged 20–35 years. The respondents were university students of different fields of study from 16 countries including Albania, Bosnia & Herzegovina, Bulgaria, Croatia, Czech Republic, Germany, Greece, Italy, Serbia, Slovakia, Slovenia, Spain and Portugal located in Europe, Argentina located in South America, Israel in West Asia and Australia.

The results of our KAP pilot survey constitute a significant contribution to the establishment and advancement of our understanding regarding young people's knowledge, perceptions and attitudes towards the potential risks and benefits associated with the use of GM forest trees in plantations. Several reasons add more value to this work. First, it is a pioneer cross-country pilot study that has brought to publicity information on issues related to public knowledge, perceptions and attitudes towards the potential risks and benefits that may accrue if GM forest tree cultivations would be authorized. Second, the study focused on young educated people, who are expected to be the future consumers, developers or policy makers, as no commercial GM forest plantations are currently grown any where in the world except China. Third, it provided the required preliminary information on the knowledge gaps of the young population regarding the potential risks and benefits of GM forest cultivations, so as to stimulate appropriate awareness efforts. The main findings of our study can be summarized as follows.

More than half of the respondents in all countries stated that they knew the meaning of GM forest trees, but they did not know whether GM forest plantations were grown commercially or whether any final GM forest, such as wood, biofuel, pulp and paper were available in the market. However in at least half of the participating in the study countries, more than 50 % of the respondents were not familiar with the prompted potential benefits of GM forest plantations and, therefore, were not able to rate their importance in their country context. Based on the ratings of the overall importance indices of the GM forest tree benefits, three potential benefits were rated as very important in all countries. These included the potential lower demand of the GM forest plantations for pesticides, the potential of GM forest trees for restoration of contaminated soils and the potential GM forest tree higher productivity.

In contrast with the general pattern recorded in the cohort regarding the knowledge of potential benefits of GM forest plantations, most of the respondents appeared to be more aware of the potential risks of GM forest plantations and able to rate the importance of the prompted potential risks in their country context.

Based on the ratings of the overall importance indices of all the GM forest tree risks, three potential risks were perceived as the most serious hazards in all countries. These involved the loss of biodiversity due to gene flow between transgenic and wild trees, the adverse effects of biotrophic processes on host ecosystems and the cultural adaptation to changing biodiversity conditions due to transgene escape.

Overall, the young educated people that took part in the KAP study appeared to approve the GM forest tree cultivation. This might become a driver for future market adoption of GM forest tree products. The majority of respondents also appeared willing to buy products from such plantations, such as wood products or pulp and paper. Over 80 % of respondents from all countries were in favour of using labelling to identify products of GM origin, while the great majority of these (over 80 %) would prefer that this labelling be legally mandatory.

The results of the study presented in the article contribute to the improvement of the scientific basis that is required in order to establish safe tree plantations and implement appropriate policy directives. However, our study was limited in scope, particularly due to financial constraints. The main limitation concerns the focus group, which included only young educated people. There is certainly a need for more elaborated and larger scale socioeconomic research that will advance further the scientific knowledge on the general public attitudes towards the GM forest tree cultivation. Moreover, due to time constraints our data analysis was confined at this stage to the identification of the general pattern of people's perceptions and attitudes towards the potential risks and benefits of GM forest plantations. Exploration of the relationships between the acceptance of GM forest tree cultivation and other factors, such as the different field of studies, the year of study attendance, the gender and age has already begun and it is planned the new findings to be brought soon to publicity.

Finally, the findings of our study suggest that more can be done to increase peoples' awareness, particularly about the potential benefits of GM forest plantations. Any effort that should be initiated to increase public awareness, most likely it will be the driving force in determining market opportunities for GM forest tree products.

References

Aguilera J, Nielsen KM, Sweet J (2013) Risk assessment of GM trees in the EU: current regulatory framework and guidance. iForest 6:127–131

Buah JN (2011) Public perception of genetically modified food in Ghana. Am J Food Technol 6 (7):541–554

Chapotin SM, Wolt JD (2007) Genetically modified crops for the bioeconomy: meeting public and regulatory expectations. Transgenic Res 16(6):675–688

Druckman JN, Kam CD (2009) Students as experimental participants: a defense of the 'Narrow data base'. Available at SSRN: http://ssrn.com/abstract=1498843 or http://dx.doi.org/10.2139/ssrn

El-Lakany MH (2004) Are genetically modified trees a threat to forests? Unasylva 55(217):45–47 1

European Commision (2010) Europeans and biotechnology in 2010, winds of change? Luxembourg, pp 172. Available at http://ec.europa.eu/public_opinion/archives/ebs_341_winds_en.pdf

FAO (2008) The potential environmental, cultural and socio-economic impacts of genetically modified trees. UNEP/CBD/SBSTTA/13/INF/6, pp17

FAO (2010) Forests and genetically modified trees. Rome, Italy, pp 235

Farnum P, Lucier A, Meilan R (2007) Ecological and population genetics research initiatives for transgenic trees. Tree Genet Genomes 3:119–133

Ferguson CA, Chan-Halbrendt C, Wieczorek A, Wen N (2002). Results from a Hawaii opinion survey on genetically modified organisms, BIO-2, CTAHR- Nov 2002

Flachowsky H, Hanke M-V, Peil A, Strauss SH, Fladung M (2009) A review on transgenic approaches to accelerate breeding of woody plants. Plant Breed 128:217–226

Fladung M, Altosaar I, Bartsch D, Baucher M, Boscaleri F, Gallardo F, Häggman H, Hoenicka H, Nielsen K, Paffetti D, Séguin A, Stotzky G, Vettori C (2012) European discussion forum on transgenic tree biosafety. Nat Biotechnol 30:37–38

Grace D, Randolph T, Affognon H, Dramane D, Diall O, Clausen P-H (2009) Characterization and validation of farmers' knowledge and practice of cattle trypanosomosis management in the cotton-zone of West Africa. Acta Trop 111:137–143

Grice J, Wener MK, Romanach LM, Paton S, Bonaventura P, Garrad S (2003) Genetically modified sugarcane: a case for alternate products. AgBioForum 6(4): 162–168, http://www.agbioforum.org/

Goutille F (2009) Knowledge, attitudes and practices for risk education: how to implement KAP surveys. Handicap International, Lyon 84 pp

Häggman H, Sutela S, Walter C, Fladung M (2014) Biosafety considerations in the context of deployment of GE trees. In: Fenning T (ed) Challenges and opportunities for the world's forests in the 21st century. Forestry Sciences 81, Springer Science + Business Media Dordrecht, pp 491–524

Häggman H, Raybould A, Borem A, Fox T, Handley L, Hertzberg M Lu M, Macdonald P, Oguchi T, Pasquali G, Pearson L, Peter G, Quemanda H, Seguin A, Tattersall K, Ulian E, Walter C, and McLean M (2013) Genetically engineered trees for plantation forests: key considerations for environmental risk assessment. Plant Biotechnol J 1–14

Häggman H, Find JM, Pilate G, Gallardo F, Ruohonen-Lehto M, Kazana V, Migliacci F, Ionita L, Sijacic-Nikolic M, Donnarumma F, Harfouche A, Biricolti S, Glandorf B, Tsourgiannis L, Minol K, Paffetti D, Fladung M, Vettori C (2012) Biosafety of genetically modified forest trees (GMTs) –COST Action FP0905- a common action of European scientists. In: 2nd international conference of the IUFRO working party 2.09.02. Mendel lectures & Plenary MLP-3, p 13

Harfouche A, Meilan R, Altman A (2011) Tree genetic engineering and applications to sustainable forestry and biomass production. Trends Biotechnol 29(1):11–17

Hinchee M, Rottman W, Mullinax L, Zhang C, Chang S, Cunningham M, Pearson L, Nehra N (2009) Short-rotation woody crops for bioenergy and biofuels applications. Vitro Cell Dev Biol Plant 45(6):619–629

Hossain F, Oryango B, Adelaja A, Schilling B, Hallman W (2002). Public perceptions of biotechnology and acceptance of genetically modified food. Food Policy Institute Publication No. WP-0602–002

Magnusson M (2004) Consumer perception of organic and genetically modified foods. Acta Universitatis Upsaliensis. Comprehensive summaries of Uppsala dissertations from the faculty of social sciences 137, 71 pp, Uppsala, ISBN: 91-554-5935-8

Pereira de Abreu DA, Rodriguez KV, Schroeder M, Mosqueda MB, Perez E (2006) GMO technology. Venezuelans' consumers perceptions: situation in Caracas. J Technol Manage Innov 1(5): 80–86

Peterson RA (2001) On the use of College students in social science research: insights from a second order Meta-analysis. J Consum Res 28(3):450–461

Peterson G, Cunningham S, Deatch L, Erickson J, Quinlan A, Raez-Luna e, Tinch R, Troell M, Woodbury P, Zens C (2000) The risks and benefits of genetically modified crops: a multidisciplinary perspective. Conserv Ecol 4(1): 13 (online) URL: http://www.consecol.org/vol4/iss1/art13/

Pew Research Center (2015) Public and scientists' views on science and society. Available at: http://www.pewresearch.org

Sedjo RA (2006) Toward commercialization of genetically engineered forests: economic and social considerations. Resources for the Future, pp 46

Shehata S, Cox LJ (2007) Attitudes of Hawaii consumers toward genetically modified fruit, BIO-7, UH-CTAHR- Apr 2007

Siwakoti S (2009) Knowledge, attitudes and practices of women and men towards recycling in North St. James Town, Toronto, St James town initiative, Wellesley Institute, http://www.welesleyinstitute.com/wp-content/uploads/2012/07/Report-20090401-Knowledge-Attitudes-and-Practices-Towards-Recyling-in-North-St.-James-Town.pdf

Tsourgiannis L, Kazana V, Iakovoglou V (2015) Exploring consumers' potential behavior towards transgenic forest products: The Greek experience. iForest (in print)

Valenzuela S, Balocchi C, Rodriguez J (2006) Transgenic trees and forest biosafety. Electron J Biotechnol 9(3) on line at http://www.ejbiotechnology.info/content/vol9/issue3/full/22/

Van Frankenhuyzen K, Beardmore T (2004) Current status and environmental impact of transgenic forest trees. Can J For Res 34:1163–1180

Vettori C, Pilate G, Häggman H, Gallardo F, Ionita L, Ruohonen-Lehto M, Glandorf B, Harfouche A, Biricolti S, Paffetti D, Kazana V, Sijacic-Nikolic M, Tsourgiannis L, Migliacci F, Donnarumma F, Minol K, Fladung M (2014) COST action FP0905: biosafety of forest transgenic trees. In: Mérillon J-M, Ahuja MR (eds) Ramawat KG tree biotechnology. CRC Press, Boca Raton, pp 112–124

Vivas A, Gelaye B, Aboset N, Kurmic A, Berhane Y, Williams MA (2010) Knowledge, attitudes, and practices (KAP) of hygiene among school children in Angolela, Ethiopia. J Prev Med Hyg 51(2):73–79

Williams CG, Davis BH (2005) Rate of transgene spread via long-distance seed dispersal in *Pinus taeda*. For Ecol Manage 21:95–102

Williams GC (2006) Opening Pandora's box: governance for genetically modified forests, ISB News Report, January 2006

A Comparative Analysis of Consumers' Potential Purchasing Behaviour Towards Transgenic-Derived Forest Products: The Greek Case

Lambros Tsourgiannis, Vassiliki Kazana and Valasia Iakovoglou

Abstract Transgenic forest products are not currently purchased in the markets mainly due to lack of transgenic forest plantations. This is partly associated with the early stage of development of GM trees and also due to biosafety concerns and the existing regulation frameworks, particularly in Europe. However, due to the increased global demand for wood products and bioenergy applications, there is an increased interest in the improved characteristics associated with transgenic forest plantations. In order to determine the future demand and the market potential for transgenic forest products there is a need for scientific information for policy makers and developers about the potential purchasing behaviour of consumers. A study was conducted of the potential purchasing behaviour of consumers to three major categories of transgenic-derived forest products: (a) paper, (b) woody biomass energy and (c) wood products. In particular the study, (i) examined the main factors influencing consumers' potential purchasing behaviour towards the three types of transgenic forest products, (ii) classified consumer groups with similar buying behaviour towards each category of transgenic forest products and (iii) profiled each consumer group according to their attitudes towards the use of biotechnology in the forestry sector. Field interviews were conducted in a randomly selected sample consisting of 418 consumers throughout Greece in 2011. Principal Components Analysis (PCA) was performed to identify the main factors that might have affected the consumers' potential purchasing behaviour towards the three main categories of products. Further, Discriminant Analysis was implemented to assess how these factors could predict cluster membership, while the Hierarchical and

L. Tsourgiannis (✉)
Directorate of Public Health and Social Care of Regional District of Xanthi, Region of Eastern Macedonia and Thrace, 67100 Xanthi, Greece
e-mail: ltsourgiannis@gmail.com

V. Kazana · V. Iakovoglou
Eastern Macedonia and Thrace Institute of Technology, Department of Forestry and Natural Environment Management, 1st km Drama-Mikrohori, 66100 Drama, Greece
e-mail: vkazana@gmail.com

V. Iakovoglou
e-mail: viakovoglou@yahoo.gr

C. Vettori et al. (eds.), *Biosafety of Forest Transgenic Trees*,
Forestry Sciences 82, DOI 10.1007/978-94-017-7531-1_13

non-hierarchical cluster techniques were employed to classify the consumers with similar behaviour for each transgenic product. This study indicated that in Greece there are potential buyers for transgenic-derived products, such as woody biomass for energy production. Marketing issues such as the attractiveness of the packaging could positively influence consumers' potential purchasing decisions towards those products. However, there were concerns such as the possible negative health issues from the transgenic-derived products. This study indicated the potential purchasing behaviour of consumers and is of value to developers and policy makers. It is recommended that the results of this pioneer study should be used to stimulate further investigation, both in Greece and other parts of Europe.

1 Introduction

The global consumption of forest products, such as wood, paper and woody biomass energy products expanded rapidly between 1965 and 2007 (UNECE 2012). The annual average growth rate for paper products consumption over this period was approximately 3 %. Despite the fact that since 2008 when the economic crisis started, an approximate decline of 2–3 % was observed in the USA, Canada and the EU countries, the use of paper products increased by 5–10 % in Russia, China and the countries of South America (Jonsson 2012; UNECE 2012). Econometric modelling indicated that demand for paper products for countries such Russia and China will keep increasing even more (Jonsson 2012). Forest wood biomass also has been used to address multiple energy needs in the form of firewood, chips, charcoal, briquettes and pellets, as well as feedstock needs for the industry biofuels (Hinchee et al. 2009; Sedjo 2010; Harfouche et al. 2011). However, long-term economic feasibility for forest wood biomass utilization for energy production depends to a great extent on its productivity rate. This should reach 8–10 dry tons/acre/year for industrial energy applications (Hinchee et al. 2009). Forest trees are in general slow growing, with low productivity rates and conventional breeding and selection is unlikely to greatly increase their productivity or utilization efficiency.

Under these circumstances, much discussion by plant breeders and the industrial private sector has focused on the potential of transgenic forest trees to help meet the projected increased worldwide demand for forest products, such as wood, paper and energy associated with woody biomass (Gartland et al. 2003; Van Frankenhuyzen and Beardmore 2004; Carman et al. 2006; Sedjo 2006, 2010; FAO 2008, 2010; Zhu and Pan 2010; Harfouche et al. 2011, www.cost-action-fp0905.eu).

Transgenic forest trees are genetically modified (GM) or genetically engineered (GE) trees that through DNA insertion from other species manifest specific traits (www.forestguild.org). Such traits mainly concern higher growth rates, reduced lignin levels and increased resistance to herbicides and/or pests (Harfouche et al. 2011). However, despite the fact that for the past 20 years much research has been conducted on field trials for transgenic forest trees, there is no current commercial production of GM trees in Europe, the USA or other parts of the world, with China the only exception (Häggman et al. 2012).

There are three main elements that are considered important for the commercialization of GM forest trees: (i) technical issues, (ii) the regulatory framework and (iii) public acceptance. Technical issues concern the required biotechnology research to improve traits, such as growth, quality, site adaptability and stress tolerance, to improve GM forest plantations productivity and use efficiency, particularly for bioenergy applications. Considerable progress has been achieved already in this direction, with a limited number of forest species (poplar, eucalyptus and pines) that have been grown as experimental crops in North and South America and some countries of Northern and Central Europe. Also, the private sector shows increased interest in funding research related to GM forest trees. For example, ArborGen, which is a private company, has developed three varieties of GM forest trees: fast growing pines (*Pinus taeda*) and freeze-tolerant *Eucalyptus* species for the southern USA and low lignin *Eucalyptus* species for plantations in South America (Farnum et al. 2007; Hinchee et al. 2009).

Regulatory frameworks are established in relation to the biosafety and risk assessment that is required prior to commercialization that differs among countries. In the European Union Directive 2001/18/EC regulates releases of GM forest trees at 2 levels: (i) releases for placing the product in the EU markets and (ii) releases for experimental purposes (field trials). In the first case, applications must include an Environmental Risk Assessment, whose structure is outlined in the Directive (Aguilera et al. 2013). Different legal documents regulate all aspects of GMO's, such as safety assessments, labelling, marketing, transport, sampling, detection and post-marketing environmental monitoring. Applications for experimental field trials require assessments only at a national level. All the GMO applications in the EU for marketing and/or cultivation are assessed by the European Food Safety Authority (EFSA). Overall, regulations in the EU are aiming at providing high levels of protection for human health and the environment, while ensuring the consumer's interests and the promotion of competitive market. However, political interference with the decision making and authorization of GM plants is the main reason that applications take a long time to be processed and licenses granted for deliberate release of GM plants. This is likely to cause considerable problems for the approval of GM trees.

Public acceptance as related to the consumers' consumption behaviour towards transgenic-derived forest products reflects the public perspective regarding the utilization of other GM plants. These perspectives involve environmental concerns, such as the potential spread of trans genes from GM trees to other species, the potential for spatial and temporal long-distance pollen spread from long-lived trees, as well as the possible adverse effects on biodiversity levels and scientific uncertainty (Gartland et al. 2003; Sedjo 2006; FAO 2008).

Empirical studies have shown that the consumers' buying attitudes towards forest products are influenced by factors, such as cost and availability (DuPuis 2000). Moreover, consumers' purchasing behaviour is dynamic and is influenced by many factors including information, social and cultural norms, beliefs, values and perceptions (Ajzen 1991; Ajzen and Fishbein 1980; Beckett and Nayak 2008).

Claims made by respected people such as opinion leaders contributed significantly to the development of consumers' attitudes towards a certain product (Ajzen 1991; DuPuis 2000). Therefore, information that consumers might receive through labeling, branding or other promotional and marketing efforts can influence their response towards forest products (Tokarczyk and Hansen 2006; Ajzen 1991).

There are numerous studies that have been carried out on the consumers' attitudes towards GMOs, but a very limited number for transgenic forest products. Some evidence comes from the USA, where a study indicated that the industrial users and processors of wood and timber were generally satisfied to use transgenic forest trees. By contrast, consumers of transgenic wood products showed higher levels of concern (Sedjo 2004). However, to our knowledge, no research has been conducted regarding the consumers' potential buying behaviour towards transgenic-derived forest products in Europe.

Although forest products from transgenic trees might not be in the market for the next 10–15 years, scientifically based information on the anticipated attitudes of consumers is extremely important both for developers and policy makers. For the developers, this type of information is very important because, without the expectation of viable markets, investment is unlikely to be forthcoming. For the policy makers this information is important, because it can help them to develop appropriate regulatory and decision-making tools.

This paper reports a pioneer survey study conducted in Greece and presents comparative results of consumers' potential buying behaviour towards three categories of forest products: paper products, wood products and woody biomass energy products of transgenic origin. Also, the study profiled the consumers according to their responses towards the future establishment of transgenic forest plantations. This study was initiated under the frame of the EU COST Action FP0905 (www.cost-action-fp0905.eu).

2 Methodology

2.1 Research Hypotheses

This study aimed to reject the following null hypotheses:

- **$H_0$1**: Consumers are not influenced by similar factors regarding their potential purchasing behaviour towards paper products, woody biomass energy products and wood products of transgenic origin.
- **$H_0$2**: Consumers cannot be classified into similar groups according to their potential purchasing behaviour towards paper products, woody biomass energy products and wood products of transgenic origin.
- **$H_0$3**: Consumers' opinions towards the establishment of transgenic forest plantations are not significant in relation to a particular buying behaviour (group of consumers) towards paper products, woody biomass energy products and wood products of transgenic origin.

2.2 Analysis Methods

A survey with face-to-face interviews was conducted throughout Greece to gather the necessary information. Cluster sampling was used to form the sample. In particular, the general population was stratified into two levels: regions and prefectures. Based on the methodology presented by Oppenheim (2000), in order to have a representative sample for the geographical area of Greece, 9 regions were randomly selected from a total of 13. In the second stage, one prefecture was randomly selected from each studied region. The sampling took place at the capital of each prefecture at highly utilized areas such as shopping malls and supermarkets. Based on a systematic random sampling, every sixth person who visited the sites was questioned (McCluskey et al. 2003). The total number of consumers who were questioned at each sampled prefecture was 50 consumers.

A total productive sample of 418 consumers came up from the survey methodology, as 32 were missing data because some respondents refused to answer all the questions mainly due to time constraints. The sample was considered representative of the total population since all the sample characteristics did not differ from those of the total population, based on Census data of 2011 (Chen 2007; Tsourgiannis et al. 2008). Prior to the main sampling, a pilot survey took place in October 2011 to test whether the research objectives could be met by the designed questionnaire. The pilot survey was performed for a total of 30 consumers based on Census data of 2011 (Chen 2007; Tsourgiannis et al. 2008). Analysis of the pilot results indicated that no modification of the questionnaire would be required and therefore the survey sample was considered as adequate in order to conduct the final survey. The main survey took place in November and December of 2011. Consumers were asked to answer questions concerning the factors that would affect their potential purchasing behaviour towards the products that could be derived from transgenic forest trees in accordance with their opinion about the establishment of transgenic forest plantations, based on the Likert scale from 1 to 5. Box 1 presents the variables used in this study.

Box 1: List of Variables Used in This Study

(A) Variables describing consumers potential purchasing behaviour towards products that could be derived from transgenic trees

I. Paper Products

1. Attractiveness of the packing
2. Advertisement
3. Production methods
4. Special characteristics of the product
5. Health safety issues
6. Possible negative environmental impact of transgenic plantations

II. Woody Biomass Energy Products

1. Attractiveness of the packing
2. Advertisement
3. Special characteristic of the woody biomass energy product
4. Quality of the product
5. Certification of the product's origin
6. Labelled as products made by wood derived from GM plantations
7. Health safety issues

III. Woody Biomass Energy Products

1. Attractiveness of the packing
2. Advertisement
3. Quality of the product
4. Special characteristic of the product
5. Brand name
6. Labelled as products made by wood derived from GM plantations
7. Certification of the product's origin

(B) Variables describing consumers' opinions towards the establishment of transgenic forest plantations

1. Transgenic forest plantations will increase job flexibility
2. Transgenic forest plantations will reduce cost production
3. Transgenic forest plantations will increase income
4. Transgenic forest plantations will reduce production losses
5. Transgenic forest plantations will have negative environmental impacts
6. Transgenic forest plantations will have a negative impact on biodiversity
7. Transgenic forest plantations will have negative impacts on non GMO products
8. Transgenic forest plantations may harm human health
9. Transgenic forest plantations are not important
10. Transgenic forest plantations will have a negative impact on climate change
11. Transgenic forest plantations will be important for biomass energy production

Multivariate analysis techniques were applied in three stages to the responses for the total of 418 consumers to reveal the key information these contained. More specifically, these were applied separately to the 205 consumers, who declared that they would buy transgenic paper products, to the 231 consumers, who were willing to buy transgenic wood products and to the 220 consumers, who were willing to buy transgenic woody biomass energy products. Principal Component Analysis (PCA) was used to identify the variables that accounted for the maximum amount of variance within the data in terms of the smallest number of uncorrelated variables (components). The anti-image correlation matrix, the Bartlett's test of sphericity and the Measure of Sampling Adequacy (MSA) were also used in order to check the fitness of the data for subsequent factor analysis. The variables with a high proportion of large absolute values of anti-image correlations and MSA less than 0.5 were removed before analysis. An orthogonal rotation (varimax method) was conducted and the standard criteria of eigenvalue equal to 1, scree test and percentage of variance were used in order to determine the factors in the first rotation (Hair et al. 1998). Different trial rotations followed, where factor interpretability was used to compare the reduced through PCA variables to a smaller set of underlying factors that related to consumers' purchase behaviour towards forestry products of transgenic origin.

These PCA scores were then subjected to cluster analysis to group consumers with similar patterns of scores into similar clusters of buying behaviour. Both hierarchical and non-hierarchical methods were used (Hair et al. 1998) in order to develop a typology of the consumers' buying behaviour. Quadratic Discriminant Analysis (QDA) was performed to assess how accurately the identified key factors could predict and discriminate cluster membership through factor analysis. Furthermore, the Friedman one way test was performed to identify the relationship between the consumers' opinion towards establishment of transgenic forest plantations and their particular buying behaviour.

3 Results and Discussion

The results of the study were analyzed initially for the factors that influenced consumers in their potential purchasing behaviour towards transgenic paper products, wood products and woody biomass energy products. Next they were analyzed for the classification of consumers into groups according to their buying behaviour towards each of these categories of forest products. Finally, the analysis explored how the buying attitudes of the consumers willing to buy transgenic forest products were affected by their opinion towards establishment of transgenic forest plantations.

3.1 The factors that influenced consumers in their purchasing behaviour towards transgenic paper products, wood products and woody biomass energy products

PCA and factor analysis was conducted to reduce the variables that addressed the consumers' potential purchasing behaviour towards paper products, wood products and woody biomass energy products to a smaller set of key factors. The latent root criterion (eigenvalues = 1), the scree plot test and the percentage of variance were used to determine the number of factors for each category of products (Tsourgiannis et al. 2013).

The analysis indicated that three factors may affect the consumers' buying behaviour towards each of the three forest product categories of transgenic origin, paper products, wood products and woody biomass energy products. These are detailed in Table 1.

Consumers' potential purchasing behaviour was influenced by two similar factors: (a) marketing issues and (b) product features, towards all transgenic forest product categories under study. In particular, attractiveness of the packing, product's advertisement and special characteristics of the product mainly affected consumers' potential buying behaviour for all three product categories. Health

Table 1 Key factors influencing the consumers' purchasing behaviour towards transgenic paper products, wood products and woody biomass energy products

Paper products		**Woody biomass energy products**		**Wood products**	
Key attitude dimensions	**Factor loading**	**Key attitude dimensions**	**Factor loading**	**Key attitude dimensions**	**Factor loading**
Marketing issues		**Marketing issues**		**Marketing issues**	
Attractiveness of the packing	0.873	Attractiveness of the packing	0.868	Attractiveness of the packing	0.819
Advertisement	0.856	Advertisement	0.826	Advertisement	0.807
Product features		**Product features**		**Product features**	
Production Methods	0.788	Special characteristics of woody biomass energy products	0.859	Quality of the product	0.802
Special characteristics of paper products	0.704	Quality of the product	0.663	Special characteristics of the wood product	0.719
		Certification of the product's origin	0.609	Brand name	0.614
Environmental impact of established transgenic plantation		**Health safety issues**		**Labelling**	
Health Safety issues	0.747	Labelled as product derived by GM plantations	0.863	Labelled as product derived by GM plantations	0.873
Possible negative environmental impact of transgenic plantations	0.741	Health safety issues	0.748	Certification of origin	0.811
KMO MSA:0.613		KMO MSA:0.614		KMO MSA:0.638	
Bartlett test of sphericity = 183.366, $P < 0.001$		*Bartlett test of sphericity* = 301.725, $P < 0.001$		*Bartlett test of sphericity* = 223.017, $P < 0.001$	

safety issues were an important factor that influenced only the potential buyers of paper or woody biomass energy products of transgenic origin, whilst certification of products' origin and labeling regarding the transgenic origin of the product had significant impact on consumers' buying preferences towards transgenic woody biomass energy and wood products. Finally, the potential purchasers of paper products of transgenic origin paid attention to the possible negative environmental impact of those plantations and the health safety issues.

Thus the research hypothesis "**$H_0$1**: Consumers are not influenced by similar factors regarding their potential purchasing behaviour towards paper products, woody biomass energy products and wood products of transgenic origin" was rejected.

3.2 Consumers' classification into groups according to their purchasing behaviour towards transgenic paper products, wood products and woody biomass energy products

Consumers of paper products of transgenic origin were classified in three groups: (a) opportunists, (b) those who were interested in marketing issues and (c) those who were concerned about the environmental impact of biotechnology in forest trees. Consumers of transgenic woody biomass energy products were also classified into three groups: (a) opportunists, (b) those who were interested in marketing issues and (c) those who were influenced by the product related health safety issues. Moreover, consumers of wood products were classified in the following three groups: (a) opportunists, (b) those who were interested in marketing issues and (c) those who were interested in products' features. Hence, most of the potential consumers of the three transgenic forest product categories under study exhibited a similar purchasing behaviour, as they were either interested in marketing issues or they were not influenced by any identified factor and therefore their buying behaviour towards those products was opportunistic (Table 2).

More particularly, the potential consumers of transgenic paper products who were interested in marketing issues were influenced in their purchasing decision by the attractiveness of the paper products packing, as well as the products' advertisement. Also, they paid attention on the products' special characteristics and the production methods employed, including the cultivation techniques, pulp and paper production. On the other hand, they were concerned of the environmental impact of the biotechnology use in forest trees. Furthermore, the potential consumers of transgenic biomass woody forest products who were interested in marketing issues would be attracted by the products' packing and advertisement. These consumers would also be influenced in their purchasing decision by the products' quality, special characteristics and brand name, but surprisingly not by the products' health safety issues. The potential consumers of transgenic wood products, who were influenced by marketing issues, would prefer the products to be labelled, so as to indicate that they were derived from GM plantations. These potential consumers were also interested in the origin of the products, the attractiveness of packing and their advertisement. On the other hand, they were not interested in the product features.

From the paper products' potential consumers 39 % seemed to pay attention to the possible negative environmental impact of the use of transgenic trees. They were very cautious regarding the establishment of such plantations and their products due to their potential negative impact on human health, the environment and biodiversity. Marketing issues and products' characteristics did not affect the purchase decision of those consumers. From the woody biomass energy products' potential consumers 40 % were influenced by the health safety issues. These consumers would like the products to be labelled as products derived from GM plantations, in order to have adequate information prior to any purchasing decision they would make. They would also be influenced to a smaller extent by the products' quality features and marketing issues. Moreover, 40 % of the wood products potential consumers were interested in products' features. In particular, they were influenced in their purchasing decision by the quality of the product, its special characteristics and brand name. They were also interested in the attractiveness of its packing and its advertisement, while they were not interested in the labelling issues (Table 2).

These findings coincide with the results of other studies related to consumers' purchasing behaviour towards non-transgenic forest products (DuPuis 2000; Tokarczyk and Hansen 2006). Furthermore, the findings of this study regarding the purchasing behaviour of the consumers who were mainly interested in the possible negative impacts that the establishment of transgenic forest plantations might have on biodiversity and the environment, support the arguments of other researchers (Gartland et al. 2003; Sedjo 2006). A summary of the cross-validation classification derived through QDA is presented in Table 3. It is evident that the attitude dimensions could accurately predict and discriminate consumers' group membership.

Therefore, the hypothesis "**$H_0$2**: Consumers cannot be classified into similar groups according to their potential buying behaviour towards products that can be derived from transgenic forest trees" was rejected.

3.3 The purchasing behaviour of consumers towards transgenic paper products, wood products and woody biomass energy products in relation to their opinion about the establishment of transgenic forest plantations

The typology of the consumers' potential buying behaviour towards the transgenic forest products under study in relation to their opinions towards the establishment of transgenic forest plantations is portrayed in Table 4. Most of the consumers who would buy paper, woody biomass energy and wood products of transgenic origin had similar attitudes towards the development of transgenic plantations. Most of the transgenic paper, woody biomass and wood potential consumers who were "interested in marketing issues" expressed the view that the development of transgenic forest plantations will reduce the production losses and production cost and will contribute to an increase of job elasticity and of farmers' income.

Table 2 Classification of consumers according to their buying behaviour towards products that could be derived from transgenic forest trees

Groups of Consumers of Paper Products				
Factors	**Opportunists**	**Consumers concerned about the environmental impact of biotechnology use in forest trees**	**Consumers interested in marketing issues**	***p-value***
Environmental impact of biotechnology use in forest trees	−1.12004	0.63788	0.46880	0.001
Marketing issues relevant to paper products	−0.12989	−0.66380	1.10833	0.001
Products characteristics	−0.12032	−0.20738	0.44450	0.001
Number of consumers (N =205)	**69**	**80**	**56**	
Groups of consumers of woody biomass energy products				
Factors	**Opportunists**	**Consumers who are influenced by products' health safety issues**	**Consumers interested in marketing issues**	***p-value***
Marketing issues	−0.82390	0.34919	0.45351	0.001
Product quality features	−0.74820	0.29285	0.44586	0.001
Products' health safety issues	−0.03854	0.80178	−1.08095	0.001
Number of consumers (N = 220)	**71**	**87**	**62**	
Groups of consumers of wood products				
Factors	**Opportunists**	**Consumers interested in product features**	**Consumers interested in marketing issues**	***p-value***
Promotion issues	−1.073330	0.56800	0.33152	0.0001
Product features	−0.4886	0.77177	−0.97946	0.0001
Labeling	−0.4994	−0.6862	0.59869	0.0001
Number of consumers (N = 231)	**70**	**92**	**69**	

Moreover, most of the paper products' potential consumers who "were concerned about the environmental impact of the biotechnology use in forest trees" had the view that the development of transgenic forest plantations might harm human health and have negative environmental impact, but it could be important for biomass energy production. Besides, most of the woody biomass energy products

Table 3 Summary of Classification with Cross-validation

Predicted groups of consumers of paper products			
Actual classification of consumers of paper products	**Opportunists**	**Consumers concerned about the environmental impact of biotechnology use in forest trees**	**Consumers interested in marketing issues**
Opportunists	68	2	0
Consumers concerned about the environmental impact of biotechnology use in forest trees	0	76	0
Consumers interested in marketing issues	1	2	56
Total N	69	80	56
Corrected *N*	68	76	56
Proportion of corrected classification	98.6 %	95 %	100 %
N = 205	*N* correct = 196	Proportion correct = 0.956	
Predicted groups of consumers of woody biomass energy products			
Actual Classification of Consumers of Woody Biomass Energy Products	**Opportunists**	**Consumers influenced by products' health safety issues**	**Consumers interested in marketing issues**
Opportunists	66	2	0
Consumers who are influenced by products' health safety issues	5	82	0
Consumers interested in marketing issues	0	3	62
Total *N*	71	87	62
Corrected *N*	66	82	62
Proportion of corrected classification	93.0 %	94.3 %	100.0 %
Predicted groups of consumers of wood products			
Actual Classification of Consumers of Wood Products	**Opportunists**	**Consumers interested in products features**	**Consumers interested in marketing issues**
Opportunists	68	0	2
Consumers interested in products features	1	70	2
Consumers interested in marketing issues	0	0	88
Total *N*	69	70	92
Corrected *N*	68	70	88
Proportion of Corrected classification	98.6 %	100.0 %	95.7 %

Table 4 Consumers' opinions towards the establishment of transgenic forest plantations

Impact of transgenic forest plantations establishment	Consumers of paper products of transgenic origin			Consumers of woody biomass energy products of transgenic origin			Consumers of wood products of transgenic origin		
Transgenic forest plantations	A^a (χ^2 = 30.877, df = 10, $p < 0.001$)	B^a (χ^2 = 52.458, df = 10, $p < 0.001$)	C^a (χ^2 = 43.267, df = 10, $p < 0.001$)	D^b (χ^2 = 60.580, df = 10, $p < 0.001$)	E^b (χ^2 = 56.158, df = 10, $p < 0.001$)	F^b (χ^2 = 31.452, df = 10, $p < 0.001$)	G^c (χ^2 = 20.677, df = 10, $p < 0.001$)	H^c (χ^2 = 61.845, df = 10, $p < 0.001$)	I^c (χ^2 = 20.677, df = 10, $p = 0.023$)
1. Will increase job flexibility	7.65	6.81	7.53	7.11	6.97	6.93	6.90	7.46	7.23
2. Will reduce cost production	6.30	5.95	7.66	7.11	6.01	6.70	7.01	6.84	6.67
3. Will increase income	5.93	6.27	7.28	6.67	6.39	6.55	6.75	6.59	6.23
4. Will reduce production losses	5.96	6.89	7.71	6.66	6.25	5.54	6.60	6.65	5.64
5. Will have negative environmental impacts	6.57	7.32	6.04	5.69	5.71	6.22	5.45	5.08	5.38
6. Will have a negative impact on biodiversity	6.59	6.72	6.02	5.91	5.68	4.96	5.69	5.68	5.79
7. Will have negative impacts on non GMO product	7.62	7.34	6.10	5.89	5.01	6.06	5.64	5.58	6.02

(continued)

Table 4 (continued)

Impact of transgenic forest plantations establishment	Consumers of paper products of transgenic origin			Consumers of woody biomass energy products of transgenic origin			Consumers of wood products of transgenic origin		
Transgenic forest plantations	*A*[a] ($\chi^2 = 30.877$, df = 10, $p < 0.001$)	*B*[a] ($\chi^2 = 52.458$, df = 10, $p < 0.001$)	*C*[a] ($\chi^2 = 43.267$, df = 10, $p < 0.001$)	*D*[b] ($\chi^2 = 60.580$, df = 10, $p < 0.001$)	*E*[b] ($\chi^2 = 56.158$, df = 10, $p < 0.001$)	*F*[b] ($\chi^2 = 31.452$, df = 10, $p < 0.001$)	*G*[c] ($\chi^2 = 20.677$, df = 10, $p < 0.001$)	*H*[c] ($\chi^2 = 61.845$, df = 10, $p < 0.001$)	*I*[c] ($\chi^2 = 20.677$, df = 10, $p = 0.023$)
8. May harm human health	6.03	6.70	5.85	4.99	7.36	6.61	5.46	5.41	5.99
9. Are not important	6.04	5.46	5.15	4.60	5.36	5.90	4.70	5.20	5.67
10. Will have a negative impact on climate change	7.28	5.19	5.50	4.89	4.84	5.70	5.32	5.30	5.73
11. Will be important for biomass energy products	6.07	7.21	6.45	6.49	6.39	4.83	6.49	6.21	5.66

[a]*A* Opportunist, *B* Consumers concerned about the environmental impact of biotechnology use in forest trees, *C* Consumers interested in marketing issues
[b]*D* Opportunist, *E* Consumers who are influenced by products' health safety issues, *F* Consumers interested in marketing issues
[c]*G* Opportunist, *H* Consumers interested in products features, *I* Consumers interested in marketing issues

potential consumers who "were influenced by health safety issues" mainly believed that forest plantations may harm human health. On the other hand, the majority of the potential consumers of wood products who were influenced by products' features thought that the establishment of transgenic forest plantations will increase job elasticity and reduce the production cost and production losses.

Hence the hypothesis, $\mathbf{H_o3}$: "The consumers' opinion about the establishment of transgenic forest plantations is not significant in relation to a particular buying behaviour (group of consumers) towards paper products, woody biomass energy products and wood products that could be derived from transgenic forest plantations" was rejected.

4 Conclusions

The results of this study contribute significantly to the understanding of the consumers' potential purchasing behaviour towards forest products that could be derived from transgenic forest plantations. In particular, the study indicated that there might be potential buyers in Greece for products originating from transgenic forest plantations, such as paper and woody biomass energy products.

Marketing issues including attractiveness of the packing, advertisement, as well as special characteristics of the products influenced consumers in their potential purchasing decisions towards paper, woody biomass energy and wood products that could be derived from transgenic forest trees. Furthermore, most of the potential consumers of transgenic woody biomass energy and wood products paid attention to the product's quality, certification and the labelling regarding the product's transgenic origin. Health safety issues influenced the potential consumers of both transgenic paper products and woody biomass energy products, whilst the production methods and the possible negative environmental impact of transgenic plantations were crucial factors for the potential buying behaviour of the transgenic paper products potential shoppers.

In general, most of the potential consumers of forest transgenic products in Greece would be willing to purchase such products based on their buying decisions on marketing issues, such as advertisement, packing, quality, labelling and branding. Taking into consideration that most of this types of products are not directly linked with human health impacts, there is a potential for a market development for such products, particularly nowadays that an economic depression is experienced. Hence, from the managerial perspective, this study shows a market segmentation of potential buyers of transgenic origin forest products.

Although forest products from transgenic plantations might not be in the market in the near future, scientifically based information on the anticipated purchasing behaviour of consumers is extremely important both for developers and policy makers. For the developers such information is important, because the expectation of viable markets will motivate investments. For the policy makers, this type of information will help them to develop appropriate decision-making positions and respond more adequately through regulation and programs. In this respect, the empirical results of this pioneer study should trigger further investigation, both in Greece and other parts of Europe.

References

Aguilera J, Nielsen KM, Sweet J (2013) Risk assessment of GM trees in the EU: current regulatory framework and guidance. iForest 6:127–131

Ajzen I (1991) The theory of planned behaviour. Organ Behav Hum Decis Process 50:179–211

Ajzen I, Fishbein M (1980) Understanding attitudes and predicting social behaviour. Prentice-Hall, Englewood Cliffs, NJ

Beckett A, Nayak A (2008) The Reflexive Consumer. Market Theor 8(3):299

Carman N, Langelle O, Perry A, Petermann A, Danna Smith JD, Tokar B (2006) Eocological and social impacts of fast growing timber plantations and genetically engineered trees, http://www.forestethics.org/downloads/GEtreeport.pdf

Chen M (2007) Consumers attitudes and purchase intentions in relation to organic foods in Taiwan: Moderating effects of food-related personality traits. Food Qual Prefer 18:1008–1021

DuPuis EM (2000) Not in my body: BGH and the rise of organic milk. Agric Hum Values 17 (3):285–295

FAO (2008) The potential environmental, cultural and socio-economic impacts of genetically modified trees. UNEP/CBD/SBSTTA/13/INF/6, p 17

FAO (2010) Forests and genetically modified trees. Rome, Italy, p 235

Farnum P, Lucier A, Meilan R (2007) Ecological and population genetics research initiatives for transgenic trees. Tree Genet Genomes 3:119–133

Gartland K, Crow R, Fenning T, Gartland J (2003) Genetically modified trees: production, properties, and potential. J Arboric 29(5):259–266

Häggman H, Find JM, Pilate G, Gallardo F, Ruohonen-Lehto M, Kazana V, Migliacci F, Ionita L, Sijacic-Nikolic M, Donnarumma F, Harfouche A, Biricolti S, Glandorf B, Tsourgiannis L, Minol K, Paffetti D, Fladung M, Vettori C (2012) Biosafety of genetically modified forest trees (GMTs)—COST action FP0905—a common action of European scientists. 2nd international conference of the IUFRO working party 2.09.02. Mendel lectures & Plenary MLP-3, p 13

Hair JF, Anderson RE, Tatham RL, Black WC (1998) Multivariate data analysis. Prentice Hall Inc, New Jersey

Harfouche A, Meilan R, Altman A (2011) Tree genetic engineering and applications to sustainable forestry and biomass production. Trends Biotechnol 29(1):11–17

Hinchee M, Rottman W, Mullinax L, Zhang C, Chang S, Cunningham M, Pearson L, Nehra N (2009) Short-rotation woody crops for bioenergy and biofuels applications. Vitro Cell Develop Biol Plant 45(6):619–629

Jonsson R (2012) Econometric modelling and projections of wood products demand, supply and trade in Europe. Geneva timber and forest discussion paper 59, UNECE/FAO, EE/TIM/DP/59. Geneva, Switzerland, p 196

McCluskey J, Grimsrud K, Ouchi H, Wahl T (2003) Consumer response to genetically modified food products in Japan. Agric Res Econ Rev 32(2):222–231

Oppenheim AN (2000) Questionnaire design, interviewing and attitude measurement. Continuum, New York

Sedjo RA (2004) Transgenic trees: implementation and outcomes of the plant protection act, April 2004, Resources for the future

Sedjo RA (2006) Toward commercialization of genetically engineered forests: economic and social considerations. Resources for the Future, p 46

Sedjo RA (2010) Transgenic trees for biomass. The effects of regulatory restrictions and court decisions on the pace off commercialization. AgBioForum 13(4):391–397

Tokarczyk J, Hansen E (2006) Creating intangible competitive advantage in the forest products industry. Forest Prod J 56(7/8):4–13

Tsourgiannis L, Eddison J, Warren M (2008) Factors affecting the marketing channel choice of sheep and goat farmers in the region of East Macedonia in Greece regarding the distribution of their milk production. Small Ruminant Res 79:87–97

Tsourgiannis L, Kazana V, Karasavvoglou A, Nikolaidis M, Florou G, Polychronidou P (2013) Exploring consumers' attitudes towards wood products that could be derived from transgenic plantations in Greece Elsevier. Procedia Technol (8):554–560

UNECE (2012) Forest products statistics 2007–2011. Timber BulletinECE/TIM/BULL/65/2. www.unece.org/forests/fpm/onlinedata.html

Van Frankenhuyzen K, Beardmore T (2004) Current status and environmental impact of transgenic forest trees. Can J For Res 34:1163–1180

Zhu JY, Pan XJ (2010) Woody biomass pre-treatment for cellulosic ethanol production: technology and energy consumption evaluation. Bioresour Technol 101:4992–5002

Socio-Economic Considerations for Decision-Making on GM Tree Cultivation

Detlef Bartsch, Achim Gathmann, Valasia Iakovoglou and Jeremy Sweet

Abstract In the European Union (EU) genetically modified (GM) trees—like all genetically modified organisms (GMO)—are regulated based on the process and product, not on the phenotype alone. The regulatory system includes a risk assessment which is science-based and concentrates on the assessment of risks rather than benefits. Lately consideration of socio-economic (SE) assessments are planned to get a more prominent role in EU decision-making. Several institutions of the EU and the Member States (MSs) are attempting to clarify the role of SE considerations in decision-making on GM plant cultivation. The SE assessment will typically be a multidisciplinary and iterative process to address all relevant costs and benefits with as little bias as possible. There is a high likelihood that there will be divergent views the methods for SE and evaluating costs versus benefits. This could lead to further delays on decisions on GM plant authorisations.

D. Bartsch (✉) · A. Gathmann
BVL: Bundesamt für Verbraucherschutz und Lebensmittelsicherheit,
Federal Office for Consumer Protection and Food Safety,
Department 4 'Genetic Engineering', 10117 Berlin, Germany
e-mail: detlef.bartsch@bvl.bund.de

A. Gathmann
e-mail: achim.gathmann@bvl.bund.de

V. Iakovoglou
Department of Forestry and Natural Environment Management, Eastern Macedonia and Thrace Institute of Technology, 1st km Drama-Mikrohori, 66100 Drama, Greece
e-mail: viakovoglou@yahoo.gr

J. Sweet
Sweet Environmental Consultants, 6 Green Street, Willingham, Cambridge CB24 5JA, UK
e-mail: jeremysweet303@aol.com

C. Vettori et al. (eds.), *Biosafety of Forest Transgenic Trees*,
Forestry Sciences 82, DOI 10.1007/978-94-017-7531-1_14

1 Introduction: EU Regulations and Socio-Economics

In the EU, GMOs are regulated based on the process and product, not solely on the phenotype of the organism or its products. The regulatory system—as an expression of early policy making at the beginning of this millennium—is laid down in Directive 2001/18/EC[1] and Regulation 1829/2003/EC.[2] The approval process is science based and concentrates on the assessment of risks rather than benefits. The final scientific risk assessment opinion for placing on the EU market is produced by the European Food Safety Authority (EFSA) in cooperation with competent authorities from the 28 MSs and some associated countries such as Norway and Switzerland. The EU Commission (COM) then takes over the process and can give an authorization unless a qualified majority of MS votes against.[3] More than 50 GM crop products can now be legally imported and used, but currently one GM plant is authorised for cultivation[4] as another authorised GM plant has been withdrawn. The current regulatory process for authorising GMOs in the EU only requires risk (to human/animal health and the environment) related information from applicants. Experience from EU funded biosafety research and the scientific literature has not shown any biologically relevant effects of approved GM products or the cultivated GM plants in comparison with existing non-GM products and crops (Sweet and Bartsch 2012). These studies have confirmed the comprehensive scientific assessment of potential risks carried out by EFSA and shown that the scientific quality of their scientific opinions is beyond reproach (Müller-Röber et al. 2013). Despite that, the authorization process for GM plant cultivation has been hampered by political conflicts that have delayed or cancelled several approval procedures.

A new discussion was started by The Netherlands (COGEM 2009) and Austria (Greiter et al. 2011; Spök 2011) to broaden the scope of GMO assessment and regulation by considering SE aspects. These considerations could provide decision-makers with more options to cope with political influences on the decision-making process. However, a series of additional requirements (e.g. risk–benefit analysis or economic modelling) for applicants may potentially make the regulatory process even more complex and cause further delays to decision-making of approvals.

---

[1]Directive 2001/18/EC of the European Parliament and of the Council of 12 March 2001 on the deliberate release into the environment of genetically modified organisms, http://eur-lex.europa.eu/LexUriServ/LexUriServ.do?uri=OJ:L:2001:106:0001:0038:EN:PDF.

[2]Regulation (EC) No 1829/2003 of the European Parliament and of the Council of 22 September 2003 on genetically modified food and feed, http://ec.europa.eu/food/food/animalnutrition/labelling/Reg_1829_2003_en.pdf.

[3]For more details see http://ec.europa.eu/food/plant/gmo/authorisation/cultivation_commercialisation_en.htm.

[4]See http://ec.europa.eu/food/dyna/gm_register/index_en.cfm.

2 The Scope of SE for GM Plants

SE evaluations concern the interaction of social and economic factors: how economy is affected and shaped by social processes and vice versa. There is as yet no clear definition of the scope (and implementation) of its possible use in GMO regulation. Since SE considerations include a wide range of subjects, a number of international activities have attempted to clarify their use.

2.1 Copenhagen EEA Conference

A workshop "Framing SE assessment in GMO and chemicals regulation" held on 6th and 7th December 2012 with 35 experts from the European Environment Agency, the COM, the European Chemicals Agency, representatives of academia and regulatory experts from MSs discussed pending questions related to the practical application of SE assessment in GMO and chemicals regulation, and specifically questioned how to frame and implement SE assessments (Eckerstorfer and Gaugitsch 2014).

The workshop aimed to provide conceptual clarity for stakeholders involved in SE assessment. A working definition for SE assessment was developed: "Other legitimate factors in the context of decision making going beyond an environmental **risk assessment** including **risks** to human health, such as (a) Environmental and human health **benefits**, (b) social risks and benefits, (c) economic risks and benefits and (d) ethical issues".

One general conclusion of the discussion was that the SE assessment should be a tool to inform, but not replace decision-making. SE assessment developed for the purpose of supporting decision-making will necessarily be restricted in time and resources, so that it provides timely inputs. SE assessment would need a high level of transparency, clear communication of the applied approaches and focussed conclusions directed towards decision-makers and the general public. However, uncertainties are also inherent to the SE assessment due to the specific approach (e.g. integrated and systematic ex-ante or post-ante analysis of economic, social and/or environmental impacts) taken and due especially to lack of data availability (e.g. due to confidential business information) to address issues regarded as important. A further general conclusion was that the task of producing an SE assessment can be highly complex leading to ambiguous results. The outcome of a SE assessment is very dependent on the methodology used and assumptions taken. However, the smallest scale or unit of SE assessment might well be a certain crop/trait combination, and not the GMO event as used in the risk assessment process. Further difficulties were identified during the conference including (a) availability of data for taking management decisions, (b) anticipated costs of coexistence, (c) foreseen overestimation of costs and underestimation of benefits and (d) clarification of 'receiving environment'. EU food law requires (GMO) safety as the basis for all further action. There is a need to draw a line between the risk-focused GMO authorization and SE assessment, the latter could have added value when used after authorization.

The SE assessment should enable a systematic analysis of the main impacts of an application or risk management measures. It should be proportionate to the problem that is addressed, i.e. scope and depth will require case-by-case considerations. The SE assessment will typically be a multidisciplinary and iterative process to address all relevant costs and benefits with as little bias as possible. It should be taken into account that SE assessment is a tool to support decision-making, thus it is very important to define the purposes of its application.

2.2 *The European GMO SEs Bureau (ESEB)*

In 2011 the COM published a report on SE implications of cultivation of GM plants describing the state of knowledge (EC 2010a). The report called for an advanced reflection at the European level, with a sound scientific basis, and the objective of:

- Defining a robust set of indicators to capture SE consequences of cultivation of GM plants from seed production to consumers across the EU
- Exploring approaches to make use of the increased understanding of these SE factors in the management of the cultivation of GM plants.

One of the initiatives towards this goal was the introduction of the European GMO SEs Bureau (ESEB).[5] The ESEB consists of scientific experts nominated by the MSs and experts from the COM. The mission of the ESEB is to organise and facilitate the exchange of technical and scientific information regarding the SE implications of the cultivation and use of GMOs between MS and the COM. The Bureau develops Reference Documents that enable a science based assessment of these impacts in the MSs and across the EU.[6] The ESEB agreed to predominantly focus on SE assessment of GM crop cultivation in the EU.

The structure of ESEB's outputs, in the form of reference documents, will be based on crop/trait combinations. Given the limited resources available and the need to establish a 2-year work programme, the work will focus on crops that are cultivated in the EU or are in the regulatory pipeline, which will currently exclude GM trees (Lusser et al. 2012). Potential crop/trait combinations are thus limited to maize, soybean, cotton and sugar beet with either herbicide tolerance or insect

[5]See https://ec.europa.eu/jrc/en/eseb.

[6]See minutes of first meeting: https://ec.europa.eu/jrc/sites/default/files/Minutes_first_ESEB.pdf.

resistance traits. Crops with other traits such as enhanced nutritional composition or drought tolerance will be considered in a later period.

The ESEB recently published the first reference document describing the framework for the socio-economic analysis of the cultivation of genetically modified crops (Kathage et al. 2015). Specific documents for different crop trait combinations are in preparation.

2.3 International Considerations: WTO, CBD and Others

The last Conference of the Parties to the Convention on Biological Diversity[7] took place in October 2014 and addressed the outcome of an expert working group for SE considerations (CBD 2014). The EU position stresses that any outcome of the discussions in which SE considerations would be incorporated in risk assessment, and/or become a compulsory and systematic element to consider in the GMO authorization process, would be incompatible with the EU legislation. However, there is little experience with the interface between risk assessment and international obligations that may be relevant to SE considerations. It is clear that the scope of SE considerations includes economic, social, ecological, cultural/traditional/religious/ethical and human health related issues. Based on the complexity of the scope proper methodology should be used in order to get the best study outcome regarding:

- Situational analysis and baseline information
- Scenario planning
- Ex-ante and/or Ex-post studies
- Quantitative and/or qualitative studies
- Public consultation and participation modalities
- Multi-criteria analysis
- SE assessments
- Valuation of biological diversity

Unfortunately a number of complex factors depending on the organism, trait and intended use will affect SE considerations in detail. The resulting uncertainty or insufficient information on SE considerations will be challenging for decision-makers. Given the contrasting views of stakeholders involved in the design of SE considerations assessment (see e.g. Eckerstorfer and Gaugitsch 2014), any weight of evidence approach might well end finally in court actions where external expert opinion and the availability of evidence (or absence of evidence) will shape the possibility to take a decision.

[7]Conference of the Parties serving as the Meeting of the Parties to the Cartagena Protocol on Biosafety, the 7th meeting, (COP-MOP 7) 29 September–3 October 2014, Pyeongchang, Republic of Korea, see http://www.cbd.int/doc/?meeting=mop-07.

More investigation is (still) required on recent World Trade Organization (WTO) case law which could provide some more guidance on what kind of SE considerations would be in line with sanitary and phytosanitary measures, technical barriers to trade, or the general agreement on tariffs and trade (Spök 2011), see also Dederer (2010). In addition, SE issues and public awareness are important for developing countries in order to harmonise their efforts towards utilisation of GM products (Malboobi 2012).

3 New Opt-Out Regulation for EU MSs in the Pipeline

In 2009, 13 MSs asked the COM to provide a sound legal basis in the related EU legal framework in order to allow MSs to restrict or prohibit cultivation of GMOs, which have been authorised or are under authorisation at the EU level, in all or part of their territory. In response the COM proposed in 2010 new rules for the authorisation of GM crops for cultivation (EC 2010b). The final proposal was adopted by the EU Parliament (EU 2014a, b) and contains a non-exhaustive list of possible grounds that can be used by MSs to restrict or prohibit cultivation of GMOs, including, notably, environmental policy objectives, SE considerations, land use, town and country planning, agricultural policy objectives and public policy issues, etc.[8] All restriction measures shall be in conformity with the EU law, reasoned, proportional and non-discriminatory and, in addition, shall be based on compelling grounds.

With the adoption of the new regulation[9] MS can now ban GMOs on policy grounds other than the risks to health and the environment already assessed by EFSA, However, it is still not clear how this proposal would work in regulatory practice. For example, there is no internationally agreed definition of what is meant by SE considerations.

According to Dederer (2010) the establishment of *general* GMO-free areas is a violation of basic EU law concerning right of property, economic or entrepreneurial freedom, free movement of goods and freedom of research. However, in his opinion the construction of *specific* GMO free areas might be possible taking into account environmental policy, nature conservation or coexistence considerations. In this respect farm level restrictions might be more legally defendable than wider prohibitions of GMO use.

It is difficult to align political intentions and objectives one-to-one with these compelling grounds. Data would need to be generated to establish specific GMO-free areas on regional/geographic levels using differentiated and GMO specific reasoning. It is economically and administratively simpler to establish and control general

[8]See http://www.europarl.europa.eu/news/en/news-room/content/20150109IPR06306/html/Parliament-backs-GMO-opt-out-for-EU-member-states.

[9]Based on the new article 26b in Directive2001/18/EC, implemented through Directive (EU) 2015/412 as regards the possibility for the Member States to restrict or prohibit the cultivation of genetically modified organisms (GMOs) in their territory

GMO free areas within greater political borders. However, basic EU and international law may be the guiding force based on reasoned arguments to cope with legally defendable arguments.

4 Outlook

Consideration of SE impacts, if implemented, are challenging as the basis for making decisions on the cultivation of GM plants, including trees. One important issue is that very little experience exists on the SE of GM trees. In China poplars expressing transgenic Cry1 and Cry3 proteins[10] providing resistance to Lepidopteran and Coleopteran pests, have been cultivated for several years for environmental restoration and timber production. Commercialisation occurred in 2002 and now well over one million insect resistant GM poplars have been planted. According to official statistics from the Chinese Forestry Academy, Bt poplars are currently being commercially grown on around 200–300 ha. There are a further 300 ha of small research sites, distributed over several provinces.[11]

These trees have improved the area where they have been established by trapping rain, reducing water runoff, and preventing soil erosion. They are also more productive as they are not defoliated by insects each year. Therefore a range of environmental, economic and social benefits can be identified as well as some environmental hazards associated with non-target effects of the cry toxins.

A GM fruit tree, the GM ring-spot virus resistant papaya, has been authorised in the US (State of Hawaii), and has been commercially grown since 1999. According to Brookes and Barfoot (2014), 77 % of the Hawaii's papaya was GM (395 ha of fruit bearing trees). There was a significant increase of yields relative to conventional varieties. Compared to the average yield in the last year before the first GM papaya cultivation in 1998, the annual average yield increase of GM papaya has been within a range of +15 % to +77 % (17 % in 2012) relative to conventional crops. The total papaya production increased by 12.7 %. Economically, the net annual impact has been an improvement per ha between $2400 and $11,400, and in 2012, this amounted to a net farm income gain of $2420/ha and an aggregate benefit across the state of $0.95 million. According to Brookes and Barfoot (2014), the cumulative farm income benefit was $23.9 million since 1999. Overall, the data presented by these authors cover basically only farm income as economic value and very few social considerations. Virus resistant papaya are also reported to have been grown in China, (4500 ha in 2012) and are being field tested in other Southeast Asian countries. However, more data on economic and environmental impacts of

[10]There is little information publically available at http://www.gmo-safety.eu/science/woody-plants/316.seeing-once-studying-thousand-times.html.

[11]http://www.gmo-safety.eu/en/wood/poplar/325.docu.html.

GM papaya cultivation are not to be yet available from China or other Southeast Asian countries (Brookes and Barfoot 2014).

Independent of this papaya example, in the EU a trade-off is foreseen between legal restrictions as consequence of broad public concerns and the freedom of public institutions or private companies to develop and use the new technology. The economic importance of GM trees is not (yet) a big issue, especially for multinational companies, but small and medium enterprises as well as public institutions, such as national forestry and tree fruit institutes and breeding programmes, would benefit from balanced SE considerations.

References

Brookes G, Barfoot P (2014) GM crops: global socio-economic and environmental impacts 1996–2012. PG Economics Ltd, Dorchester, UK, 189 p. http://www.pgeconomics.co.uk/pdf/2014globalimpactstudyfinalreport.pdf

CBD (2014) Ad hoc technical expert group on socio economic considerations. (Article 26 of the Cartagena Protocol on Biosafety) First meeting report. http://www.cbd.int/doc/meetings/bs/bs-ahteg-sec-01/official/bs-ahteg-sec-01-03-en.pdf

COGEM (2009) Socio-economic aspects of GMO's. Building blocks for an EU sustainability assessment of genetically modified crops (090929-01). Report CGM/090929-01, 94 p

Dederer HG (2010) Weiterentwicklung des Gentechnikrechts - GVO-freie Zonen und sozioökonomische Kriterien für die GVO-Zulassung: eine Untersuchung der Regelungsspielräume und ihrer europa- und welthandelsrechtlichen Grenzen. LIT Verlag Münster, 259 p

EC (2010a) Report from the Commission to the European Parliament and the Council on socio-economic implications of GMO cultivation on the basis of Member States contributions, as requested by the Conclusions of the Environment Council of December 2008, 11 p. http://ec.europa.eu/food/plant/gmo/reports_studies/docs/socio_economic_report_gmo_en.pdf

EC (2010b) Communication from the Commission to the European Parliament, the Council, the Economic and Social Committee and the Committee of the Regions on the freedom for Member States to decide on the cultivation of genetically modified crops. COM 375 final, 9 p. http://ec.europa.eu/food/plant/gmo/legislation/docs/communication_en.pdf

Eckerstorfer M, Gaugitsch H (2014) Framing socio-economic assessment in GMO & chemicals regulation. Workshop report, 17 p. http://www.umweltbundesamt.at/fileadmin/site/umweltthemen/gentechnik/EEA-SEA_Dec2012/Summary-report_EEA-Workshop-SEA_6-7Dec2012.pdf

EU (2014a) Council of the European Union—Proposal for a Directive of the European Parliament and of the Council amending Directive 2001/18/EC as regards the possibility for the Member States to restrict or prohibit the cultivation of genetically modified organisms (GMOs) in their territory. http://register.consilium.europa.eu/doc/srv?l=EN&f=ST%2010972%202014%20INIT

EU (2014b) Committee on the Environment, Public Health and Food Safety of the European Parliament: Final compromise text as agreed at the COREPER on 10 December 2014 concerning the Proposal for a Directive of the European Parliament and of the Council amending Directive 2001/18/EC. http://www.europarl.europa.eu/meetdocs/2014_2019/documents/envi/dv/gmo_final_crp10122014_/gmo_final_crp10122014_en.pdf

Greiter A, Miklau M, Heissenberger A, Gaugitsch H (2011) Socio-economic aspects in the assessment of GMOS-options for action. Umweltbundesamt Reports, Vienna 52 p

Kathage J, Gómez-Barbero M, Rodríguez-Cerezo E (2015) Framework for the socio-economic analysis of the cultivation of genetically modified crops. JRC Science and Policy Reports, 28 p

Lusser M, Raney T, Tillie P, Dillen K, Rodríguez-Cerezo E (2012) International workshop on socio-economic impacts of genetically modified crops co-organised by JRC-IPTS and FAO—workshop proceedings. Publications Office of the European Union and the Food and Agriculture Organization of the United Nations, 134 p. http://ftp.jrc.es/EURdoc/JRC69363.pdf
Malboobi MA (2012) GM crops: the socio-economic impacts. J Plant Mole Breeding (JPMB) 1:1–9
Müller-Röber B, Boysen M, Marx-Stölting L, Osterheider A (Hrsg.) (2013) Agricultural biotechnology. Current scientific, economic and societal developments. Summary—supplement of the interdisciplinary research group 'Gene Technology Report'. Berlin-Brandenburg Academy of Sciences and Humanities (BBAW)
Spök A (2011) Assessing socio-economic impacts of GMOs, issues to consider for policy development—final report. Federal Ministry of Health; Federal Ministry for Agriculture, Forestry, Environment, and Water Management, Vienna, 127 p. http://bmg.gv.at/cms/home/attachments/5/0/0/CH1050/CMS1291038713992/assessing_socio-economic_impacts_of_gmos,_band_2_20101.pdf
Sweet JB, Bartsch D (2012) Synthesis and overview studies to evaluate existing research and knowledge on biological issues on GM plants of relevance to swiss environments, NFP 59: review of international literature. Swiss National Research Program 59, vdf Hochschulverlag AG, ETH Zuerich, 194 p. http://dx.doi.org/10.3929/ethz-a-007350187

Part IV
Value Communication in the Field of GM Trees

Value Communication in the Field of Agri-Food

Guido Nicolosi

Abstract European surveys show that Europeans are not technophobic. At the same time for some Europeans, the jury is still out on biotechnology. All the surveys carried out to measure public perception on the topic of technoscientific innovations demonstrate a certain ambivalence in the judgment of public opinion. It is generally positive with respect to medical applications (diagnosis, prevention and cure of disease). On the contrary, biotechnological innovation in the agricultural and food sector is a source of anxiety and considerable fear. The case of agri-food is emblematic. It is an entity that is at the crossroads between territory and technique and it highlights well the need we have to update the 'toolbox' for the appraisal of environmental and development problems. This is because agri-food products are perceived as substances of physical–biological maintenance, pharmacological remedy and 'cultural object' able to channel important symbolic meanings. The cultural and symbolical value of agri-food practices implies that agri-food and communication are strongly intertwined. Presenting a theoretical background and some evidences emerging from empirical research, this chapter suggests some guidelines for a value-based system for risk and benefit communication in agri-food applications.

1 Introduction

1.1 Public Opinion on Agro-Biotechnology: Fears, Anxiety and Symbolic Value in the Agri-Food Sector

European surveys (for example Eurobarometer) show that Europeans are not technophobic. The greater majority thinks that telecommunications, computers and IT, the Internet, solar and wind energy and mobile phones will improve our way of

G. Nicolosi (✉)
Department of Social and Political Sciences, University of Catania,
Via Vittorio Emanuele 49, 95131 Catania, Italy
e-mail: gnicolos@unict.it

C. Vettori et al. (eds.), *Biosafety of Forest Transgenic Trees*,
Forestry Sciences 82, DOI 10.1007/978-94-017-7531-1_15

life over the next 20 years. Trend data since 1991 show little change in this optimism for telecommunications, computers and information technology.

At the same time for some Europeans, the jury is still out on biotechnology. Around 20 % said 'Don't know' when asked whether it would improve our way of life or not. By contrast amongst those who expressed an opinion, 53 % of Europeans were optimistic and 20 % pessimistic about biotechnology (Eurobarometer 2012).

In the period 1999–2002, optimism has increased to the level seen in the early 1990s after a decade of continuously declining optimism in biotechnology. An index of optimism shows an appreciable change from the declining trend of the years 1991–1999. This rise in optimism holds for all the EU Member States with the exception of Germany and the Netherlands, where such a rise was only observed between 1996 and 1999.

1.1.1 Attitudes to Medical, Industrial and Agri-Food Applications of Biotechnology

All the surveys carried out to measure public perception on the topic of techno-scientific innovations demonstrate a certain ambivalence in the judgment of public opinion. An attitude that has been defined by some commentators as '*utilitarian*' (Cerroni et al. 2002) seems to govern opinion formation. In consequence, medical applications (diagnosis, prevention and cure of disease) are generally appreciated. On the contrary, biotechnological innovation in applications in the agricultural and food sector is the source of anxiety and considerable fear.

Judgments about six applications of biotechnology—genetic testing for hereditary diseases, cloning human cells and tissues, GM enzymes for soaps, transgenic animals for xenotransplantation, GM crops and GM foods show that Europeans continue to distinguish between different types of applications, particularly medical in contrast to *agri-food applications*. Genetic testing for hereditary diseases is seen as useful, morally acceptable and worth to be encouraged (supported), the same holds for cloning human cells and tissues, even though this application is also seen as a risk. These two applications are supported in all the EU member states.

A majority of Europeans do not support GM foods. These are judged not to be useful and to pose a risk for society. For GM crops, support is lukewarm, while they are judged to be moderately useful they are seen as almost as risky as GM foods. While GM crops are supported in Spain, Portugal, Ireland, Belgium, UK, Finland, Germany and the Netherlands, with the exception of Belgium, all the countries which called for the extension of the de facto moratorium on the commercial exploitation of GM crops (France, Italy, Greece, Denmark, Austria and Luxembourg) have publics that are, on average, opposed to GM foods (Eurobarometer 2003). Why are GM technology applications with an emphasis on agri-food so differently perceived? We will suggest some cultural and symbolical explanations requiring specific communicative strategies for the different fields of biotechnology.

For example, the main fears of Italians, according to the findings of Censis (Social Investments Study Centre) are those where foods play a primary role (Censis 2001). Closely correlated to these data, according to Censis, is the exponential increment in the growth of the market share of biological products and of quality alimentary products certified by PDO (Protected Denomination of Origin) and PGIs (Protected Geographical Indications) which are estimated at 7 million euro. The consumption of biological fruit and vegetable products is around 5 % of the market. The anxieties shown by Italian consumers reflect those found in a broader European context, i.e. that fear and pessimism are not connected to biotechnologies in general, but are mainly focused on agri-foods.

In the agri-food sector the more anxiety-inducing aspects practically coincide with the fear of contaminating one's own body. An important aspect emerging is that biotechnology innovation in the agri-food sector becomes more acceptable if the 'contaminating' products (for example GMOs) are accompanied by informative labels. Indeed, 61 % of those interviewed stated that they were 'favourable to the introduction on the market of biotechnological products if the relative information were indicated on the label' (Cerroni et al. 2002). Highly interesting is the interpretation that Cerroni gives to this data. In his view, labelling does not so much perform an informative role (as the supporters of the cognitive approach would have believed), but rather realizes an essentially psychological task. Cerroni cites various empirical researches to show that the value of labelling does not consist in assuring an effective increase in the power of control over technology by the consumer, because in the background persist:

(a) a profound *informative asymmetry* between producer and consumer;
(b) a fundamental emotional and symbolic dimension that cannot be resolved in a mechanistic manner (or hydraulic) through the mere 'pumping' of information.

According to these researchers, citizens perceive science and technology as hazardous for health and the environment due to their inability to analyze the risks and thus develop an understanding of the technology. The profane, therefore, would react in a 'non-scientifically rational' way because they are ignorant of the technology. According to this view, there is an information deficit (deficit model) at the roots of the unwarranted anxieties of contemporary society. It has been suggested that increasing the amount of information available to people would resolve this problem of the widespread resistance to change and progress (Lupton 2003). However, this point is challenged by Wynne (1994).

The greater propensity of the consumer to GMOs in the presence of labelling is, instead, explicable by a *control illusion*. Interpreting the work of Slovic, Cerroni asserts that through labelling it is possible to reduce 'the subjective component of unknown risk'. The existence of such a component is demonstrated, *a contrario*, by the disavowal or the scant importance attributed to the danger of day-to day hazardous behaviours (domestic incidents, cars, etc.). In particular, Cerroni quotes the renowned distinction between *dread risk* and *unknown risk*: the former definable as a fear of unimaginable effects; the latter owing to the fear of 'the hidden nature of the process'.

The interpretation proposed in the following argument is that this special attention towards biotechnological applications in foods may be explained by recognizing that food continues to hold, even in complex contemporary societies, a decisive symbolic weight. Unfortunately, science (in particular medicine) obstinately continues to underrate this aspect.

For such a reason, doctors and scientists are often vexed by the broad spread of alimentary fears; and they are critical of the behaviour of consumers which is defined 'irrational'. In the magazine *Salute* (weekly insert magazine of the daily paper 'La Repubblica') 16 February 2006, n° 479, Riccardo Crebelli, renowned toxicologist of the Higher Institute of Health, in an interview on the theme of banning Teflon and Pfoa (with particular reference to so-called 'non-stick frying pans'), emblematically, asserted:

> [...] Pfoa have been found everywhere: in 92% of American non- exposed citizens [...] they enter into the atmosphere, they are concentrated, they end up in the alimentary chain, the aquifers and become ubiquitous and persistent [...] as a toxicologist I always ask myself why people's attention is pathologically concentrated on food, while one never thinks of air or water. We breathe approximately 15 kilos of air a day, drink 2 litres of water, we only eat 1 and a half kilos of food, yet our main worry...

Prof. Crebelli's question expresses the mechanistic and rationalist perspective so widespread in the medical–scientific field. The main thesis developed within this chapter takes a contrasting view: for or against, we will never understand the widespread fears and consequent resistance to biotechnological innovations in the food if not by starting off from a serious reflection on the symbolic importance of the agri-food sector.

2 Theoretical Background

2.1 Technological Innovation and the Crisis of Trust: The Case of Agri-Food Applications and the Role of Communication

2.1.1 Technique, Territory and Culture

The public debate on environmental conditions is to a large extent founded on judgments, appraisals, forecasts and desires that concern two key dimensions of the life of Man: *technique* and *territory*. The disagreeable sensation pervading some of those following this debate is that it is marred by a serious error of perspective. I refer, in particular, to the consolidated tendency to think of technique and territory as neutral and objective entities; and to the consequential choice of leaving them to be dealt with by only those with specific technical–scientific expertise (engineers, geologists, etc.).

Territory and technique are socially rooted entities and a serious discussion should always include a deep reflection on the sociohistorical conditions that determine their definition. Territory is socially rooted to the extent in which it is defined in a continuous and mutual exchange between ecological, environmental and sociocultural conditions of human groups that inhabit it. It is technique that is the main form of mediation implied in such exchange. But far from neutral, technique is itself produced by Man. By means of technique, Man projects himself, his culture, to the outside, in time, creating an objective and *meaningful* world, in which he lives.

Such a relationship is not determinist, neither unidirectional: by means of technique Man also expresses the way in which the natural world conditions and constrains him (Fig. 1).

The relationship with the world needs mediation and this can be done with material objects (forks, umbrellas), but also with immaterial or symbolic objects (beliefs, codes…). But the same material objects incorporate symbolic systems and cultural practices and by mediating Man's relationship with the world carry out an important communicative function.

But technique is not only the medium of Man's relationship with the non-human and the non-social. Technical action, namely 'a kind of acting that creates artefacts', is also one of the fundamental conditions that determine the very social nature of Man. Indeed, there are three modalities of technical acting: (a) for the purpose of *employment*; (b) directed towards the *modification* of what is found; (c) and *productive* (Popitz 1990). But to employ, modify and produce imply, next to a subject–object relationship, also a subject–subject relationship:

In this sense, technique is 'a form of social relationship and production'. Rather than the social consequences of technique, we should discuss 'sociotechnical systems', namely: '*networks of human and technological elements operating jointly in order to achieve a determined purpose*' (Gallino 1998; Latour 1995). The case of agri-food is one of the most emblematic. It is an entity that is at the crossroads between territory and technique, highlighting the need that we have to update the 'toolbox' for the appraisal of environmental and development problems.

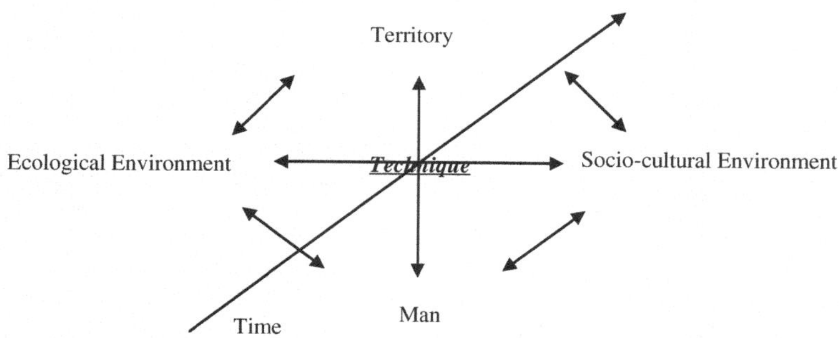

Fig. 1 Model of ecocultural evolution (Nicolosi 2008)

2.1.2 Agri-Food Between Nature and Culture

To consider nutrition a merely biological fact is reductive. If it is true that Man, like any other animal, must ingest nourishing substances to live, it is equally true that he does so only after having transformed them into *aliments*: natural elements culturally *elaborated* and consumed within the framework of social codified practices (Poulain 2002).

The food chain is characterized by three or four fundamental stages: harvest, (transport), transformation and consumption. The transformation stage is that which phylogenetically distinguishes the greatest discontinuity between Man and other animal species. In fact, if some physical–chemical transformations are also done by other living species, humans are unique in practicing cooking and combining foods. For this reason, for Man feeding means using natural products from a specific territory and transforming them through culinary techniques. That is, if the alimentary act unites Man with the animal, the culinary act (which is a technocultural act), an experience that is exclusive to the human species (Perlès 1979, p. 5). Already in this, Man shows his specificity: to crossbreed nature and culture with a techno-transformative action. In this sense, I propose to consider the human condition as 'physiologically' *biotechnical* and biocultural (and therefore biosocial, biopolitical, etc.): from his very beginning, Man has always lived in a biotechnical environment, in which nature and culture are continually interwoven.

There is a complex relationship between culture and nutrition. Survival is guaranteed by satisfying alimentary needs; but, in the case of Man, this is achieved in different and at times hostile natural environments. This is possible (beyond that of a specific omnivorous physiology) because Man is a social animal endowed with culture and symbolic ability to construct more favourable environmental conditions. For this reason, in the case of the human animal, many anthropologists rightly speak of biocultural evolution.

Moreover, because of the *principle of incorporation*[1] (Fischler 1988), there is a complex but essential food–body nexus. There are three dimensions able to specify such complexity. Food is: substance of physical–biological maintenance; pharmacological remedy[2]; 'cultural object' able to channel important symbolic meanings.

These three dimensions, in effect, are traceable to two fundamental spheres: a material (life, health, energy, etc.) and another immaterial (symbolic projection).

---

[1]Founding principle of the relationship between man and food, consists in the banal fact that eating means incorporating the external world within, breaking the barrier between the interior and exterior. This implies the fact that, both on a material as well as symbolic level, incorporating food means incorporating a part or all of its qualities (positive or negative). Much of primitive thinking is based on such a 'belief'.

[2]Indeed, food as of being the main material with which our bodies are made, has always been, in every society, the first pharmacological remedy (Hippocrates invited making his medicine into food). But if in archaic societies this fact further augmented its symbolic importance, in modern western societies it has increased its rationalist reification (nutritional science, etc.).

Moreover, both may be tinted in an ambivalent way (and in a synchronic or diachronic sense) by conservative or innovative instances: protect the body (physical identity), the Self (symbolic identity) or innovate both. Such fundamental ambivalence is, in the last instance, traceable to the so-called *paradox of the omnivore* (Fischler 1998), namely that the human being, in so far as an omnivorous animal, lives out an elementary anthropological contradiction. On one hand, he inhabits the reign of freedom, to the extent he is not limited to one food alone, but can choose from a more or less limitless variety of possibilities. This renders him particularly flexible, creative and ready to adapt to the changes in the environment in which he lives (*neophilia*). On the other hand, as always happens in the reign of the freedom, he experiences the distressing constraint of choice. He must choose and decide among infinite possibilities, some of which are irremediably toxic, corruptive or lethal (*neophobia*).

For all these reasons, Man does not only feed on food, but also on culture and must satisfy the two levels that compose him: the material (corporeal) and immaterial (symbolic). It may happen that symbolic–cultural requirements provoke disastrous endemic deficiencies of a nutritional kind.[3] In compliance with the matrix of the anthropological charter of human nature (in which biological, cultural, individual and social are *fused* inseparably), the *homo sapiens'* relationship with food is complex and cannot be reduced to one-dimensional readings. If we add the fact that various cultures respond to alimentary needs in different ways, we can easily see that in the case of cultural processes it can be misleading to speak of 'evolution'.[4] It would be more correct to admit a good dose of symbolic *arbitrariness*.

But it would be wrong for this reason to remove the cultural dimension from the context of what must be scientifically analyzed. At times, genes adapt more readily than human cultures to changes. **Any kind of new intervention aimed at the introduction of new practices, technologies or agri-food production must be able to deal with this aspect.**

[3]It is no accident that the phase of weaning (passage from 'natural' to 'cultural' nutrition) is the riskiest moment for babies' survival.

[4]For example, with the advent of the Neolithic age the agricultural revolution had the effect of increasing enormously the amount of food available and the possibility of storing foods, but at the expense of a slow but significant narrowing of the qualitative spectrum of alimentary consumption (Gaulin 1979, p. 48). In effect, agriculture marks a paradox: an animal that bases its evolutionary successes on the condition of omnivore, 'chooses' to practice a relative but important alimentary specialization. Such a choice is evidently cultural and to some extent can be read as a veritable 'regression': 'Agricultural societies, by partially reducing the fluctuation in resources, or at least, the irregularity of alimentary cycles, introduced the risk of crises with catastrophic consequences' (Fischler 1979, p. 197). Nevertheless, what from the point of view of mere biological adaptation might be considered dysfunctional behaviour, from the historical point of view has had exceptional implications: the affirmation of the earliest kinds of private property by means of production, construction of cities, foundation of empires, until arriving at the modern industrial era.

2.1.3 Food and Communication: An Inseparable Binomial

The cultural and symbolical value of agri-food practices implies also that food and communication are strongly intertwined. Approximately 10,000 years ago, the introduction of the products of animal origin meant the passage from the condition of gathering to that of hunting (after a probable stage of necrophagy). Such a change most likely contributed to make feeding become a collective and sexually differentiated action. In particular, since Man was not endowed with suitable 'natural instruments' (tusks, claws, etc.), there was the need to develop an action of collective hunting with an advanced level of coordination.[5] This implied, besides lengthening the alimentary process (between need and its satisfaction), also a meaningful development of the system of communication between the players involved (Perlés 1979).

The practice according to which feeding became first of all a social action founded on the division of labour, exchange and reciprocity was consolidated. Most likely, from this moment on the consuming of meals was to be done in a non-individualized way, in a shared place and time, becoming, therefore, a fundamental practice of community integration. Probably, in this particular moment of the hominization process food takes on a privileged function of *medium* of communication, acquiring in such a way an exceptional symbolic charge.[6]

Many authors have found this charge stimulating for their own reflections; for example Lévi-Strauss, in the context of his structuralist theory of culture as communication (2002–2004). As is known, moreover, for Lévi-Strauss a society's cuisine can be analyzed, exactly like a language, according to structures of opposition and correlation and based on specific *gustemes*.

Mary Douglas has often stated that the alimentary dimension is able to go beyond the border between the sphere of the natural and that of the cultural. Needs of the body and needs of society are, in food, inextricably interwoven, so much so that 'it is culture that creates among men the *communication system* that defines the edible, the toxic, satiety' (Douglas 1979).

With reference to the alimentary problems of development, Douglas emphasizes how in communities (often rural ones that feed following the close ties with seasonal correspondences) it is utopian to think of introducing important changes in food practices without deeply affecting their fragile inner equilibrium and what binds them to the environment. In general, the overly abrupt introduction of new foods may even provoke serious forms of widespread anxiety. **Equally, the forced realization of a plan of agricultural development foreign to local ethno-gastronomic traditions can have the same negative effect as the compulsory introduction of a foreign language**.

---

[5]Also the harvest of vegetable products may not be exclusively individual, but it implies a decidedly reduced coordination and division of labour.

[6]For example, meat, the product of hunting becomes a male food par excellence. The distribution of the venison among the members of the group, moreover follows a logic of social importance based on roles, hierarchies, exchange and allegiance relationships.

For Igor De Garine, the food–communication nexus is also central. It is verified empirically that human groups living in very similar environmental conditions develop alimentary behaviours and give value, from nutritional and symbolic points of view, also highly varied foods. This fact demonstrates that feeding represents one of the main communicative registers that allow asserting the inner cohesion of a society (De Garine 1979). Alimentary behaviours, in such a perspective, carry out a *demarcative* function aimed at boosting cultural heterogeneity. Moreover, according to De Garine (1979), valued foods serve to explicate the *differential gaps* between groups and categories that cohabit in each culture. The interpretation given by the French anthropologist is highly evocative. The differential gaps are not found at a biological level, but at a cultural one that performs a function similar to that carried out in the field of sexuality by the taboo of incest: by differentiating alimentary behaviours, the inner cohesion of the original groups, as well as the social heterogeneity between individuals, categories, social groups and cultures is strengthened at the same time. This makes communication and exchange possible, rather the main 'instruments' by means which human societies are formed and persist.

In conclusion, I have briefly shown how communication has great importance in the definition of alimentary processes. These are embedded in a social–cultural matrix with a twofold valence: on one hand, nutrition (in all its stages: harvest, transformation, circulation, etc.) always implies a social exchange of a communicative kind; on the other, food may be interpreted as a symbolic system of communication in itself. Both levels have been radically overturned by the advent of alimentary modernity.

2.1.4 Modernity and Communication: The Agri-Food System as Expert System Generates Alimentary Fears

Our age is characterized by a broad diffusion of alimentary fears. I suggest that they do not represent a secondary or contingent phenomenon. **The case of bovine spongiform encephalopathy (BSE), the problem of the GMO, the avian flu crisis, the diffidence towards biotechnologies, etc. are all manifestations of a much deeper structural phenomenon**. The thesis proposed here is that they are the manifestation of the social affirmation of a rather significant anxiety-inducing syndrome that expresses the particular symbolic relationship we have established with food and the body in the contemporary scenario. I describe such a relationship by defining our society as an *orthorexic society*.[7]

[7] *Orthorexia nervosa* is a psychocultural syndrome, whose definition is the work of Steve Bratman, that may summarily be described as the obsession for healthy (opportune) feeding. Although the clinical picture has still not been definitively established, it represents a condition that affects an increasing proportion of the western population and, here, serves as a metaphoric representation of an 'epoch-making' condition. The orthorexic spends an important part of his time in search of dietetic perfection, organizing, researching and selecting food.

Such a syndrome can be understood with three processes set in motion by modernity: the distancing, in the agri-food production sector, of the producer from the consumer ('opacity' of food), the erosion of the constraining aspect of norms regulating a correct diet (the culinary Order) and the closure in an individualist frame of the human body. Particularly, here, is the first process we have to underline.

Modernity has inaugurated a series of radical transformations in the production processes of food. Domestic cooking has been replaced by an industrial conception, and the new agro-industrial system has rendered food processing opaque. Its identification, namely in the new order of alimentary production, is increasingly difficult and an ever greater part of the population consumes food whose history and origins it does not know. At the same time, the olfactory/taste consistency of a food has increasingly less of the very nature that identifies it. Its 'sensible' characteristics are increasingly an issue of *sign* appearance (marketing, packaging, etc.) exposed to technological sophistication (artificial flavour, additives, preservation techniques, etc.). Also natural and traditional products are subject to imitation or industrial elaboration.

In the framework of such epochal transformations (individualisation and untransparency of the alimentary act), **insecurity and alimentary uncertainty represent an ever more decisive aspect of contemporary society**. Only in such a way can the ever greater demand for informative labelling of food products be understood (ingredients, nutritional tables, traceability, marks of quality and origin, etc.).

Alimentary modernity and the affirmation of the long globalized chains of production create a process of *uprooting* of the alimentary act from the local and specific 'eco-biocultural' ('techno-territorial') context. Through the *taylorization* of production, transformation and distribution[8] food 'takes off' from the territory and it becomes 'indifferent' to the characterizations that in the past[9] rendered it unique. In this way, an *abstract* alimentary system is created. Namely, this is the concrete application to food of those radical transformations that Anthony Giddens has described with reference to modernity in general (1994): *space-time distanciation*, *disembedding* and *reflexivity*.

For Giddens, modernity has lifted out social relationships from the 'local contexts of interaction', assuring at the same time their restructuring 'across indefinite spans of time and space' (Giddens 1994). The most important examples of such uprooting processes are the so-called *expert systems*.[10] In my view, the modern alimentary system must be studied and understood, in its features and social consequences, precisely as an expert system.

---

[8]To this should be added the processes of rationalizing alimentary consumption through: (a) the substantial participation of nutritional science; (b) the taylorist–fordist organization of the food offer [see the *Mcdonaldization* of society by Ritzer (1993)].

[9]Thanks to the earth, climate, seasons and material and symbolic practices.

[10]Giddens defines systems of technical realization or professional competence that organize wide areas in material and social ambient in which we live today (*ibidem*, p. 37).

The *disembedding* of the expert system from the local context of interaction implies that 'trust' becomes essential to its workings. That is because social expectations are created in space-time distanced contexts. As Giddens often states, such trust is at the same time an article of faith and what Simmel called 'weak inductive knowledge'. Because of this dual matrix the confidence entrusted to expert systems is, at the same time, a quasi-magical component (of symbolic type) and a pragmatic one (tied to observation and experience). We have confidence in the expert systems because they are constructed on specialist *knowledge* that is inaccessible and mysterious to us, but which some elected few possess (professionals, technicians). These are considered the clergymen of the churches (with their own technical and legal apparatuses and, above all, with their own methods of monopolistic certification). But we also have faith because we know by experience that such systems generally work and satisfy our expectations. This dual nature can sometimes provoke important dyscrasia (at times when airplanes crash or cows go mad) that fracture the confidence that is generally granted to expert systems, creating a climate of anxiety.

The expert system is prevalently disembodied, founded on functions, roles, flows, etc. Uprooted from local contexts of human interaction, it works as a professional structure that sets aside sympathetic relations. Nevertheless, the same Giddens recognizes that trust given is not only tied to abstract emblems (anonymous); an important part of this trust is conveyed by particular '*facework commitments*'.[11] In this sense, a fundamental component of the expert systems is the so-called *access points*: points in which the abstract systems open up to the external world, guaranteeing contact between its representatives and the profane. They are very important to ensure the maintenance of trust, precisely so that common people *physically* meet the operators of the system. These *openings* to the profane world represent the interface of the system; in fact, they are literally, spaces of shared interaction. For this reason, the access point constitutes, for certain aspects, the most delicate point of the system. The place in which great efforts must be concentrated in order to limit the onset of 'friction' that may throw the entire system into crisis. A crisis that takes on the guise of 'crisis of confidence'.

Now, is important, for our scope, to remember that much of this effort is *communicative*. **All the force or weakness of the expert system depends on the workings of the communicative processes that drive it**. And, equally important, to a large extent the success of such processes depends on the quality of the non-verbal dimension of the communication.

This aspect is determinant for understanding the role and influence that mass media and the ICT (Information and Communication Technologies) have had in provoking the current crisis of confidence in the expert alimentary system. Electronic and digital media are able to influence the 'definition of the situation'

---

[11]Facework commitments are defined by Giddens as relationships of trust supported by or expressed in social bonds instituted in circumstances of simultaneous presence (Giddens, *ibidem* p. 85).

circumventing the material limits imposed by physical structures and geographical and temporal distances; and to break down the barriers constructed to maintain the separation between stage and backstage. To what extent did the images of the staggering cows broadcast by the media during the 'mad cow crisis' influence the propagation of the food crisis? In our view, very greatly indeed. They managed to circumvent the many official reassurances that, in those troubled times, were issued by various official operators of the system (veterinary surgeons, ministers, retailers, breeders, etc.).

This aspect shows the determining role that the third founding characteristic of modernity plays: *reflexivity*, that is, 'the regularized use of knowledge about circumstances of social life as a constitutive element in its organization and transformation' (Giddens 1991). And the media performs a key function in diffusing, accelerating and reinforcing modern reflexivity. Here reflexivity is understood not so much as an enlightened project of cumulative growth of knowledge at the expense of the arbitrariness of tradition. It is seen, on the contrary, as a systematic affirmation of a radical principle of methodological 'doubting'. With reference to food, the example of the media coverage of the so-called 'mad cow' or BSE crisis was emblematic.

It is for this reason that the importance of communicative processes must be stressed. In fact, what we are witnessing, at least since the BSE scandal, is a ferocious *communication war* with still uncertain results. The protagonists of this communication war are various: environmentalist and consumer protection groups, the assorted *stakeholders* are the powerful food industry, the public powers, and in the middle, the media.

In particular, the food industry and the public powers have been creating a communicative strategy aimed at reducing anxieties heightened by the media representation of the crisis in food safety. They have had to touch on what many sociologists and anthropologists have been saying for a long time: eating unknown manufactured products, without a past and social roots can mean losing the last sense of oneself; and eating 'against nature' (Genetically Modified Organisms) can mean symbolically denaturalizing oneself. Therefore, in order to face the current of modern alimentary anxieties, two paths have been undertaken to restore identity to food: labelling (traceability, guarantee of origin, etc.) and advertising narration. Both realize a new artificial form of '*communicative embedding*'.[12]

It is interesting to notice that advertising strategies carry out a function in many aspects analogous to that carried out by traceability. In both cases this concerns, through *narration*, restoring an identity to 'objects' that have become 'opaque' and unrecognizable over time since they are culturally and socially uprooted. Naturally, they involve different forms of narration. Traceability seeks to tell an 'objective'

[12]At the Department of Analysis of Political, Social and Institutional Processes of the University of Catania, we have been making an empirical diachronic investigation into the narrative strategies adopted by food advertising (press) that seems to confirm this (see Nicolosi 2007).

history, a biographical history of a determined food. Publicity, through *discourse* and *branding*, often tells a mythical or fantastic story (stories), so that the consumer can newly incorporate meanings with the food.

3 Communication, Narration and Symbolic Values: An Empirical Analysis

The role of narration in order to restore identity of agri-food products has been shown by research we carried out in 2006 on communication strategies in food advertising in the Mediterranean area. We used a method elaborated on the basis of a four quadrants model by Ferraro (1998).

This model is the result of two axes intersecting each other. The horizontal and vertical axes cover values and issues, respectively, from relative to absolute and from subjective to objective. 'Relative' means that reality is seen as socially constructed; 'absolute' means that reality is seen as nature, that is independent from human intervention. On the vertical axis, we have 'subjective' which means that individuals are seen as the most important, and 'objective', which means that society (or the group) is seen as the most important. The horizontal and vertical axes intersect resulting in four quadrants (Fig. 2). The four quadrants define (clockwise) four different *narrative strategies* or *regimes: causal, positional,*

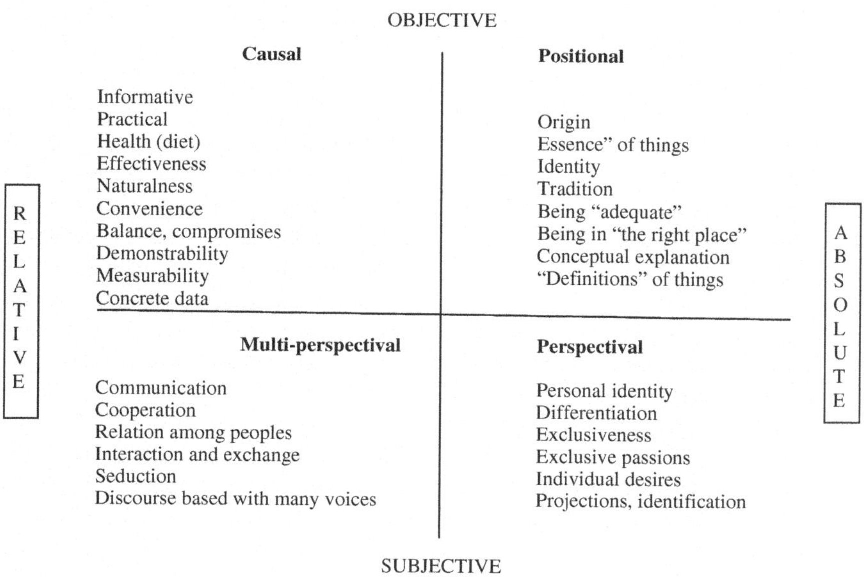

Fig. 2 Four quadrants of narrative strategies and regimes based on Douglas (1970, 1996) and Ferraro (1999) showing the narrative forms and values involved

perspectival and *multiperspectival*. Together, the four quadrants classify in a fairly general and universal manner orientations in societies and cultures. According to the above-mentioned authors, these four orientations or strategies mutually exclude each other, which means for example, that it is impossible to use simultaneously the causal and the positional perspective. The orientations cannot be compromised because they are incompatible organizing principles, as Douglas (1996) states: 'Any choice which is made in favour of one is at the same time a choice against the others'.

The *causal* regime is defined by the objective and relative axes: this strategy is structured in a series of logical steps based on cause and effect relations. The approach tends towards objectivity, observability and rationality. The central values of this model are *information, effectiveness, force, energy* and *power*. What counts are the concrete facts and not values associated with images. Advertising in this field tend to promote the product as a convenient and effective solution capable of solving even complex problems. In texts belonging to this quadrant, a central role is played by compromises, that is, the attractiveness of 'taking into account' different and apparently irreconcilable needs like taste and physical fitness, quality and savings. As Ferraro writes 'it is not a perfect world that is discussed, but a somewhat improved world, a more rational, significant, comfortable, world, in which we have more resources and more energy to face the problems that constantly arise' (Ferraro 1999). He calls this type of advertising 'informative'. The advertising texts that belong to this quadrant emphasize the *practical, measurable* and *objective* qualities of food (like the nutritional value and calories.). *Rules* and *measurements* are central aspects and *diet* (in a broad sense of living healthy with food) is the texts' central message. Nature does not dictate what humans should eat, but scientific (human) knowledge about what is healthy and good for us.

The *positional* regime is located in the objective/absolute quadrant; here the fundamental value is the intrinsic and 'objective' qualities of the product. The narrative formula used in texts belonging to this quadrant emphasizes the value of the *essence* of things, their true nature. For this reason, the food products have according to this perspective static qualities, referring to their traditional origin. *Authenticity* holds a supreme, *absolute* value (Ferraro 1999). Food advertising belonging to this regime constantly evokes *tradition, nature* and the value of traditional things. Ferraro calls the actual advertising associated with this regime 'identity advertising'; the message is based on 'the expression of the full adequacy of a product to its essence' (Ferraro 1999), an essence that is usually original, but can also be the final result of a process.

The *perspectival* regime is defined by the subjective/absolute axes. This regime is presented as coming from a specific subject or from a 'system of values' expressing a specific (and usually exclusive) 'world vision' (Ferraro 1999). The central aspects of this quadrant are *exclusiveness* and *uniqueness*. Products are associated with prestige and luxury. They are presented as powerfully captivating, irresistible, and as the fulfilment of irrepressible desire. There is often use of irony and smile. Typical objects of this narrative strategy are clothing, jewels and perfume. The *exclusiveness* and *uniqueness* that define this quadrant are expressed in the sector of food by the centrality of 'gourmandize', the consumer is perceived as a 'gourmet'. The characteristics of food are not defined as objective elements in the real world, but presented as dependent on the taste of an exclusive and unique subject, the gourmet, i.e. the expert and connoisseur of good food.

Finally, the *multiperspectival* regime of the subjective/relative quadrant is characterized by a strategy that presents the interaction and integration of different perspectives as various systems of values are playfully interacting. The wealth of diversity of food in the world is seen as one continuous conversation between cultures. 'Food adventurers' enjoy the rich variety of cuisines. Food advertising according to this perspective centres on the deregulation of traditions and associates with creativity, the 'de-ritualizing' of meals, and with cooking innovations, like new dishes, new origins and new combinations.

3.1 The Empirical Research

In our empirical research we analyzed food advertisements that appeared in 2004 in 3 weekly periodicals (one Italian and two Spanish). The research was based on two different counting systems. First, we counted how many ads belonged to each narrative regime. In this case, each ad was counted only once and therefore repeating appearances of an ad are not counted. The resulting counting shows how much the advertising world uses the four regimes with a given text (we called this text 'models' in Table 1). Second, we also sought to evaluate the 'diachronical impact' on the readers of the journals, by counting the total number of times the same ad expressing a given narrative regime, was published (the 'frequency' in the table). Results were analyzed separately for each publication. Third, we analyzed some representative advertisements more in depth, to highlight some interesting features of the dominant regime.

Table 1 Models and frequency—D di Repubblica

Regime	Models	Frequency
Causal	8	13
Positional	33	96
Perspectival	24	86
Multiperspectival	–	–
Total	65	195

3.2 Data from Italy: D di Repubblica

We analyzed a weekly magazine called *D di Repubblica: La Repubblica delle Donne*,[13] distributed on Saturdays along with the newspaper *Repubblica*, and targeted at a mainly public female of middle-upper class (the 'D' in the title is for *donna*, woman). We examined the issues published between 14 March 2004 (n° 392) to 30 April 2005 (n° 447). Research therefore included a detailed analysis of 54 issues[14] and 195 total ads. The results were shown in Table 1.

The majority of observed food advertisements belonged to the quadrant we called 'positional'. More specifically, as shown in Table 1, 33 out of 65 ads with a given text (sometimes the same brand and/or product is presented through different texts) belong to this quadrant. Out of the total 195 texts surveyed, a total of 96 (49 %) belong to this quadrant.[15]

Quantitative analysis, however, produced an extremely interesting datum: a good deal of the ads belonging to the positional quadrant also used the typical tools of *causal* advertising to suggest the health qualities of the products. There are references to the ingredients of the product, often presented in terms of percentages references to the process of production, or information on the nutritional values (like minerals and fibres). Specifically, 22 out of 33 advertisements belonging to the positional narrative regime adopt rational and informative elements in support of their 'argument'. If we consider the 'diachronical impact' (frequency, counting also repetitions) this approach is found in 49 out of 96 appearances (51 %). It seems as if the narrative strategies are not so exclusive vis-à-vis each other as theoreticians want us to believe!

Another interesting aspect is the frequent use of brands and symbols (such as logos of consortia for the protection of consumers) that guarantee that the product has been constantly monitored and its quality and safety have been certified. This aspect is present in 36 appearances out of 96 belonging to the positional strategy (38 %). Only one case from the causal regime (13) appeals to brands and symbols for the protection of consumers.

---

[13]This study was done in collaboration with Dr. Venera Trepiccione, University of Catania, Italy.

[14]The only gap is issue 421 which we were unable to obtain.

[15]In reality, we are dealing with a rather uncommon discursive 'polarization' between positional regime (48 % of the cases) and perspectival regime (43 % of the cases). The high presence of ads belonging to the perspectival regime can probably be explained in terms of the social characteristics of the audience of the magazine: tendentially high social status and high level of education. This audience looks for highly distinctive, even exclusive, consumer products. This is confirmed by the analysis of the textual content of the magazine's columns and articles (fashion, design, architecture, etc.). It is worth remembering also that many ads belonging to the perspectival quadrant advertise ice cream and coffee.

3.3 Discussion of Two Advertisements

The first advertisement, which we consider particularly significant, was that of 'Grana Padano' (Fig. 3).

A central theme of the advertisement is the relation of the product to milk—but not to the cow. By only talking of milk, and not alluding to cows, the ad distracts the consumers' attention from recent anxieties like the 'mad cow'-disease. This metaphorical 'slippage' is evident in the *visual headline*, where the rational and linear argument is supported and reinforced by the photo collage (cheese as a cup of milk). Besides, milk always functions very well as a symbolical synecdoche. It represents nature and purity and refers directly, through another considerable metaphorical slippage, to the idea of maternal love. And moreover, what could be more reassuring than milk? Milk is our first meal as newborn and it makes up the only diet in the first months of our lives.

The ad reassures stressing that the milk is particularly safe insofar as it is Italian (instead of English, for example). Furthermore, the authenticity of the product is confirmed by the origin (space) and by tradition (time in the sense of history); this last one is guaranteed with 50 years on the side of quality ("*50 anni dalla parte*

Fig. 3 Image of 'Grana Padano' in 'D di Repubblica'newspaper n° 397, 17 April 2004

Fig. 4 Image of 'Parmigiano Reggiano' in 'D di Repubblica'newspaper n° 397, 17 April 2004

della qualità"). There are other devices that refer to the authenticity and quality of the product, like the reference to the 'Grana Padano Protection Consortium' that supervises cheese production through 'continuous and rigorous verifications'. Reference is also made to its PDO certification (Protected Denomination of Origin), which is an additional guarantee of quality. This certification and the existence of a consortium serve to reassure consumers that this is a 'safe cheese'. But also a cheese that respects regional traditions and therefore contributes to the continuation of the local identity—and even to the life stories of the consumers.

A second highly significant example of territorial and cultural association of food and location is found in the ad for 'Parmigiano Reggiano' (Fig. 4). In the foreground there is a whole Parmesan cheese with the mark of the brand name clearly visible on the cheese itself. The cheese is surmounted by photos of five important monuments of five well-known cities of the Emilia region.[16] This collage of photos underlines the 'monumental' status of the product. Like the depicted monuments, the cheese also functions as a symbol of a geographical area (due to its history and

[16]Note, incidentally, that Parma is presently the European seat of the Agency for Food Safety. Parma is well known in Italy and abroad for the quality of its food products though its reputation recently received a blow because of a financial scandal involving Parmalat, the city's largest food company. The Italian government lobbied heavily for Parma to be the seat of the European agency.

culture) famous for its food products. For this reason, the product is presented as 'unique as its origin' (see subheadline). The first part of the subheadline in the upper half suggests the long tradition behind the product ('70 years') stressing the fact that 'this extraordinary example of nature and wisdom' is a 'gift' from 'generations of the place of origin'. Tradition, territory and nature are merged in a single message.

The idea of the product as traditional and natural is reinforced by another element: the phrase in the payoff: 'Parmigiano Reggiano. You don't build it, you make it'. The product is presented as the result of a tradition, which stays clear of industrial locations and procedures. It is 'prepared solely with milk from cows of the area of origin', as specified in the *bodycopy*. The latter also states that the product is 'without additives', reassuring customers about possible chemical contaminations. The Consortium vouches for all these qualities, ensuring that the product is 'controlled unit after unit, day after day'.

3.4 Hybridization of Advertising Strategies

Tradition, nature, geographical and cultural origins are almost obsessively present in the ads examined. Everything seems to revolve around the 'quasi-mythical' celebration of these elements, presented as actual and fixed absolute entities. In a certain sense these advertisements do not express a genuine story since they do not show a historical development starting with a point of departure through a process to a final goal. The readers are immersed in a realm of immutable elements and are drawn in a movement of eternal return of the same elements again and again. The idea of authentic purity acquires an almost divine quality and the authenticity of food is its most marked expression. Eating and food consumption is presented as a sacral act in which the essential things in life are intimately preserved and maintained. However, apparently in contradiction to these absolute items, **the texts of the advertisements do refer to human and scientific items, like information on the safety and health of the product**.

The narratives used in actual advertising show rarely one strategy in its pure sense. In particular, ads belonging to the 'objective' (or upper) side of the four quadrants (causal and positional) tend to influence each other. In the Italian case it is evident that positional advertising, which is the most common, often shows elements typical of causal advertising (objective characteristic encounters relative characteristic).

These two features give these texts a certain ambivalent or hybrid character. The hybridizing consists in the frequent use of a technique which could be defined as 'information support'. It results in a significant type of advertising halfway between identity and informative advertising, aimed at providing rational reassurance. Measurability, reference to concrete data and the need to (scientifically) demonstrate the positional strategy are the most significant aspects we have detected. In a certain sense science is needed to support the cultural (positional) elements: science and cultural myths of origin and authenticity fuse in one new way of advertising.

Seen from this perspective of hybridization, the examined food advertisements seem often to break, somewhat unpredictably, the traditional distinction between mythical and scientific thought. Lévi-Strauss (1978) has studied the differences and similarities between science and cultural myths and indicates that both have the function of ordering elements of reality. Science and cultural myths both refer to concepts of 'order' and 'rule' (code). However, they differ in their method of ordering reality. Scientific thought and reductionism initially fragmentize problems into as many parts as necessary with the aim to order and to solve them afterwards, it reduces complex questions into single problems that can be analyzed and explained, whereas mythical thought is a totalizing form of thought, which gives the illusion of understanding the world while incapable of providing great control over it (Lévi-Strauss 1978). The ads we have analyzed seem to try to overcome this distinction through a persuasive act based on both forms of thought. They sometimes 'incorporate' both thoughts in attempting to persuade as many consumers as possible.

In several cases discussed by Nicolosi and Korthals (2008), the most common narrative form is the causal one (informative advertising). Consumers are confronted with a text that refers to 'demonstrable' data and provides measurements with the purpose of presenting and legitimating the truth of the product. But in a significant number of cases, informative advertising takes on a hybrid character that is remindful of the typical form of identity advertising. But more than tending towards a relative configuration (like comparison and improvement) the texts appear to present some values as absolute and ethical, which is typical of the positional quadrant, like respect for the body. The advertisements address not the improvement of imperfect bodies but rather represent the achievement of a perfect and healthy body through a complex process of giving analytical information and an authentic value. The healthy and perfect body is considered as something of sacral value, and health is sanctified in sanctity (compare Douglas 1996; Fischler 1996) refers to the same phenomena with his concepts of sanitas and sanctitas). The moralistic aspect of these advertisements consists not so much in the perfection of the product as in the perfection of a lifestyle that one can achieve through this product. It expresses a late modern ideal of a scientifically and rationally achievable 'sanctity'. Perfection is not given, it can be acquired: this could be the meta-headline of all these advertising texts. Perfection does not belong to an ahistorical and atemporal non-place independent of human actions and volition (as in the case of the positional model). The texts suggest that we are not born gods, but that we can become one; and everyone can do it, so long as he or she follows the indications and respects the rules (the diet), the principle of a 'correct' and healthy life, a life founded on correct and healthy food, i.e. food interpreted according to the doctrine of natural law: the divine law expressed as a law of nature.

Moreover, a research study carried out in Italy shows some important aspects. The analysis of weekly newspapers put in evidence that the narrative advertising

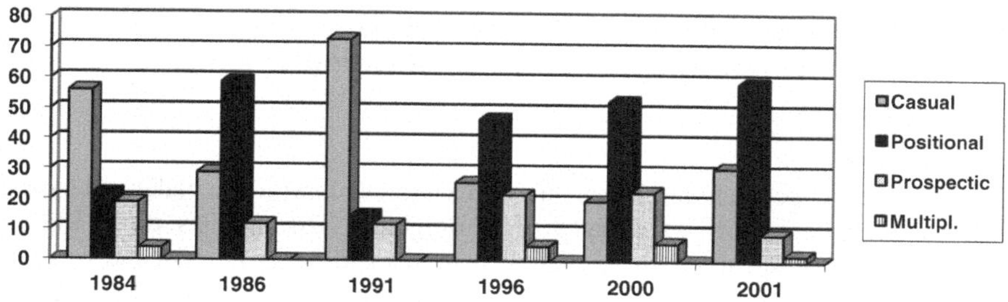

Fig. 5 Narrative advertising (%) weekly newspaper 'Gente'

seems to be exposed to relevant changes in accordance to some important chronological steps (for example the BSE scandal) (Fig. 5).

Diachronic analysis showed:

(a) in 1984 (before the BSE scandal) informative advertising is prevalent: it is the development of *energetic* power of food which is the most important benefit pointed out;
(b) but in 1986 (after the first case of 'mad cow' in England) we verify a dramatic slump of informative advertising and the growth of positional advertising;
(c) a new balance between causal and positional advertising is restored for next 5 years of 'relative calm';
(d) a new dramatic change in 1996, with a new trend of growth for the positional;
(e) this trend is reinforced in 2001, alongside a new revival of the causal but, as we will see, supporting new 'values' (the new benefit is now *health/safety*).

Those who followed the 'mad cow' case knows that 1996 and 2001 are special years. In spring 1996, Europe 'trembles' as British authorities admit, for the very first time, that the disease can be transmitted to humans. In January 2001, infected animals have also been found in Italy, considered safe for years by national health authorities.

What seems to have happened is that the media explosion of BSE scandal induced a strategic change within narrative advertising from a rational, informative and comparative narration to a symbolic, absolute and in some cases mythological (nature, purity, etc.) one. Moreover, the content analysis put in evidence that:

- positional advertising is more and more hybrid, that is, symbolical communication (tradition, geographical location, etc.) is supported by rational elements of measurability (labelling, data, certification, etc.);
- causal advertising abandons, in time, the priority acknowledgement of the benefit *energy* (prevailing in 1984) for focusing on the benefit of *health/safety* (prevailing in 2001).

4 A Value-Based System for Risk and Benefit Communication in Agri-Food Applications: Some Guidelines

To communicate coherent messages on agri-food related issues is challenged by the broad range and variety of sources of information. This applies especially to risk/benefit communication where citizens perceive and interpret risks and benefits in different ways. Therefore, the topic on risk/benefit perception and communication in agri-food applications is a subject of major importance.

It is considered highly relevant to develop a communication model valorizing risk/benefit communication in the agri-food sector and enhance European wide communication with consumers, by integrating a crucial, hitherto widely ignored aspect: The significant importance of values in risk and benefit perception.

In order to do this, it is highly recommended to:

(a) Develop a risk/benefit taxonomy and correlating communications strategies for specific risk types;
(b) Classify the dominant values in risk and benefit perception;
(c) Identify technical and value-laden communication barriers in risk/benefit communication;
(d) Develop different tools designing communication strategies precisely tailored to population groups, with shared value horizons on the basis of objectives 1–3.

The EU has set in place policies that allow Europe to play the leading role in agri-food safety issues and has developed a consistent, science-based and transparent system in the agri-food sector whilst constantly further developing the established standards. Basically, our food has never been safer. Nevertheless, in a globalized agri-food economy, risk/benefit communication is even more important than ever before. New strategies can no longer be postponed. Strategies taking account of:

- a **BROADER CONCEPT OF RISK** that considers the risk and benefit assessment of citizens;
- an **APPROPRIATE MODEL OF COMMUNICATION** that takes into account target audience as actively constructing information;
- an **UPGRADED CONCEPT OF RISK AND BENEFIT COMMUNICATION** that is informed by the recipients' value horizons.

Today, one of the vital challenges of contemporary European policy is not only to safeguard high standards in agri-food but to also be prepared to communicate effectively and give public guidance. Risk/benefit communication faces at least the following fundamental challenges:

- **BROAD RANGE OF INFORMATION**: European citizens receive a broad range of information from a variety of sources that may be unbalanced or interpreted as conflicting and/or confusing;

- **VULNERABILITY OF PUBLIC TRUST**: Scandals in recent years proved the vulnerability of public trust: Trust is created slowly, but can be destroyed in an instant, by just one single mistake. Consequently, it cannot be taken for granted that European citizens trust the communicator of risks and benefits;
- **INFORMATION OVERKILL**: Messages about agri-food issues run the risk of drowning in a flood of information, failing to attract citizens' attention.

For all these reasons, following answers are required:

(1) **CONSIDERING VALUES**: Regarding risk communication, the exclusive provision of scientific facts is not sufficient for all types of risks. The European Commission justly pointed out, that consumers need to be provided with easily accessible and understandable information relating not only on scientific opinions. Tailoring information precisely to the values and worldviews of recipients enhances the efficiency and probability of successful communication;
(2) **BUILDING TRUST**: Building trust implies transparency concerning the values. Therefore, integrating the value-laden perception of agri-food issues is necessary to understand consumer behaviour and gain trust as an important aspect of risk and benefit communication;
(3) **TAILORING MESSAGES**: Knowing the value-laden structure of consumers' perception allows the message to be adjusted according to values that play a vital role in citizens' lives.

The common approach of risk communication addresses the quantifiable aspects of risks and benefits. At this point communication failed in the past since only the quantifiable aspects of risks were communicated and did not address the wide range of values at stake. Therefore, we suggest to go beyond the common approach, by addressing the existing plurality of value dimensions which are at stake. For example, 'Healthy Food', 'Animal Welfare', 'Organic', 'Sustainable Rural Life', etc.

The opposing and successful strategy is to communicate shared worries that indicate value similarity, which is a vital element for any trust or trust building process.

Last but not least, it is needed to give relevance to:

(a) 'Containers'

  - the **FUNDAMENTAL ROLE OF NEW MEDIA**: Achieving the objective of developing a value-based communication support system means also to analyze various new communication channels and their advantages and disadvantages compared to traditional channels in risk/benefit communication (web, mobiles, apps, etc.).
  - the use of **TRANS-MEDIA TECHNIQUES**: that is, telling a single story or story experience across multiple platforms and formats using current **digital technologies**. Trans-media storytelling involves creating content that

engages audience using various techniques to permeate their daily lives. In order to achieve this engagement, a trans-media production will develop stories across multiple forms of media in order to deliver unique pieces of content in each channel.

I suggest we validate user-friendly, web-based tools that advise and design agri-food related risk/benefit communication strategies, precisely tailored to the target audience. Informed by experiences made during former agri-food scares, these tools will explicitly address the tremendous influence of individual value horizons on risk/benefit perception. The tool users (such as European policy makers, national and local decision makers and other responsible actors) can design and adjust their communication strategies precisely to target audience value horizons, in order to communicate risks and benefits effectively and appropriately. These tools would be designed with detailed input from consumer organizations and media channels to ensure that they meet user needs and expectations.

(b) 'Content'

- the **RELEVANCE OF NARRATION**: Narrative (or storytelling) is the systematic recitation of an event or a series of event. The narrative transport is a crucial part of social cognition. Stories have a powerful and universal appeal and storytelling is a fundamental component of effective public speaking;
- the **IMPORTANCE OF HYBRIDIZATION STRATEGIES**: Narration can be supported by measurability, reference to concrete data and the need to (scientifically) *demonstrate* the positional strategy. Science and cultural myths of origin and authenticity fuse in one new way of communicating.

References

Censis (2001) XXXV Rapporto annuale sulla situazione sociale del Paese, Roma

Cerroni A et al (2002) Biotecnologie e opinione pubblica. Una ricerca sulla percezione della scienza in Italia. Sociologia e ricerca sociale, 67:117–140

Douglas M (1970) Purity and danger. An analysis of concepts of pollution and Taboo. Penguin Books, Harmondsworth

Douglas M (1979) Les structures du culinaire. Communications 31:145–170

Douglas M (1996) Thought styles, critical essays on good taste. Sage, London

De Garine I (1979) Culture et nutrition. Communications 31:70–92

Eurobarometer (2012) Biotechnology Bruxelles

Eurobarometer (2003) Europeans and Biotechnology in 2002, Bruxelles

Ferraro G (1998) 'Senso e passione'. In: Brugo I, Ferraro G, Schiavon C, Tartari M, Al sangue o ben cotto. Miti e riti intorno al cibo, Meltemi, Roma, pp 11–51

Ferraro G (1999) La pubblicità nell'era di internet. Meltemi, Roma

Fischler C (1979) Gastro-nomie et gastro-anomie. Communications 31:189–210

Fischler C (1988) Food, self and identity. Soc Sci Inf 27(2):275–292

Fischler C (1996) Pensée magique et alimentation aujourd'hui, Les Cahiers de l'OCHA, Paris
Gallino L (1998) Critica della ragione tecnologica. Valutazione, governo, responsabilità dei sistemi sociotecnici. In: Ceri P, Borgna P (eds) La tecnologia per il XXI secolo, 5–24. Einaudi, Torino
Gaulin SJC (1979) Choix des aliments et évolution. Communications 31:33–52
Giddens A (1991) Modernity and self-identity: self and society in the late modern age. Polity Press, Cambridge
Giddens A (1994) Le conseguenze della modernità. Il Mulino, Bologna
Latour B (1995) Non siamo mai stati moderni. Saggio di antropologia simmetrica. Elèuthera, Milano
Lévi-Strauss C (1978) Myth and meaning. University of Toronto Press, Toronto
Lévi-Strauss C (2002) Antropologia strutturale. Il Saggiatore, Milano
Lévi-Strauss, C. (2004) Il crudo e il cotto, Milano : Il Saggiatore
Lupton D (2003) Il rischio. Percezione, simboli e culture, Il Mulino, Bologna
Nicolosi G (2007) Lost food. Comunicazione e cibo nella società ortoressica, Ed.it, Catania
Nicolosi G (2008) Communic(e)ating. Communication and the social embedding of food. In: Ruivenkamp G, Hisano S, Jongerden J (eds) Reconstructing biotechnologies: critical social analyses, Wageningen Academic Publishers, Wageningen, pp 229–242
Nicolosi G, Korthals M (2008) Narrative strategies in food advertising. In: Coff C, Barling D, Korthals M, Nielsen T (eds) Ethical traceability and communicating food. Springer, Dordrecth, pp 63–78
Perlès C (1979) Les origines de la cuisine. L'acte alimentaire dans l'histoire de l'homme, Communications 31:4–14
Popitz H (1990) Fenomenologia del potere. Il Mulino, Bologna
Poulain JP (2002) Sociologies de l'alimentation. Puf, Paris
Ritzer G (1993) The Mcdonaldization of society. Sage, London
Wynne B (1994) Scientific knowledge and the global environment. In: Redclift M, Benton T (eds) Social theory and the global environment. Routledge, London, pp 169–189

The COST Action FP0905 Experiences on the Web: Web 2.0 and Scientific Dissemination

Fabio Migliacci, Donatella Paffetti, Jeremy Sweet and Cristina Vettori

Abstract Forms of communication have evolved over time with increasing speeds reaching the present form of the dynamic Web 2.0 with the emergence of communities, social networks (i.e., Facebook, Twitter), wikis, etc. The internet is no longer just a means of information and use of content for daily use, but it has become part of everyday life, changing our behavior and consequently the way we think. For spreading information relevant to this COST Action FP0905, the website (http://www.cost-action-fp0905.eu/) was set up to provide communication between participants of the Action, and to disseminate the activities, knowledge, and technology produced beyond the COST Action network. The principal aims of the website were to provide a database of the main information on genetically modified trees (GMTs) and to update the website with science based information of public interest on GMTs related to the activities within the Action. Therefore, Web 2.0 technologies were used to support the aims of the COST Action, with specific positive (YouTube and Twitter) but sometimes also negative communication experiences (i.e., hacker's attacks) of the Action, which are reported.

F. Migliacci
National Research Council (CNR), Institute of Chemistry of Organometallic Compound (ICCOM), Via Madonna del Piano 10, 50019 Sesto Fiorentino, FI, Italy
e-mail: fabio.migliacci@iccom.cnr.it

D. Paffetti
Department of Agrifood Production and Environmental Science, University of Florence, P. le delle Cascine 18, 50144 Florence, Italy
e-mail: donatella.paffetti@unifi.it

J. Sweet
Sweet Environmental Consultants, 6 Green Street, Willingham, Cambridge CB24 5JA, UK
e-mail: jeremysweet303@aol.com

C. Vettori (✉)
National Research Council (CNR), Institute of Biosciences and Bioresources (IBBR)—UOS Firenze, Via Madonna del Piano 10, 50019 Sesto Fiorentino, FI, Italy
e-mail: cristina.vettori@cnr.it

C. Vettori et al. (eds.), *Biosafety of Forest Transgenic Trees*,
Forestry Sciences 82, DOI 10.1007/978-94-017-7531-1_16

1 Introduction

In 1967, Paul Watzlawick, Austrian-American psychologist and philosopher, together with Beavin Bavelas and Jackson researchers of School of Palo Alto (California), published the book "Pragmatics of human communication: a study of interactional patterns, pathologies and paradoxes", which defines the basic axioms of communication (see Box 1) to understand the dynamics that occur in interpersonal communications and permit the obtaining of two important results:

(i) formulate effective communication;
(ii) avoid misunderstandings and inconsistencies.

Box 1 Basic Axioms of Communication (Watzlawick et al. 1967)
First axiom: "One cannot not communicate"
Second axiom: "Every communication has a content and relationship aspect such that the latter classifies the former and is therefore a metacommunication"
Third axiom: "The nature of a relationship is dependent on the punctuation of the partners communication procedures"
Fourth axiom: "Human communication involves both digital and analog modalities"
Fifth axiom: "Inter-human communication procedures are either symmetric or complementary"

Therefore, according to Watzlawick's theory on communication, these axioms are necessary to have a functioning communication between two individuals. If one of these axioms is somehow disturbed, communication might fail.

Communication is a set of processes through which we share and transfer information. However, effective communication requires that the issuer (i.e., the one who initiates communication) takes into account the various elements in order to ensure the effective transfer of the message to the recipient (either an individual or a group of people).

Often the term effective communication is used as a synonym of persuasive communication, but this is incorrect, since the two terms imply a different goal. A persuasive communication has the aim to produce a change in behavior, attitudes, or values of the recipient, and an effective communication has to successfully transfer the contents of the message.

Fattorello (1961) defined the so-called "5 Golden Rules" for effective communication (see Box 2).

Box 2 Rules for Effective Communication (Fattorello 1961)

1. *Listening*, seems a paradox but a good communicator must know how to listen, not just with your ears, but also with the eyes and heart. Active listening is the first requirement to better understand the real meaning of a message.
2. *Completeness*, i.e., the message must contain all the necessary information for the recipient to evaluate an offer or a situation, or to solve a problem.
3. *Conciseness*, be concise means to communicate all information relevant to the content of the message, without adding unnecessary details or redundant.
4. *Fairness*, the absence of grammatical or syntactic errors improves the clarity of the message and increases its impact, positively influencing also the listener.
5. *Clarity*, means focusing on one goal, thus emphasizing its importance and making an easier assimilation of the message by the receiver.

There can be no communication if you do not keep in mind (either unconsciously or automatically) these basic concepts of common speech and without you having to constantly consider whether the ultimate goal of the act is taking place, i.e., make yourself understood.

These considerations always apply to any version of communication: whether it is Web 1.0, Web 2.0, or those that follow in the near future.

The term Web 2.0 refers to a set of all those online applications that allow a high level of interactions between the website and the user, such as blogs, forums, chats, wikis, media sharing platforms like Flickr, YouTube, Vimeo, and social networks like Facebook, Myspace, Twitter, Google+, Linkedin, Foursquare, etc., typically obtained through appropriate Web programming techniques (O'Reilly 2005; de Judicibus 2008).

In this chapter, we look at how the complex world of the Web is transforming communication and training in scientific dissemination, through the experience of four years of the COST Action FP0905, in order to draw useful insights for communication activities of scientific themes on the Web for researchers, students, and the public. The following contents are drawn from studies, research, and experiences collected online, considering the Web as a primary source of information. The case histories examined refer to communication theories, the Web in its evolution, social media, international scientific journals, etc. This work is relevant to anyone who wishes to combine the social media Web 2.0 information with scientific dissemination.

Factual science-based information is being communicated through the website of the COST Action FP0905 to provide science based information on technical, socio-economic, and environmental aspects of Genetically Modified (GM) forest trees to the general public.

In particular, we refer to those online applications used within COST Action FP0905.

2 Web 2.0 and Scientific Dissemination

The **marriage between Web 2.0 and scientific dissemination** has every reason to take place, since concepts such as **sharing, collaboration, and interaction** are very relevant to the area of communication and science. Specific terms have been identified as "Science 2.0" or "Health 2.0" or "Medicine 2.0", in order to indicate the application of new technologies in the field of science or communication and to evaluate their impact on the dissemination and training/updating of researchers and the public (Sandars and Schroter 2007; Hesse et al. 2011).

Readers of science journals and websites of scientific societies have been able to observe the new features introduced and been invited to participate. The distribution of the content can be done by RSS feeds, podcasts, blogs, social bookmark, and tagging systems to classify and share content portals, and to activate special channels on major social media (especially Facebook, Twitter, and YouTube).

In science, the recipients of these new tools of aggregation and socialization are students and researchers. However, of increasing importance is the role of the general public in the online community, relying on their social networks and systems that enable them to create and share scientific information on particular topics (sustainability, biodiversity, deforestation, pollution, GMOs, etc.) based on available data and information (Eysenbach 2008).

At the global level, **through the Google Trends** search engine (https://www.google.com/trends/), you can analyze in time how interesting is a research topic, such as the one linked to genetically modified organisms (GMOs). The results are in a graph that shows the term's popularity over time in (nearly) real time for a period from January 2004 to July 2015 (Fig. 1).

If you hover your mouse over the graph (Fig. 1a), the numbers that appear show total searches for the term (GMO) relative to the total number of searches done on Google over time. They do not represent absolute values of search volume, because the data are normalized and presented on a scale from 0 to 100. The same survey shows the most scientific information related to the term GMO that has been subjected to search on the web (Fig. 1b). Moreover, the countries and cities (Fig. 1c, d) with higher search volumes are reported.

The use of the Web 2.0 tools to obtain this information for several topics is increasing. A recent survey conducted in the United States has shown that **about 44 % of Americans use the tools of social media and social networks to search for information regarding medical information** (White and Horvitz 2009).

Researchers are also increasingly using participating tools of Web 2.0. In the United States, the American Institute Manhattan Research reported in 2010 that

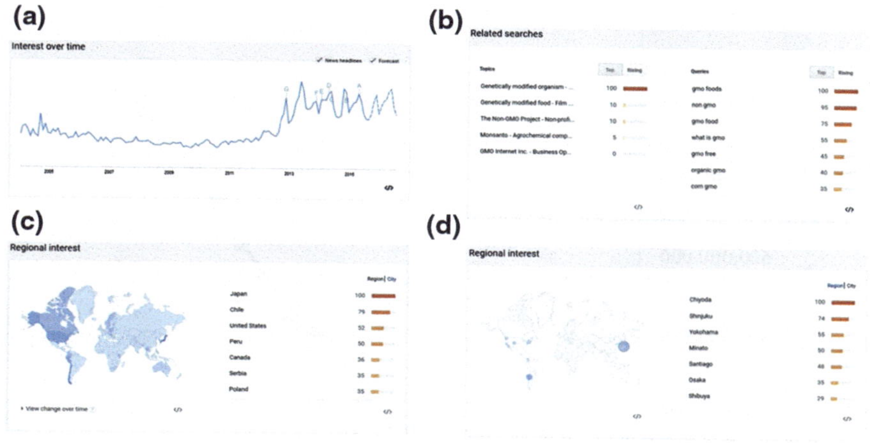

Fig. 1 Searches for GMOs on Google Trends over time with forecast included (**a**), interest for "related searches" (**b**), for region (**c**), and for city (**d**)

over 35 % of medical researchers read Web 2.0 content from blogs, chat, online forums and social networks, and in addition 20 % regularly contribute to create new ones.

3 Web 2.0 and Public Opinion: The Results

The most interesting aspect is that the general public, and not just scientists or researchers, make use of blogs, social networks, and other tools on the net for topics covering scientific themes. These people contribute to public opinion in debates and discussions and they are more inclined to have relationships with those who are interested in a specific topic than members of the scientific community. The general public have greatly enhanced their ability to connect to the Internet (Fig. 2) and, when online, they increasingly use blogs and other participatory tools available on the net to discuss their problems and arguments principally related to their future and their health (Bach 2008).

Another survey showed that, in the case of matters relating to aspects of health, as can be the case of GMOs, the public have active participation in blogs and forums on topics which arouse emotions and these emotions can be shared (Hurley and Smith 2007). Thus, the collaborative tools of the Internet, and those provided by the Web 2.0, are a means to remain "connected" with the world, especially when new events or new findings are disseminated in your country or involve aspects of daily life (Santoro 2007).

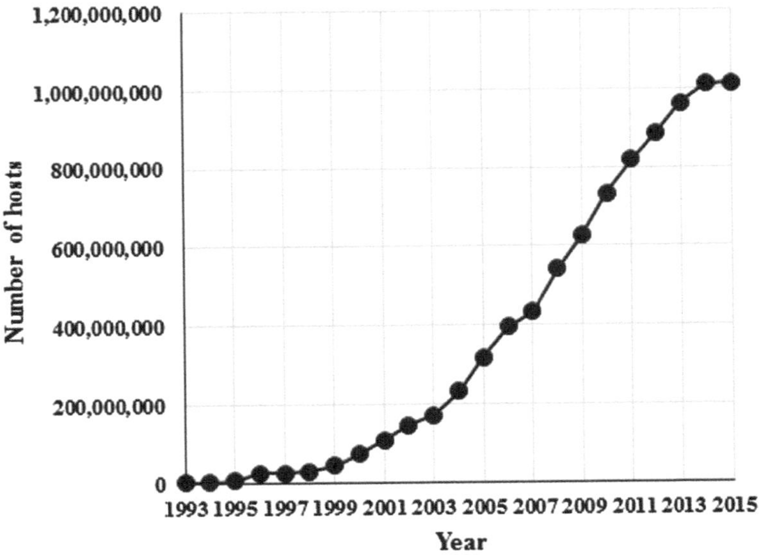

Fig. 2 Host internet 1993–2015. Data host count worldwide based on statistics published on the Internet Domain Survey (http://ftp.isc.org/www/survey/reports/current/) by Internet Systems Consortium (http://www.isc.org/). The data reported are referring to January of each year. The latest update of the series was published in January 2015

For the reasons stated above, it is important to note that the Web 2.0 technologies have been very appropriate and useful for spreading the information and technology produced by the COST Action FP0905 network to the general public in order to inform public opinion.

4 "RSS Feeds"

One aspect that characterizes the technology of Web 2.0, and why in 2011 it was chosen to be used by COST Action FP0905, is the ability to receive and automatically organize in one place the Web content relevant to the Action. It was considered that the previous practice of using "favourite" sites to monitor for new posts or articles in both the Website of the Action and in other websites was obsolete.

RSS (Really Simple Syndication) technology is an XML-based format (an extension of HTML, hypertext) that many sites use to communicate their upgraded content to Internet users. The distribution of the contents in this format is done through special files called "**RSS feeds**". These define a structure composed of fields (for example, containing the name of the authors of a scientific paper, the title, the summary, and the Internet address where to find insights or access the full article) which is updated every time that the website publishes news in this format.

RSS technology and RSS feeds have been used first by the websites of scientific journals for distribution of the contents of new issues to subscribers to appropriate services. Among the information provided in the content are usually the title, the authors and the summary of new articles, and a link to the website of the journal in order to be able to read the full text or further explore the item on line. Over time, websites and scientific journals have extended the use of RSS feeds to provide alerts for the most recent articles and those published in previews, or simply to report new articles published on blogs managed by them on specific topics.

Some typical examples are the RSS feed "Three most recent issues" offered by the American Medical Association journals, the feed "Recent articles" of BMJ, and the published in preview feed "Online First" by the Lancet.

This RSS system was fully integrated into the website of COST Action. Therefore, any participant registered to the site of COST Action, if he/she became a member of an RSS feed and of each of the websites of interest, would have benefited from the new content published on the website in a single window, through a news aggregator (feed reader), without having to visit the home page of each site.

The RSS system, integrated into the website of COST Action, permitted the reading of full articles to subscribers to the website, unless the policy of the Action specifically restricted access to that article.

In particular, the website of the Cost Action attempted to develop a service that consisted of a page of the site dedicated to the topics of the Action for each working group (WG). The user could subscribe to any channel of each WG and, in addition to the functions described above, was allowed to receive internal reports that belonged to one of the four WGs or articles reported by members including links to the journals that contained them.

In addition, the **scientific portals** were organized to distribute, through specific RSS feeds, published information. They ranged from simple RSS feeds to distribute the most recent news, to a multi-channel distribution (Frost and Massagli 2008).

RSS technology and RSS feeds have become **valuable tools for the professional development of scientists and operators**. RSS technology and RSS feeds also offer enormous potential that could be exploited in the fields of communication, especially scientific and research. An useful example is the scientific portal developed by the "Società Italiana di Medicina Generale" developed with "ASCO (Aggiornamento Scientifico Online)" project (http://www.progettoasco.it/), in which each medical area has a specific RSS feed.

5 Podcast

A podcast is an audio file (also called audiocast) **or video** (also known as video podcast or vodcast) usually available in MP3 or MP4 format, which can be downloaded manually or automatically from a server located on the Internet to anyone who subscribes to a particular service of periodic distributions.

The peculiarity of the podcast (and vodcast) is that it can be heard or seen by anyone, at anytime, and in every place. In fact, once downloaded, they can be played on the computer (for example, through the Windows Media Player software), on iPod or any other MP3 player. This feature makes them different from stream systems (such as for example YouTube) in which the audio and video files may be heard or seen at the user's request but provided continuously by an activated Internet connection with the servers on which they reside.

This is an important difference from stream systems, since **anybody can automatically receive files from multiple sources** and organize their "schedule" to consult on site in the manner and within the most opportune times. The success of the Podcast was due to its portability, combined with the opportunity of enjoying the podcast even in off-line mode, or in conditions of mobility. This tool, originally used to create new forms of Internet radio stations, has become a new means of distribution of information and media to revolutionize methods of communication.

The idea of using the podcast technology as a means of dissemination of scientific information is of questionable origin, but it is reported that Grayson Wheatley, American cardiovascular surgeon, was the first to think to exploit the iPod and MP3-player craze in the medical field. He reported in an interview in 2006: "*The idea came to me while I traveled by train to a conference and to defeat boredom listening to my iPod. I thought that this would be a wonderful opportunity for patients to obtain information on their health. Just got home from the conference, I set to work to give life to this new service*" (Murray 2006).

Currently, there are websites that collect and classify in special "directory" the podcasting systems available on the network, simplifying the job of finding and registering by internet users. In addition to iTunes Store, "PodcastDirectory.com" and "Podcast.com" are the most complete services that you can use.

Experimentation with the podcast, as happened (and happens) for many new technologies, has seen **the medical and scientific journals as first users** (New England Journal of Medicine, Nature, Science, etc.). Today, many of these journals regularly distribute audio podcasts through their own websites. These files can be manually removed from the site of the journal (where there is an archive of podcasts already published), or you can receive them automatically after registering to these services. To facilitate listening, most journals also offer the **transcript of the podcast**. Science and Nature offer channels of podcasts on specific topics, including food science and genetics, which are very similar.

Due to content provided by many journals published by Nature Publishing Group, their example was immediately followed by numerous other journals (Johnson and Grayden 2006; Savel et al. 2007), but also by a growing number of blogs and online sites specializing in issues of interest, such as GMOs. For example, in searches for the tags #GMO to select and filter content on spreaker.-com, the users themselves are producers of freely available content and programs that are then reported via RSS feed and podcast.

Experiences of podcasting have been developed by several international scientific societies to distribute (often with the aid of slides) the main contributions presented at annual conferences and comments of leading experts on the results of research presented in the context of these events.

The distribution of congressional events in this form is part of a broader strategy adopted by scientific societies for training and updating. Some organizations are pioneers of using podcasting as a means of delivery of training credits.

Today, many scientific societies, universities, and associations distribute their published articles in audiocast and vodcast format, as well as their reports submitted to annual conferences, comments, and interviews with "leaders" operating in the scientific area.

Podcasts are one of the tools that the Action FP0905 has decided to use to spread and convey the collected information to the research community and the public. This instrument was used to make available files of the COST-related issues on GMO and related meetings and workshops, unless the policy of the Action specifically restricted access to that document.

Clearly, this tool can be very useful to communicate scientific subjects in a correct and balanced way in order to inform discussions at the public level.

6 COST FP0905 and "Wiki"

The applications of "**wiki**" in science are becoming quite common. Although this is a generalist tool, with only hundreds of entries devoted to GMO, Wikipedia represented one of the most complete wiki in this new field. Nevertheless, specific Wikipedia projects on GMOs are missing, which are available in more generalist fields such as medicine, as evidenced by the section called "**Portal: Medicine**" available in English language (Santoro 2009).

The entries in Wikipedia and its portals related to GMOs are handled by more than **200 researchers, journalists, activists, and others** with different interests and degrees of knowledge on the subject. Unfortunately, Wikipedia is an initiative without specific direction or objectives which would allow the development of specific information lines using evidence-based science (Tapscott and Williams 2006).

Following the example of the Wikipedia, Working Group 4 of the COST Action FP0905 has implemented within the Website of the Action a portal using the component Wiki Mediawiki licensed under GPLv2 by Harold Prins. The philosophy is based on the assumption that a participant of the COST Action should provide the answer to a question to his/her colleagues by publishing it on the inside pages, thus contributing to the "collective intelligence" that characterizes the phenomenon of Web 2.0 (Surowiecki 2005; Keim 2007; Hesse et al. 2011).

The goal of the realization of WIKI in COST FP0905 was to provide useful information to understand the topic of GM trees in forestry for all those who do not have the possibility to access scientific journals, setting the stage for an encyclopedia that should evolve over time by integrating video, audio, and image. Further work needs to be done, but this first initiative in this specific sector demonstrated the great interest by the public which should be continued in the near future.

For example, WikiDoc is the best medical encyclopedia written by doctors and containing more than 70,000 items that are managed in an environment that is the home of social networking tools, blogs, and forums. This is a good model to be adopted in forest biotechnology.

7 Microblogging

Twitter, a microblogging system, was created by Jock Dorsey in 2006. This system enables anyone to **create a personal page on which to exchange fast and frequent messages** answering to the simple question: "What's new?". The messages (called tweets) must have **a maximum of 140 characters** and can be updated through the website, via SMS, email, or using other applications offered by the website. Often the tweets include links to websites for enlarging the news items (text, images, video, etc.).

Twitter owes its success to the tremendous ability of news to spread through its network of followers. It has an average of 460,000 new users per day and 200 million total users (over 4 million in Italy). However, Twitter does not provide the number of registered and active users but the answer can be found in the study conducted by Cornell University and "Yahoo! Research": they reported that 50 % of the 140 million tweets per day are generated by 20,000 users (only 0.05 % of the total). Therefore, Twitter does not succeed in making a transition from "registered" users to "active" participants.

Another useful feature of Twitter allows us **to be updated in real time on what is happening with respect to any matter to the world**. For example, one or more participants, experts in different areas among those at a conference, send tweets to users subscribed to the service, giving news of events of particular interest and worth investigating.

In order to trace a tweet that refers to a given topic through the search engine of Twitter, it is given a hashtag to the message, i.e., a specific keyword preceded by the symbol "#". This function is used particularly by organizers of congresses, symposia, and conferences so that all conversations related to a specific conference can be monitored. Organizers of conferences can identify the reference hashtag in advance and invite those interested in participating in the online discussion to include it in their tweet.

With regard to the COST Action FP0905, the account "@ COSTActionFP0905" has contributed to publicizing the COST meetings and allowing users to launch hashtags on topics of most interest.

8 Video

The use of images produced by the "new" tools to visualize research and discovery has been a feature of science since the eighteenth century.

To date, the video, also thanks to the spread of smartphones which can easily produce video, has filled every space-time dimension of our lives, and it is extensively used in research and scientific communication. Surfing on the network reveals the huge amount of video content of science with research institutes, museums, science centers, universities, amateur groups, and individual researchers having video sections on their websites or on a YouTube video channel.

According to the report "Cybercitizen Health v9.0" of Manhattan Research, 30 % of all Internet users watch online videos on "science", 43 % of which are received directly on YouTube (Khamsi 2007). The "emotional" component of a video and the ease of dissemination have convinced industry, and some important medical and scientific centers to open official channels on YouTube. Videos are a means of communication that cannot be manipulated after posting on a website and therefore can be a useful tool to communicate scientific information.

COST Action FP0905 considered that a collection of videos would more easily explain the nature of the work done by the COST Action FP0905 and started a YouTube channel (http://www.youtube.com/user/CAFP0905) with the aim of disseminating reports of events and meetings and to publicize the objectives and activities of the Action. The use of the YouTube channel increased significantly the visibility of the Action website and the activities of the Action on GM trees more effectively than was expected.

The use of the YouTube channel was specifically intended for communicating with society and the public, with the goal to transfer information on GM trees and their usefulness for society; a very sensitive issue. In particular, two longer videos were made at the Euroscience Open Forum (ESOF) 2012 and at the Training School held in Tromsø (Norway), on May 2013.

Videos are a means of communication that cannot be manipulated after posting on a website, and are a useful tool to communicate information from the COST Action activity on GM trees. However, because of the contentious nature of this information, it created a reaction from groups opposed to GMOs.

Anti-GMO activists manipulated some written information by extrapolating certain sentences which were then placed in a different context. This took place at the COST Action FP0905 in February 2012, when activists of the "Global Justice Ecology Project" extrapolated sentences from the paper published in Nature Biotechnology (Fladung et al. 2012), and reported in the COST Office website (http://www.cost.eu/media/cost_stories/Transgenic_Trees). In their website they attacked the activities of the COST Action FP0905 stating: "*Please support the STOP GE Trees Campaign with a donation to help us win our campaign for a global ban on these dangerous GMO trees. Make a donation today*" (http://globaljusticeecology.org/cost-promoting-dangerous-ge-trees-in-europe/).

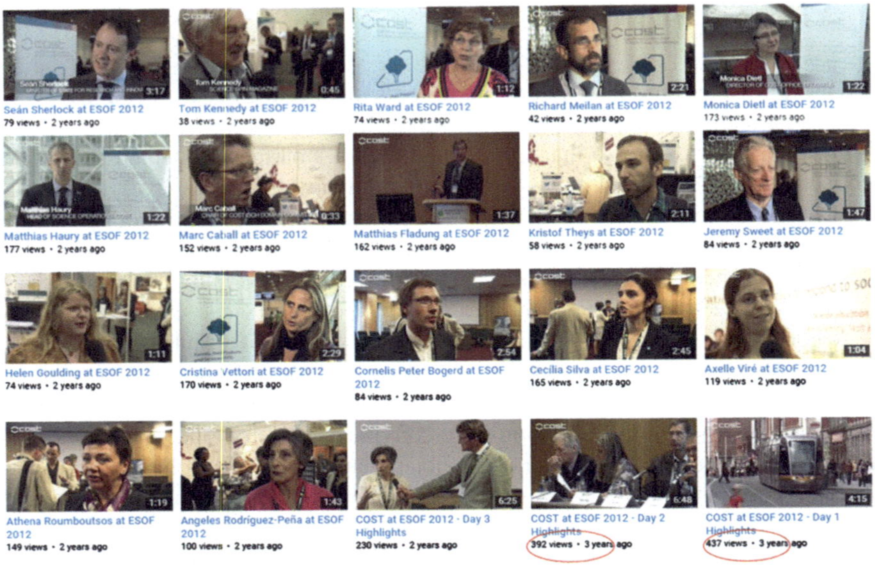

Fig. 3 Videos produced by the COST Office at ESOF 2012 (https://www.youtube.com/user/COSTOffice/videos)

Following a discussion with experts on Communication at the COST Office, it was decided to respond with a video in which scientists of COST Action FP0905 explained their activities in the context of this Action.

The opportunity came at the Euroscience Open Forum (ESOF) 2012, Europe's largest general science meeting, which took place in Dublin (Ireland) from 11 to 15 July 2012. ESOF's main aim is to highlight the latest advances in science and technology, and to promote dialog on the role of science and technology in society and public policy.

COST Action FP0905 took part in ESOF 2012 by organizing a workshop on GM trees and used the occasion to produce a video (https://www.youtube.com/watch?v=1BdToqlTB1g). The video, available on YouTube, is among the most viewed of the events of the COST ESOF 2012 with 392 views, making it second after the video presenting COST at ESOF (Fig. 3). It is interesting that, to date (July 2015), there have been no negative comments regarding the issues described in this video.

9 Facebook and COST Action FP0905

A growing number of researchers and scientists have active Facebook accounts. However, this presence on Facebook does not translate into opportunities to meet with other researchers or with people interested in research, since the current use is restricted to the private sphere. Findings from a survey published in some authoritative newspapers document how **8 out of 10 scientists would reject any request of "friendship" on popular social networks coming from users interested in their research**. This report is in relation to a study, conducted by French researchers on 405 researchers of universities, with the objective to survey their willingness to accept the demands of "friendship" that come from requests to establish a channel of virtual communication.

The main reasons for this rejection are to be found in the desire to keep their distance from the general public (98 %), the desire to protect their personal information (98 %), and the belief that this interaction would not be correct because it would lack the basic knowledge (88 %). Only 15 % would be willing to accept a virtual communication but only on a case by case basis (Ghassan et al. 2011).

Facebook is a communication channel that can give opportunities for providing information updates and this has encouraged numerous magazines, scientific societies, universities, and research organizations in the field of science to increasingly spread their news across pages of Facebook.

The COST Action FP0905 initially began an approach on Facebook, but the activity was terminated since the release of the information had to be moderated by all participants and not by an individual participant in a Working Group of the Action.

Some organizations are not allowing their employers to use Facebook and some of the participants in COST Action FP0905 were not allowed to post their contribution in Facebook. One of the main reasons for this was the article: "Experimental evidence of massive scale emotional contagion through social networks" by Kramer et al. (2014). This resulted in an Editorial Expression of Concern by The Editor-in-Chief of PNAS, Inder M. Verma, specifying that "*this paper represents an important and emerging area of social science research that needs to be approached with sensitivity and with vigilance regarding personal privacy issues*" (Verma 2014).

The study reported in PNAS was conducted with 689,003 (of the estimated 1 billion) users who have expressed their real experiences and opinions. This study showed that people are subject to emotional responses to information on the network through manipulation of the algorithm by which Facebook provides the "News Feed". Some people were exposed to posts of mainly positive emotional content for a week, and other people to negative content posts. A control group was subjected to random posts. The results of the study indicated evidence that emotional contagion on a massive scale can be generated via social networks in response to positive or negative posts.

For these reasons, some of the participants in COST Action FP0905 were not allowed by their organizations to post their contributions in Facebook. Thus Facebook was not considered appropriate for disseminating scientific information from the COST Action FP0905.

10 Visibility of COST Action FP0905

The technologies of Web 2.0, described above, have allowed us to increase the visibility of the activities of the COST Action FP0905 to an extent that would not have been possible with free traditional media (e.g., newspapers, Web 1.0). This, combined with a GMO topic of special interest and discussion around the world, has led to higher than expected numbers of visits, considering that these are scientific topics and not topics such as sport or music that have a wider interest to the general public.

The activation of the YouTube channel has been the most important tool of the Web 2.0 in determining the increase in visibility and therefore of visits to the website. The activation of the YouTube channel (March 2011) increased the visits to the simple channel with a few contents from 4,500 (average of visits from the start date of website in April 2010) to 25,000. This further increased to 75,000 with the publication of the video to promote the participation of the Action at the IUFRO Tree Biotechnology Conference (July 2011). The number of visits has increased significantly successively with higher peaks corresponding to events announced in the COST Action FP0905 website and in YouTube (Fig. 4).

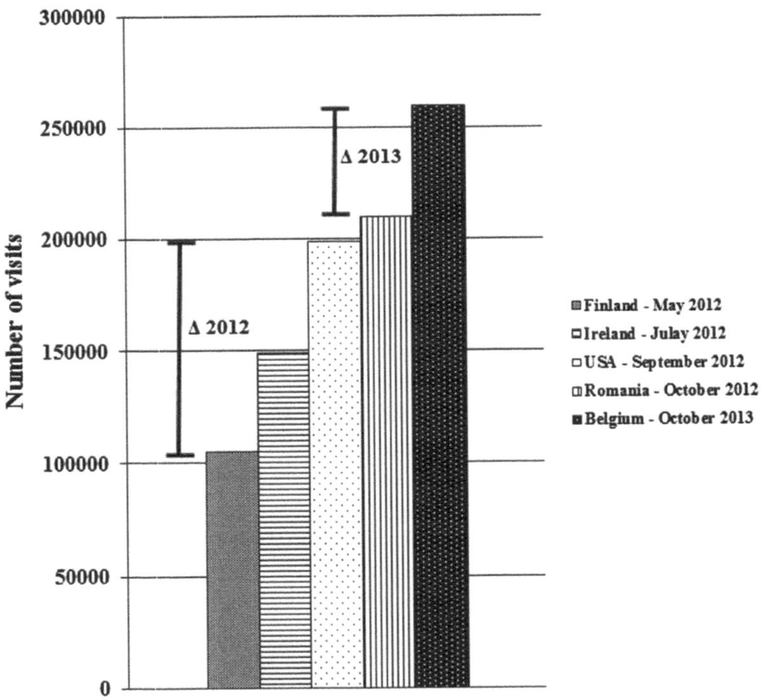

Fig. 4 Accessing statistics of COST Action FP0905 website at events of the action in the period 2012–2013. The *delta* indicates the increment of visits between 2012 and 2013

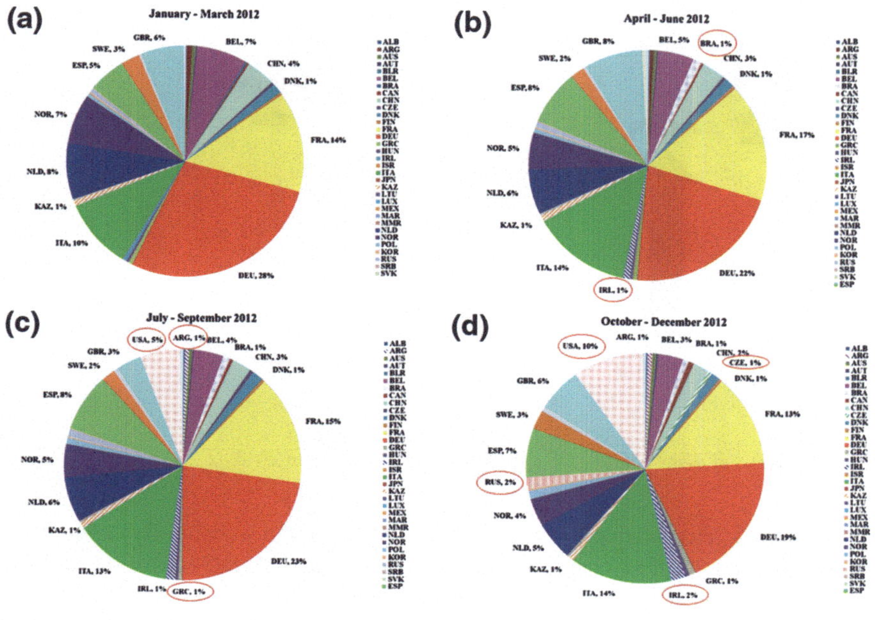

Fig. 5 Trends of the visits from EU and non-EU countries in the period 2012 expressed as % of the total. Only the visits above 1 % are shown: **a** Q1 period: no particular events; **b** Q2 period: COST Action Meeting (8–11 May 2012, Helsinki, Finland), **c** Q3 period: ESOF 2012 conference (July 10–14, 2012, Dublin, Ireland), and ISBGMO 12 conference (September 18–20, 2012, Buenos Aires, Argentina); **d** Q4 period: COST Action Meeting (October 26–28, 2012, Brasov, Romania)

Moreover, it is interesting to see the country of origin of the visits and that an increase of visits and new visits from some countries occurred after participation of the Action FP0905 in internal meetings or at important international conferences (Fig. 5).

In particular, we can see that the number of visits from USA was 5 % after the ESOF 2012 (Fig. 5c) and increased to 10 % after the ISBGMO 12 Conference and COST meeting (Fig. 5d). Moreover, some visits appeared for the first time after some events, such as from Argentina and Greece (Fig. 5c), Russia and Czech Republic (Fig. 5d) countries. Other countries showed a more or less constant number of visits and always above 1 %, e.g., United Kingdom (3–8 %), Germany (19–28 %), Belgium (3–7 %), Italy (10–14 %), Spain (5–8 %), and France (13–17 %).

An additional increase in visits occurred after the opening of a Twitter channel (#COSTActionFP0905), in April 2013, which promoted the website content, especially in those countries in which this service is highly used such as USA, Canada, Brazil, and India (Fig. 6).

This greater visibility of the website increased its popularity, which, in combination with the critical interest in the subject, generated some additional problems.

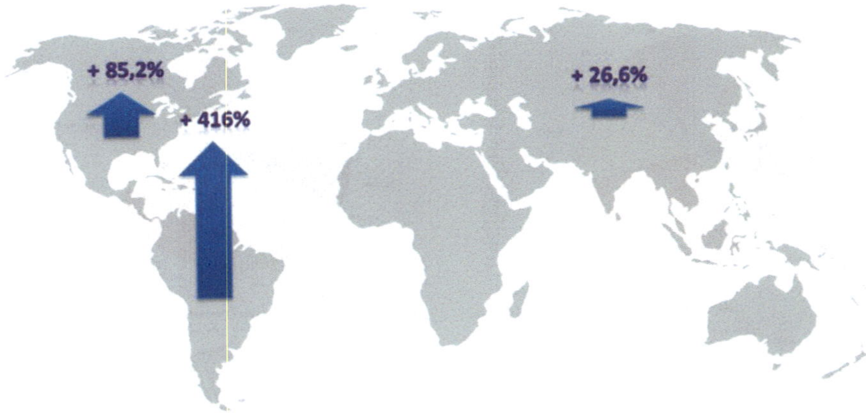

Fig. 6 Increase of visits to the COST Action FP0905 website in 2013

Anti-GMO activists made several attempts to block the website. By April 2014, the website had 260,916 visits, and 211 hacker's attacks which caused the website to go off-line for 100 days and causing complete paralysis of some activities for several weeks. In particular, two of these attacks are worthy of note.

The first big attack happened just a few days before the COST Action meeting held at Orleans (France) in September 7–8, 2012. In particular, after the publication of the information and video of this new Action meeting, fake registrations to the website increased by more than 24,000 per month, occurring every 2 min, and in total more than 60,000 (cumulative) email alerts were recorded in a year. For this reason, the Web Hosting Provider asked us to suspend the registration to the website and the wiki subsystem until it improved the security of the hosting server.

The other attack was larger and directed toward many of the most visible websites in the world. We received a message from the Web Hosting Provider saying: "*Dear Customer, at around 4 a.m., our system administration team identified a website defacement attack affecting a large number of customers .....*"

These attacks started at 4:00 a.m. on May 3, 2012, and stopped 6 min later. During this short period, our website had more than 170,000 fake users, more than 35,000 fake submissions/minute, and more than 16,000 entries in MediaWiki per second. Consequently, the website crashed and the hacked website had the homepage replaced by the words "Server HackeD."

The website of the Action was built by Joomla, and only 2 % of its typology were hacked, whereas 78 % of the self-made websites were hacked (Fig. 7).

This attack was done at world level, and more than 70,000 websites were hacked particularly the most visible ones. The attack started from a South Asian anonymous proxy server and all the attacks used anonymous proxy servers and the North European network TOR.

At the time of this attack, the website of the Action was in the 7878 position (data on May 1, 2012) in the ranking of the most visible websites in the world.

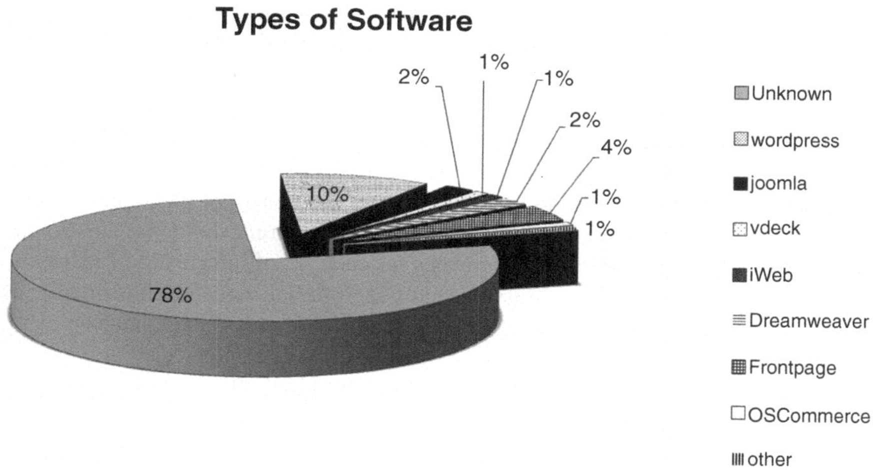

Fig. 7 Distribution of the types of software that can be used to build a website and that have been hacked

11 Genetically Modified Trees (GMTs) Database

An important goal of the COST Action FP0905 was to provide an accessible database on GMTs containing the main information on forest GMTs present in other databases or published in international Journals. This was particular important since, after 15 years of experimentation with GMTs, there is considerable amount of information but it is scattered. Therefore, it was a primary objective of the COST Action to create a database on GMTs which was freely accessible to the worldwide scientific community.

At the beginning of the Action, the database was located directly in the website with a web interface for entering the data. However, for security and safety reasons following the attacks on the website, a database for internal use with a web interface on a proprietary platform (not open source), called Filemaker X1, was set up in 2012. The private access was secured by various security levels:

- only registered users to COST Action website,
- first access through the web page of the COST Action,
- IPFilter (by country and IP institution),
- username and password for Filemaker application,
- right policy, for administration and management of access rights.

The web interface of FileMaker is also accessible from any mobile device (such as notebooks, smartphone, tablets, etc.) allowing the incorporation and management of each entry in different cards updated and verified in real time.

On the website of the COST Action, the information on the external server was replicated and updated at a predetermined time using PHP technology and a Mysql database.

The GMT database is composed of 257 entries corresponding to 9 genera (*Betula* spp., *Castanea* spp., *Eucalyptus* spp., *Larix* spp., *Liquidambar* spp., *Picea* spp., *Pinus* spp., *Populus* spp., and *Ulmus* spp.).

For each entry the information available is on the construct used and the GM trait of the GMTs (name of the construct, name of the trait, donor organism, and function of the gene introduced). References for each set of information (institution or private/industry, code of permission, etc.) are present, and this information is linked with literature and methods developed for risk assessment, management, gene flow, and the containment strategies) if available (Fig. 8).

A search tool allows surfing within the database to look for specific information and this search can be done by genus, GM trait, gene, country, and location.

The information present in the GMT database is the result of the work done in WG 1 and reported in Part I of this book and with the contribution of two early-stage researchers (ESR), Ms. Maja Popovic (Division of Genetics, Forest Tree Breeding and Seed Husbandry, Croatian Forest Research Institute, Jastre-barsko, Croatia) and Mr. José Manuel Ramos Sánchez (Universidad Politécnica De Madrid, Madrid, Spain) who received a COST Short Term Scientific Mission to work at

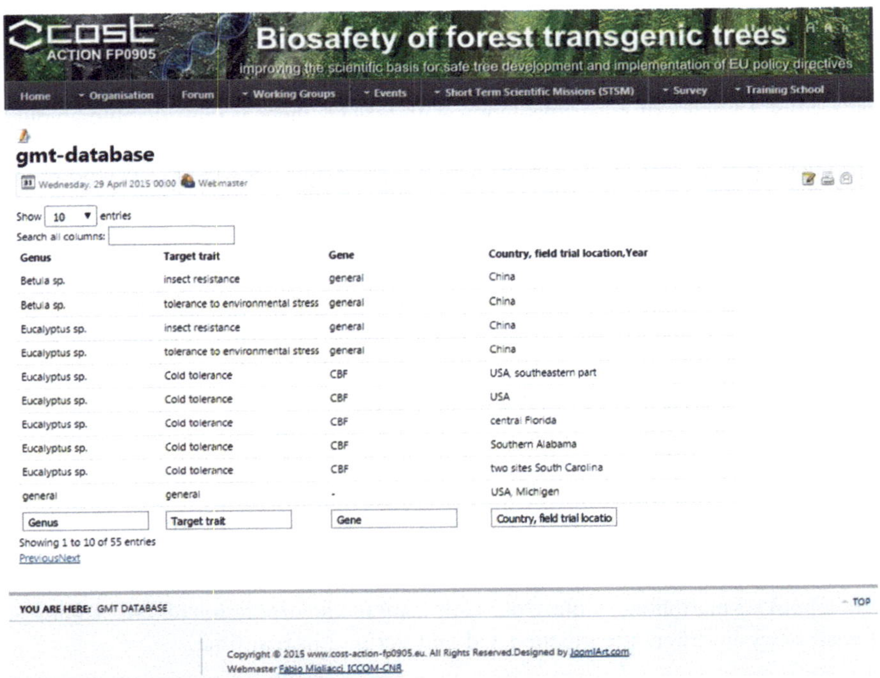

Fig. 8 GMT database of COST Action FP0905

INRA (Orleans, France) financed by this Action and under the supervision of Dr. Gilles Pilate (Leader of WG 1).

At present (July 2015), the transfer of the entire database in an open source platform (Myslq) is planned to permit exchanges with other data banks.

12 Conclusion

Dissemination of scientific information is itself a scientific technique which uses languages specific to each medium of communication, and develops new elements for directing the scientific information to the contingent interests of the public.

In addition to the Mass Media (i.e., Discovery Channel) and Scientific Journals (e.g., Nature), Web 2.0 tools through social networks (e.g., ResearchGate, blog, wiki, YouTube, Twitter, etc.) can be a valuable way to communicate scientific information to a wider non-specialist public.

Therefore, the use of the Web 2.0 for scientific dissemination can promote the production and dissemination of scientific knowledge to various social groups.

In effect, Web 2.0 and in our specific case, Science 2.0, can increase the collaboration between research groups and encourage participation in collective undertakings for the advancement of knowledge and contributing to the "collective intelligence".

The relationship between scientists, science communicators, and journalists in the Web 2.0 can change the relationship between media sources and public. The new specialist sources provide a pronounced advancement from "traditional journalism" due to an enrichment of the "ecosystem of information" through scientific bloggers, which can also support scientific journals as tools for dissemination.

Social networking and collaborative tools on the Web have spread among researchers. A survey in the US estimated that the percentage of American scientists who collaborate online through blogs, chat, and social network is about 69 % (Dimov et al. 2007). Between 18 and 25 % of respondents claimed to regularly participate in social networks limited to their field, with the aim to find relevant experiences by colleagues, for example, physicians sharing information on drugs. For other scientists information on methods of analysis, on equipment, or on possible collaborative projects is shared. In the medium future, likely development will be social networks "dedicated" to meet the requirements of young researchers working at universities, who see in these instruments a natural meeting place where they can find answers to their questions and update their information.

Nevertheless, some difficulties and impediments in the dissemination of scientific content cannot be underestimated. The absence of a moderator and/or a control in social networks (i.e., Facebook) could also result in an excess of simplification, trivialization, dramatization, and distortion of the scientific content.

The 4 years of experience of COST Action FP0905 in communicating about GM trees to the public can be considered positive. The use of the Web 2.0 tools, and in particular of YouTube and Twitter, has permitted wider spread of scientific information on a contentious topic. Nevertheless, a lot of work has to be done specifically in relation to the communication of biotechnology and GMOs. We have only had the possibility to perceive the intrinsic potential of Web 2.0 and social media technologies in enhancing the traditional communication and scientific dissemination. Scientists should understand this potential and increase the flow of scientific information freely available in the web, following the example of the medical encyclopedia "WikiDoc," the scientific portal developed with "ASCO" project in medicine field, and ResearchGate.

Web 2.0 and social media technologies are connecting scientists with public opinion directly, but there is a need to educate scientists on the most simple scientific dissemination procedures and on the ability of people to understand scientific principles.

An attempt has been made with this COST Action, and some of the information gained is still available in the website (http://www.cost-action-fp0905.eu/), even though the Action ended in April 2014. In particular, the GMT database is freely accessible in the website and is an important resource for researchers and for organizations involved in GMO risk assessment and management. In addition, it is also available to the general public and can be consulted by any person interested in gaining knowledge in particular areas related to GMTs.

We apologize if you sometimes find that the website is down due to hacking attacks, and if some documents are not available due to the necessity to upload them again after an attack. This is the risk of having the high visibility provided by Web 2.0 technologies combined with the sensitivity of the topic of GMOs.

References

Bach L (2008) Blogging during terminal care: communication, color schemes, and creating a community. J Clin Oncol 26(27):4504–4506. doi:10.1200/JCO.2007.15.4443

de Judicibus D (2008) Web 2.0 Orientation Map. L'Indipendente (The independent). Available via http://www.lindipendente.eu/wp/it/2008/02/08/web-2-0-orientation-map/

Dimov V, Fares W, Schwimmer J et al (2007) The blog website: a new educational tool for cardiology. J Am Coll Cardiol 49:275A–276A

Eysenbach G (2008) Medicine 2.0: social networking, collaboration, participation, apomediation, and openness. J Med Internet Res 10(3):e22. doi:10.2196/jmir.1030

Fattorello F (1961) Introduzione alla tecnica sociale dell'informazione. Roma, Istitutto Italiano del Pubblicismo

Fladung M, Altosaar I, Bartsch D, Baucher M, Boscaleri F, Gallardo F, Häggman H, Hoenicka H, Nielsen K, Paffetti D, Séguin A, Stotzky G, Vettori C (2012) European discussion forum on transgenic tree biosafety. Nat Biotech 30(1):37–38

Frost Jeana H, Massagli Michael P (2008) Social uses of personal health information within PatientsLikeMe, an online patient community: what can happen when patients have access to one another's data. J Med Internet Res 10(3):e15. doi:10.2196/jmir.1053

Ghassan M, Guiot A, Benhamou Y, Benhamou A, Hariri S (2011) Global medical ethics: Facebook activity of residents and fellows and its impact on the doctor–patient relationship. J Med Ethics 37:101–104. doi:10.1136/jme.2010.036293

Hesse BW, O'Connell M, Augustson EM, Chou W-YS, Shaikh AR, Finney Rutten LJ (2011) Realizing the promise of Web 2.0: engaging community intelligence. J Health Commun 16(S1):10–31. doi:10.1080/10810730.2011.589882
Hurley M, Smith C (2007) Patients' blogs—do doctors have anything to fear? BMJ 335:645–646. doi:10.1136/bmj.39343.478403.68
Johnson L, Grayden S (2006) Podcasts—an emerging form of digital publishing. Int J Comput Dent 9:205–218
Keim B (2007) WikiMedia. Nat Med 13:231–233. doi:10.1038/nm0307-231
Khamsi R (2007) Is YouTube just what the doctor ordered? In New Scientist. https://www.newscientist.com/article/dn10859-is-youtube-just-what-the-doctor-ordered/. Accessed 5 Jan 2016
Kramer ADI, Guillory JE, Hancock JT (2014) Experimental evidence of massive-scale emotional contagion through social networks. PNAS 111:8788–8790. doi:10.1073/pnas.1320040111
Murray RB (2006) Web 2.0 and podcasting: implications for health care. Population Health Matters (Formerly Health Policy Newsletter) 19(4):Article 5. Available at http://jdc.jefferson.edu/hpn/vol19/iss4/5
O'Reilly T (2005) What is Web 2.0? Design patterns and businnes models for the next generation of software. http://www.oreilly.com/pub/a/Web2/archive/what-is-Web-20.html. Accessed 30 Jan 2015
Sandars J, Schroter S (2007) Web 2.0 technologies for undergraduate and postgraduate medical education: an online survey. Postgrad Med J 83:759–762. doi:10.1136/pgmj.2007.063123
Santoro E (2007) Il web 2.0: dalla partecipazione alla in-formazione. Ricerca & Prati 138:266–269
Santoro E (2009) Web 2.0 e medicina, come social network, podcast, wiki e blog trasformano la comunicazione, l'assistenza e la formazione in sanità. Il Pensiero Scientifico Editore. Roma
Savel RH, Goldstein EB, Perencevich EN, Angood PB (2007) The iCritical care Podcast: a novel medium for critical care communication and education. Am Med Inf Assoc 14(94):99. doi:10.1197/jamia.M2205
Surowiecki J (2005) The wisdom of crowds. New York, Anchor
Tapscott D, Williams AD (2006) Wikinomics: how mass collaboration changes everything. Penguin Group (USA), New York, p 320
Verma IM (2014) Psychological and cognitive sciences. Editorial Exp Concern PNAS. doi:10.1073/pnas.1412469111
Watzlawick P, Beavin Bavelas J, Jackson DD (1967) Pragmatics of human communication. A study of interactional patterns, pathologies and paradoxes. New York, NY: Norton
White RW, Horvitz E (2009) Cyberchondria: studies of the escalation of medical concerns in Web search. ACM Trans Inf Syst 27(4):Article 23, 1–37. doi:10.1145/1629096.1629101

Concluding Remarks

The COST Action FP0905 "Biosafety of forest transgenic trees: improving the scientific basis for safe tree development and implementation of EU policy directives" was a 4 year project (2010–2014) combining scientists from 27 European countries as well as 7 non-European countries. The aim of the action was to gather information on genetically modified (GM) forest trees to strengthen the scientific basis for the implementation of the EU policy directives related to GE trees intended for cultivation and to evaluate the relevance of scientific information for biosafety protocols. The intercontinental co-operation was of great value allowing us to compare forest tree cultivation practices (semi natural forests vs. plantation forestry) and GM-related regulatory issues from the global perspective, taking into account that forest certifications, forestry industries and trade are global. The Action considered scientific perspectives on the present state of art of biological characterization, environmental impact assessment, monitoring, and also socio-economic considerations and the value of communication on GM trees in order to achieve a comprehensive view of GM trees in Europe and globally. After 4 years of intensive work, studies and discussions, we presented our main results in a book which we hope is more than its individual chapters and provides readers with a comprehensive, up-to-date view on these scientific and socio-economic aspects of GM forest trees.

Part I: Biological Characterization of Genetically Modified Trees (GM trees)

In Part I of this book, two of the chapters (*Alburquerque et al.* and *Häggman et al.*) are focused on genetic transformation technologies. *Alburquerque et al.* present new technologies being employed as alternatives to transgenic modification of trees. They include the most novel and powerful methods, including genome editing technologies based on homologous recombination. These are of great interest for the future because homologous recombination and variation between transformants will be negligible. Mixing of genetic material between species that cannot hybridize by natural means is one of the major criticisms of transgenic crops.

C. Vettori et al. (eds.), *Biosafety of Forest Transgenic Trees*,
Forestry Sciences 82, DOI 10.1007/978-94-017-7531-1

Therefore, transfer of genes from species sharing the same gene pool as the receptor species (known as cisgenesis or intragenesis), could be an interesting choice to increase public acceptance.

The paper by *Häggman et al.* is focused on the role of genetic engineering (GE) in forest tree breeding efforts. They point out the manifold roles of forests that are not covered in monoclonal forest plantations, but GM forest trees can play a significant role in highly-productive plantation culture and would be beneficial for specific purposes. The role of GE technology in forest tree breeding is compared to traditional tree breeding and discussed in the context of climate change. Moreover, the importance of the accumulating information on tree genome sequences in the perspective of future forestry, genetic engineering and forest tree breeding are is contemplated.

The chapters by *Pilate et al.* and *Strauss et al.* are both focused on GM tree field trials in Europe and USA, respectively. *Pilate et al.* underline the situation in Europe where the number of GM tree field trials is limited and it is also becoming increasingly difficult to obtain authorization for GM tree field trials. From the scientific point of view, this is very surprising for the following reasons: (1) phenotypic effects resulting from transgene expression in GM trees grown in the field appear to be stable, albeit variable, (2) most field studies have validated earlier observations made under greenhouse conditions, although in some cases the modification of target traits was less pronounced in field environments, and in a few cases, GM trees had severe growth and developmental penalties, and (3) non-target effects were consistently within the range of natural variation. *Pilate et al.* conclude that European GM tree field trials have failed to identify any significant tangible risks. Based on this evidence, it seems appropriate that Europe should now move forward beyond small confined trials to larger scale experiments better fitted to a broader context of evaluation and environmental assessment.

Strauss et al. indicate that the situation is totally different in the USA. Under USDAAPHIS federal regulatory notifications and permits, they have planted nearly 20,000 trees, mostly *Populus*, derived from approximately 100 different constructs and several different traits in more than two-dozen different field experiments. The most significant lessons from these trials are: (1) Visual abnormalities, i.e. somaclonal variants have been observed in several experiments, but are extremely rare, (2) gene expression and RNAi-induced gene suppression have been highly stable, (3) the regulatory process has largely been efficient and workable, though it imposes significant biological constraints, costs, and risks that are very difficult for an academic laboratory to bear when trials span several years. It is especially problematic when flowering is the essential research goal for many projects. (4) Field environments invoke complex and largely unpredictable changes to expression and associated phenotypes when studying wood chemistry and physiology-modifying transgenes, suggesting the need to study several field sites, genetic backgrounds, and gene insertion events over many years similar to common practices of conventional breeding. (5) Collaborative field trials with industry have shown that common transgenic traits, such as herbicide and insect resistance, can have large productivity benefits in near-operational plantation conditions suggesting that it could be highly beneficial to incorporate transgenic traits into production programs. *Strauss et al.* suggest that regulatory reforms that are based on product benefits as well as risks, and not primarily on the use of recombinant DNA methods, are urgently needed if the power of transgenic technology is to provide significant social benefits in forestry.

Part II: Environmental Impact Assessment and Monitoring of GM Trees

The chapters by *Ruohonen-Lehto et al., Glandorf and Breyer,* and *Biricolti et al.* all deal with environmental considerations for GM plants/trees and guidance on their risk assessment and monitoring but from different points of view. The chapter by *Ruohonen-Lehto et al.* describes different environmental risk assessment frameworks. They point out that the basis of any risk assessment is hazard identification, estimation of potential consequences, estimation of the likelihood of these consequences to occur, estimation of the overall risk, and deciding about risk management measures. Risk assessment usually includes a problem formulation step, definition of assessment endpoints from the broad protection goals, and identification of those attributes to be measured. As examples, the EFSA guidance on risk assessment of GM plants and the Cartagena Biosafety Protocol guidance of GM trees are described. Moreover, both the comparative and *per se* risk assessment approaches are discussed.

The chapter by *Glandorf and Breyer* describes the Netherland system of step-by-step approach for the categorization of GM field trials. Under this approach, the confinement of GM plants in a field trial can be gradually decreased and the scale of the trial increased in a step-wise manner at the same time that knowledge on the GM plant and its environmental interactions increases. They propose that a formal step-by-step approach, applicable also for GM trees, may help to facilitate the approval process for GM field trials and the collection of relevant data/material for the ERA without compromising the environmental safety.

Biricolti et al. focused on impacts of GM trees on the environment and the role/use of plant "omics" strategies in environmental risk assessment. They used a questionnaire approach targeted at the participants of the action but also at other scientists. The results indicated that important biosafety issues can only be addressed by conducting field releases of transgenic trees. The use of "omics" strategies in environmental risk assessment were extensively discussed among the participants of the questionnaire, and the conclusion was that large-scale methods that can be internationally certified and accepted are so far not available.

As several GM tree studies dealing with diverse traits have addressed potential impacts on the receiving environment, Part II also includes two case studies. *Gallardo et al.* focus on potential soil effects on GM trees and *Robischon* deals with potential environmental impacts of insect-resistant GM trees.

Gallardo et al. report that most published studies have considered effects on sensitive indicators such as mycorrhizal fungi communities which can also be indicative of possible alterations in soil ecosystems. However, no significant changes in fungal communities or in mycorrhizal colonization have been reported in published papers of GM trees to date. The authors emphasize that the lack of data about on indicator behaviour in conventional agricultural and silvicultural practices limits the evaluation of the ecological relevance of the observed changes. In the few example studies reported, plant stage, type of soil and other environmental factors may have a greater influence on indicators than on the effect of GMTs.

The chapter by *Robischon* points out that at present the resistance to phytophagous insects is a trait that is highly desirable in commercially grown forest trees. There is also a growing number of studies available focusing on non-target or wider ecosystem effects of such GM trees in field trials. Based on the literature review, *Robischon* points out that, until now, no coherent overall picture emerges, yet a few instances of unexpected outcomes of the exposure of insect resistant GM trees to biocenoses have emerged. Thus, longer exposure of transgenic trees in field studies will also produce further results, observations and information.

Part III: Socio-economic Considerations for GM Trees

The chapter by *Bartsch et al.* deals with socio-economic considerations for the decision making on GM tree cultivation. *Bartsch et al.* point out that in EU GM trees (as all GMOs) are regulated based on the process of production, not on the phenotype and the science-based risk assessment concentrates on the assessment of risks rather than benefits. Lately, consideration of socio-economic (SE) assessments are planned to get a more prominent role in EU decision making. The SE assessment will typically be a multidisciplinary and iterative process to address all relevant costs and benefits with as little bias as possible. *Bartsch et al.* conclude that there is a high likelihood of divergent views of the methods for SE assessments and evaluating costs and benefits, which may lead to further delays on decisions for GM plant authorisationsauthorizations.

Kazana et al. made a cross-country pilot "Knowledge, Attitude, Practice" survey (KAP) to better understand public knowledge and perception of safety issues towards the use of GM forest trees. The survey was conducted using university students of different disciplines as sampling subjects. In total, 1920 completed questionnaires from 16 European and non-European countries were evaluated. According to the survey, the majority of the respondents (>60 %) approved the use of GM forest trees for commercial plantations and the majority also appeared willing to buy products (e.g. wood, pulp and paper) from such plantations. Over 80 % of the respondents from all countries were in favour of using labeling to identify products of GM origin, while more than 80 % of those would prefer the labeling to be legally mandated. The top three potential benefits were lower demand of the GM forest plantations for pesticides, the potential of GM forest trees for restoration of contaminated soils and the potential higher GM forest tree productivity. The top three GM forest tree risks included the loss of biodiversity due to gene flow between transgenic and wild trees, the adverse effects of bio-trophic processes on host ecosystems and the cultural adaptation to changing biodiversity conditions. Overall, lack of knowledge regarding the benefits and potential risks of the cultivation of GM forest trees was observed in almost all surveyed countries.

Socio-economic considerations of GM forest tree biosafety in addition to environmental risk assessment are reported in another chapter by Kazana et al. They used a "Cost-Benefit Analysis" (CBA) approach targeted at highly experienced scientists in their respective fields of expertise representing 17 member countries from across the COST Action FP0905 network. This chapter presents an overview of available socio-economic approaches and tools for assessment of GM forest crops and presents options for their implementation. This analysis has allowed the identification of the most important positive and negative externalities of GM forest plantations in connection to modified traits.

The CBA approach can make a significant contribution to a more rational decision-making process in respect of the potential release of GM forest trees, as it would add an assessment of potential contributions to social welfare.

The chapter by *Tsourgiannis et al.* makes a comparative analysis of Greek consumers' potential purchasing behaviour concerning transgenic products, i.e. paper, wood biomass for energy production and wood products. The focus of the analysis was to study the main factors influencing consumers' potential purchasing behaviour. Field interviews were conducted in a randomly selected sample consisting of 418 consumers throughout Greece in 2011. The results indicated that in Greece there are potential buyers for transgenic products, e.g. wood biomass for energy production. Marketing issues such as the attractiveness of the packaging could positively influence consumers' potential purchasing decisions towards those products. However, there were also concerns about the negative health perceptions of the transgenic product.

Part IV: Value Communication in the Field of GM Trees

The chapter by *Nicolosi* deals with value communication of GM plants as example in the field of agri-food. *Nicolosi* points out that all the surveys carried out so far to measure public perception on the topic of techno-scientific innovations demonstrate a certain ambivalence in the judgment of public opinion. It is generally positive with respect to medical applications but innovation in biotechnological applications in the agricultural and food sector, is the source of anxiety and considerable fear. The cultural and symbolical value of agri-food practices implies that agri-food and communication are strongly intertwined. Presenting a theoretical background and some evidences emerging from empirical research, *Nicolosi* suggests some guidelines for a value-based system for risk and benefit communication in agri-food applications, which might also be of value for the GM tree sector.

The chapter by *Migliacci et al.* considers the experiences of using the world wide website as a value communication and scientific dissemination tool for COST action FP0905. During the 4 year period of the COST action, the web pages have been in active use enabling establishment of an interactive database including information on GM forest trees, GM field trials, traits and gene constructs used. The database is user friendly allowing specific searches, e.g. on species, gene construct or trait level and it will be certainly appreciated by scientists and specialists of the field. Moreover, the value communication between scientists and the public was ongoing on throughout the 4 years. The web site was also an important information source for events arranged by the action such as training schools, short-term scientific missions conferences and for producing video presentations. Some of the functionalities of the database (e.g. the use of the database) will be maintained after the end of the Action. COST Action FP0905 used a range of the most innovative tools of web communication to spread scientific information to the community and for interactions among scientists and the public. Throughout the 4 year project, severe problems were created by the high frequency of hacker attacks resulting in shutting down of the COST Action FP0905 website in some instances. All the innovative tools of web communication have been made possible and the hacker attack problems solved by the excellent web master Fabio—Grazie!

Index

C. Vettori et al. (eds.), *Biosafety of Forest Transgenic Trees*,
Forestry Sciences 82, DOI 10.1007/978-94-017-7531-1

The manufacturer's authorised representative in the EU is Springer Nature Customer Service Centre GmbH, Europaplatz 3, 69115 Heidelberg, Germany. If you have any concerns regarding our products, please contact ProductSafety@springernature.com

Printed and bound by CPI Group (UK) Ltd, Croydon, CR0 4YY
22/09/2026
02221057-0016